Michele Pagano (Ed.)

Cell Cycle Control

With 17 Figures

Springer

Dr. Michele Pagano
New York University School of Medicine
Department of Pathology, MSB 548
550 First Avenue
New York, NY 10016
USA

ISBN 978-3-662-21695-8 ISBN 978-3-540-69686-5(e-Book)
DOI 10.1007/978-3-540-69686-5

Library of Congress Cataloging-in-Publication Data

Cell cycle control/Michele Pagano, ed. p. cm. —(Results and problems in cell differentiation; 22)
Includes bibliographical references and index. ISBN 3-540-64031-2
1. Cell cycle. 2. Cellular control mechanisms. I. Pagano, M. (Michele), 1961- . II. Series.
QH607.R4 vol. 22 [QH605] 571.8'35 s—dc21 [571.8'44] 98-10581

Originally published by Springer-Verlag Berlin Heidelberg New York in 1998
Softcover reprint of the hardcover 1st edition 1998

Cover Design: Meta Design, Berlin
Typesetting: perform k + s textdesign GmbH, Heidelberg

SPIN 10567999 39/3136 - 5 4 3 2 1 0 - Printed on acid-free paper

Results and Problems in Cell Differentiation

Series Editors:
W. Hennig, L. Nover, U. Scheer

22

Springer-Verlag Berlin Heidelberg GmbH

Preface

Leafing through a journal of biology will almost certainly turn up several articles on some aspect of the eukaryotic cell division cycle. The progress in understanding the modus operandi of the cell cycle machinery and its central role in controlling cell proliferation has awakened a great deal of interest in many biologists. However, the cell cycle field often seems unnecessarily complicated, perhaps because data are acquired in bits and pieces and almost all simultaneously from studies using different systems and organisms. The aim of this book is to make the background of the cell cycle field as unambiguous and clear as possible by distinguishing what is commonly accepted in the field from what are the unsolved problems. Therefore, this book is intended to serve a large audience: graduate and postgraduate students, basic and clinical scientists involved in the study of cell growth, differentiation, senescence, apoptosis, and cancer. My hope is that this book will stimulate further interest in the fascinating field of the cell cycle and help scientists find connections between this and their fields of interest.

This book brings together experts to review all major aspects of the eukaryotic cell division cycle. The cell cycle is a coordinated and highly regulated series of events during which the cell replicates its DNA, and then divides. Formally, it has been divided into four phases: G1 (gap1), S (DNA synthesis), G2 (gap2) and M (mitosis). During G1, the cell monitors its environment and size before committing to enter S phase, during which it replicates its DNA. During G2, the cell ensures that DNA replication has been faithfully completed. In mitosis, the cell first undergoes nuclear division and subsequently divides into two daughter cells. An overview of the molecular mechanisms controlling the G1 phase is presented in the first chapter by Robert Sheaff and Jim Roberts, while the chapter by Anindya Dutta considers the regulation of DNA replication. The regulation of the G2 to M transition is analyzed by Jonathan Pines in Chapter 3.

It is now known that the sequential activation of different *c*yclin-*d*ependent *k*inases (CDKs) plays a central role in controlling the passage through the different phases of the cell cycle. CDKs are mainly controlled by phosphorylation/dephosphorylation events and by the association with inhibitory subunits. These two important aspects of CDK regulation are described in the chapters by Mark Solomon and Philipp Kaldis, and Andy Koff et al., respectively. Finally, the cellular abundance of both activating and inhibitory CDK subunits is primarily regulated at the levels of transcription and degradation (via the ubiquitin pathway). Joyce Slingerland and I have reviewed the latter mechanism of cell cycle control in Chapter 6. Important aspects of transcriptional control of the cell cycle is reviewed in the last three chapters (see below).

The cell cycle is not only the story of CDKs, but also of tumor suppressors and proto-oncogenes. Tumor suppressors and proto-oncogenes regulate cell cycle progression and are often regulated by CDKs. In addition, some cell cycle regulators (e.g., some G1 cyclins) are proto-oncogenes and others (e.g., some CDK inhibitors) are tumor suppressors/anti-oncogenes. Thus, the stories of the cell cycle and cancer are so interwoven that they cannot be dissociated. The reader will find in every chapter a reference to oncogenic events connected with the cell cycle. However, three chapters specifically cover this connection. The first one, by Mark Ewen, reviews the prototype of a tumor suppressor that regulates the cell cycle, namely the retinoblastoma gene product, pRb. The next chapter, by Martin Eilers et al., describes the regulation of cell proliferation by the proto-oncogene c-Myc, which, among all proto-oncogenes, is the one most intimately implicated with cell cycle control. Finally, the third chapter by Lili Yamasaki discusses the transcription factor E2F which appears very intriguingly to be a bona fide oncogene and a tumor suppressor at the same time. Interestingly, pRb, c-Myc, and E2F are all involved in the transcriptional control of cell cycle regulators.

As with all books, there are many people who deserve thanks. I am most grateful to the authors for their contributions and would like to thank my friend and colleague, Massimo Loda, for encouraging me to undertake this task.

New York, November 1997 Michele Pagano

Contents

Regulation of G1 Phase
R. J. Sheaff and J. M. Roberts

Regulation of S Phase
A. Dutta

Regulation of CDKs by Phosphorylation

M. J. Solomon and P. Kaldis

Regulation of the Cell Cycle by CDK Inhibitors

T. J. Soos, M. Park, H. Kiyokawa, and A. Koff

Regulation of the Cell Cycle by the Ubiquitin Pathway
J. Slingerland and M. Pagano

Regulation of the Cell Cycle by the Rb Tumor Suppressor Family
M.E. Ewen

Control of Cell Proliferation by Myc Proteins
A. Bürgin, C. Bouchard and M. Eilers

Growth Regulation by the E2F and DP Transcription Factor Families
L. Yamasaki

Regulation of G1 Phase

R. J. Sheaff and J. M. Roberts[1]

1
Introduction

The cell cycle describes the ordered process by which a cell executes the four essential tasks required to produce a duplicate of itself: growth (increase in mass), chromosomal duplication, chromosomal segregation, and cellular division (Mitchison 1971). The components comprising the cell cycle engine should be distinguished from the downstream events they control. The engine is responsible for periodic cycling and is composed of biochemical circuitry controlling transitions between stages of the cell cycle. This machinery acts on itself and on components directly responsible for DNA replication or cell division, thereby permitting a greater degree of regulation than would otherwise be possible.

An underlying tenet of cell cycle theory is that the processes required for duplication must be ordered; i.e., the production of genetically identical and viable daughter cells requires that chromosomal replication and cellular division be coordinated with each other (Murray and Hunt 1993). Thus cells replicate their DNA before dividing, and then divide before re-replicating their DNA (Prescott 1976). If the sole purpose of cells were simply to proliferate, then the role of the cell cycle machinery would likely be reduced to that of a simple internal oscillator coordinating DNA replication with cell division.

However, proliferation is not the only option open to cells, and in fact is oftentimes highly undesirable. In complex multicellular organisms, cells must also differentiate, senesce, undergo apoptosis (die), and in general subordinate their replicative capacity for the greater good of the organism (Alberts et al. 1989). In order to decide between disparate fates, the typical somatic cell must integrate a complex array of both intra and extracellular signals to ascertain whether proliferation is desirable and feasible. The wrong decision can have catastrophic consequences for genomic integrity and ultimately the viability of the cell and organism. Thus, while the basic milestones of cell duplication are easily identified, understanding how the cell decides to implement the mitotic program is exceedingly complex.

[1] Division of Basic Sciences, Fred Hutchinson Cancer Research Center, 1100 Fairview Ave. N. Mail-stop A3-023, Seattle, Washington 98109-1024, USA, E-mail: jroberts@fred.fhcrc.org

Even in a simple unicellular organism like yeast, the cell must still coordinate its proliferative capabilities with those of other cells and in response to environmental cues (Hartwell 1991). Therefore, while the cell cycle machinery maintains the temporal order of the four main tasks required for duplication, it also serves as a biochemical barometer linking these processes to the outside world. The purpose of this chapter is two-fold – to describe why the mitotic program devotes a significant fraction of its time to a period known as gap phase 1 (G1), and to describe how the cell cycle machinery is organized to create the circuitry which both controls G1 progression and links the mitotic program to the extracellular environment.

The first goal harks back to early studies on cell physiology – obtaining information about how cells replicate by observing cell growth and proliferation in both normal and abnormal (tumorogenic) cell populations (Mitchison 1971). This early work laid the foundation for cell cycle theory and its subsequent fruition within the fields of genetics and molecular biology. The resulting explosive growth in identifying molecules involved in cell cycle progression has led to the central conclusion that the basic components of the cell cycle machinery have been conserved throughout evolution (Murray and Hunt 1993). Thus, it is possible to offer a detailed description of cell cycle progression through G1, and be reasonably assured that for most cells it conveys an accurate picture of this critical period in the mitotic cell cycle.

2
Why Have G1?

Gap phase one (G1) is defined operationally as that period after the cell has undergone cytokinesis (cellular division, or M phase) to the initiation of DNA replication (S phase), and normally represents the longest phase of the animal cell cycle (Murray and Hunt 1993). The typical animal cell spends fully half of its time in G1, so it may be surprising to learn that specialized situations exist in which gap phases are entirely absent. An examination of the advantages and disadvantages which accrue from such an organization is highly informative in revealing why most cells spend much of the mitotic cycle in G1. The short answer is straightforward – after division cells need time to increase their mass before beginning another mitotic cycle. Explaining how cells coordinate growth with proliferation and decide whether or not to continue cycling is a greater challenge.

2.1
Division and Growth-Limited Cycles

The early embryonic cell cycle represents the most austere manifestation of the mitotic program, streamlined to maximize the rate of synchronized cellular division. Gap phases are absent, and the cell cycle is pared down to alternating processes of chromosomal replication and cellular division (Alberts et al. 1989). The embryonic egg is essentially a large cell approximately 50 times the diameter of a typical somatic cell. Upon fertilization the egg rapidly proceeds through alternating periods of DNA replication and mitosis in a synchronized fashion to create a hollow ball of cells called the blastula (Fig. 1; Newport and Kirschner 1984). The first cell cycle lasts 75 min, while the next 11 only last 30 min each (Gerhart et al. 1984). Such rapid multiplication is possible because the egg contains maternal stores of all the nutrients and enzymes required for these cycles of replication and division. Thus, the initial large size of the embryonic cell and its maternal storehouse allow the cell to bypass the requirement for growth during each cell cycle. The net result is that the embryo is subdivided into many smaller cells, after which cell multiplication slows, cycles become more asynchronous, and gap phases make their appearance (Newport and Kirschner 1982). This type of proliferative program is called "division limited", because the rate of duplication is determined by the rates of DNA replication and division.

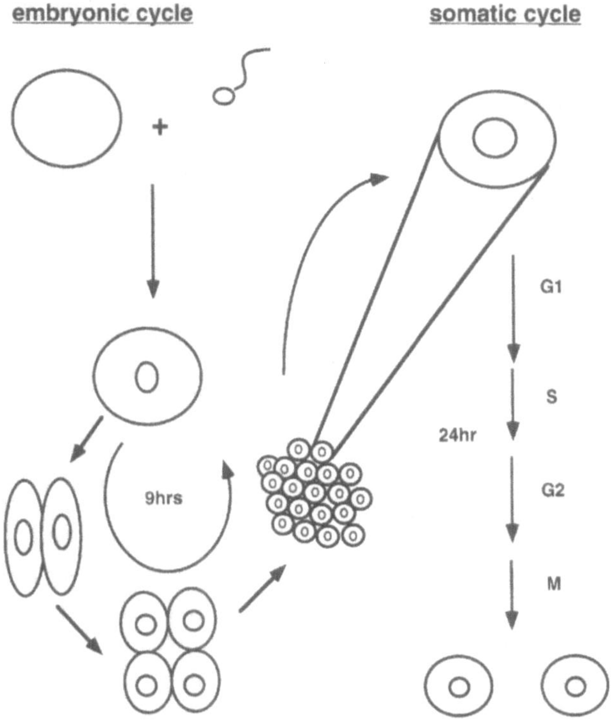

Fig. 1. Comparison of early embryogenesis and the later somatic cell cycle. Fertilization of the egg gives a zygote, which then undergoes 12 rapid, alternating cycles of DNA replication and division to produce the blastula (*left*). As this early synchrony breaks down, cells acquire gap phases and a concomitant lengthening of the cell cycle (*right*)

In contrast, the somatic cell cycle is said to be growth limited because the rate-limiting step in duplication is growth; hence somatic cells typically require 24 h to make a single copy of themselves (Fig.1). The tasks of DNA replication and mitosis are carried out in discrete periods of the cell cycle separated by gap phases, but growth is a continuous process. The smaller somatic cell spends a large fraction of its time in G1 because it does not have the luxury of a maternal pantry, and hence must coordinate cell growth with cell proliferation in order to maintain a similar cell size throughout generations (balanced growth). This is particularly important since it is clear that cells can replicate and segregate their DNA faster than they can double their mass. The complex relationship between cell growth and proliferation is examined in more detail in the following sections.

2.2 Growth Factors

Growing cells increase mass by ensuring protein synthesis exceeds degradation, while in nongrowing cells these processes are balanced. Stimulation of quiescent cells activates cell growth by stimulating production of rRNA and tRNA, ribosomes assemble into polysomes, translation factors are activated, and protein synthesis increases (Alberts et al. 1989). The rate of growth is directly proportional to the rate of protein accumulation, which is determined by the translation rate. A mere 50 % reduction in protein synthesis rates is sufficient to cause cell cycle withdrawal, suggesting tight coordination of these processes (Brooks 1977).

Growth factors are proteins which function as extracellular mitogenic signals, stimulating cell growth and influencing progression through the cell cycle (Baserga 1985). The effect of starving for growth factors, depriving cells of amino acids, or simply inhibiting protein synthesis is a reduction in protein synthesis rates leading to cell cycle arrest in mid-G1 at the restriction point (Pardee 1989). Thus there is a link between removal of growth factors, which results in decreased protein synthesis, and withdrawal from the cell cycle.

There are two main classes of growth factors; broad range, like platelet derived growth factor (PDGF), which effect many cell types, and specialized, which are specific for particular types of cells (Alberts et al. 1989). Growth factors interact with receptors in the plasma membrane, activating a signal transduction pathway which transmits information about the cellular environment to the cell cycle machinery inside the nucleus. This information (commonly transmitted by protein phosphorylation) is ultimately converted into an effect on the transcription of genes required to implement the desired response to the initial extracellular signal.

A comparison of mRNA from serum-starved and serum-treated cells revealed the induction of early and delayed genes (Murray and Hunt 1993). Early

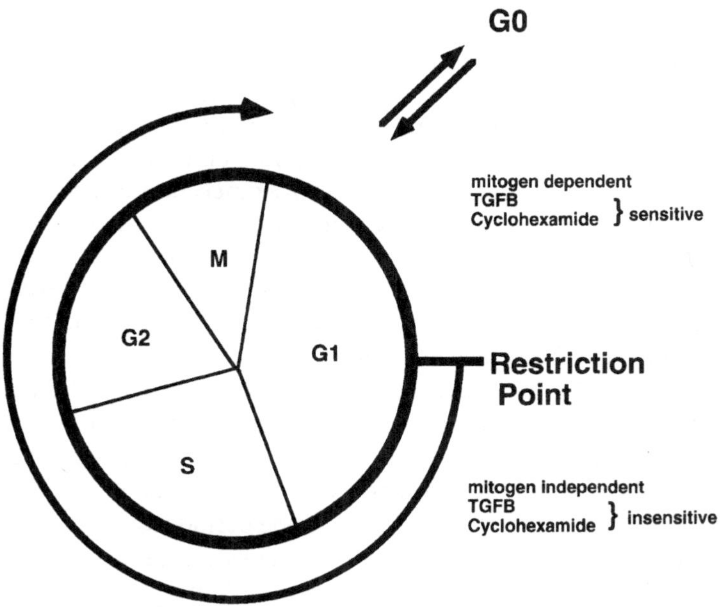

Fig. 2. The somatic cell cycle. Shown are the four phases of the cell cycle and the restriction point. Details in the text

response genes are undetectable in serum-starved cells, are rapidly induced within 30 min after refeeding, and then decline to lower levels which are maintained in the presence of growth factors. These mRNAs are not inhibited by protein synthesis inhibitors. In contrast, delayed response genes come up more slowly and are inhibited by protein synthesis inhibitors, suggesting they depend on the protein products of the early response genes (Murray and Hunt 1993). Indeed, some of the early response genes encode transcription factors specific for inducing the transcription of delayed response genes. Because the mRNA and protein of early response genes is unstable, removal of growth factors results in a rapid decrease in the delayed response genes and hence cell cycle arrest. Adding back growth factors requires that cells restore the delayed response gene transcription before continuing the cell cycle. Delayed response genes include members of the cell cycle machinery such as the G1 cyclins or components required for initiating the processes of DNA replication (Sherr 1994b).

2.3
Identifying the Restriction Point

The decision in G1 to continue the mitotic cycle or opt for an alternative fate is called the restriction point (R point), and is analogous to Start in yeast (Pardee 1974; Hartwell et al. 1973). The R point is the period in G1 at which the cell evaluates the conditions required for continued proliferation. If those conditions are not sufficient to complete the rest of the cell cycle, the cell cycle is halted and the cell withdraws into the G0 resting state (Fig. 2). The act of evaluating cell cycle progression is called a checkpoint, and encompasses any mechanism whereby the cell monitors the progress and fidelity of a cellular process to ensure that it is completed before continuing on to the next stage (Hartwell and Weinert 1989). Failure to finish a task in a timely fashion activates a biochemical checkpoint which halts cell cycle progression until the event is properly completed.

The restriction checkpoint was localized to a position several hours before the initiation of DNA replication, based on observations of the morphology of cells through the cycle before and after a period of starvation (Zetterberg and Larsson 1985). Cells starved before the R point exited to G0 and were delayed in the cell cycle in which starvation occurred. In contrast, removal of serum after the R point allowed cells to complete the current cell cycle without delay. This indicated that the R point occurred approximately 4 h after mitosis in this cell type. While most cells completed the period from mitosis to commitment within a narrow time range, there was considerable variability between cells in proceeding from the R point to S phase (Zetterberg and Larsson 1985). For this reason, it has been proposed that after committing to the mitotic cycle, cells still have the capacity to decide when to initiate DNA synthesis. This decision may be dependent on factors such as the accumulation of total cellular protein, or perhaps the levels of a particular protein involved in cell cycle progression (Zetterberg and Larsson 1991).

These results led to an operational definition of the R point in which cell cycle progression is dependent on mitogen stimulation from G1 to the R point; thereafter cell cycle progression is mitogen independent (Fig. 2). Subsequent experiments revealed that throughout most of G1 up to the R point, cell cycle progression can be stopped by mild exposure to the protein synthesis inhibitor cyclohexamide (Pardee 1989). This observation suggested that a critical protein(or proteins) was required for traversing the R point, and that this protein was most likely unstable. After the R point, however, cell cycle progression was resistant to the same levels of cyclohexamide, suggesting all the necessary components had been synthesized. Years later a third criterion was identified for cells in culture, in which prior to the R point the cytokine transforming growth factor beta(TGFB) leads to cell cycle arrest, and afterwards it does not (Laiho et al. 1990; Koff et al. 1993).

These early experiments, without invoking molecular mechanisms of cell cycle control, led to the hypothesis that prior to the R point the cell receives and integrates a variety of positive and negative information (Pardee 1989). At the R point in midG1 the cell evaluates the available information and decides whether or not to continue with the mitotic cycle. The cell can advance through the R point and commit to completing the cycle, or it can withdraw from the mitotic cycle into the quiescent G0 state.

It was speculated that the molecular basis for the R point switch involved the accumulation of a critical regulator above a pre-determined level, that this restriction(R) protein must be synthesized in G1, be unstable, and a target of mutation in cancer cells (Pardee 1989). Although this putative R protein has not yet been shown to exist, the R point hypothesis has proven remarkably useful as a framework for ordering the complex events involved in G1 progression.

2.4 Linking Cell Growth with Cell Cycle Progression

Since it is clear that during G1 cells increase significantly in size, it is reasonable to ask whether the cell maintains a checkpoint in G1 which ensures that the cell reaches a minimal size before dividing. The answer appears to be "yes" in yeast, where well-fed cells divide asymmetrically about every 90 min to produce a large mother cell and smaller daughter cell. Mothers are larger and bud faster than daughter cells, which must first reach a minimum size before budding. This is accomplished by postponing DNA replication and lengthening the G1 phase of the cell cycle to allow a longer period for growth (Hartwell and Unger 1977).

In somatic cells the answer is less straightforward. Early work indicated a requirement for cellular enlargement during G1 in order to proceed with the cell cycle (Killander and Zetterberg 1965a,b). Thus, while it seems likely that cell size and control of cell division are interconnected in somatic cells, elucidating the molecular mechanisms linking the two has proven frustrating. Furthermore, evidence suggests that the growth cycle can be separated from the proliferative cycle in some cases (Das et al. 1983; Baserga 1984). Cell physiology experiments revealed that protein accumulation, RNA synthesis, and growth factors are all normally required for DNA replication (Auer et al. 1970; Mercer et al. 1984). However, DNA replication could be initiated under certain conditions which failed to elicit the growth response (unbalanced growth; Zetterberg et al. 1982, 1984; Ronning and Pettersen 1984). Furthermore, removing growth factors from cells rapidly decreased the rate of protein synthesis regardless of their location in the cell cycle (Zetterberg and Larsson 1991). This reduction in protein synthesis in turn correlated with a small but significant reduction in cell size due to inhibition of cell growth. Nevertheless, cells past the R point continued through the cycle with normal kinetics even though their growth rate (and hence size) was compromised (Larsson et al. 1986).

These results suggested that the effect of growth factors on cell cycle progression could be separated from their effects on proliferation. Insulin was quite effective in restoring protein synthesis rates to their pre-starved level, but was unable to prevent mitotic delay. Thus, an overall increase in protein synthesis is not sufficient to prevent cell cycle exit in the absence of other growth factors (Zetterberg and Larsson 1991). In contrast, the growth factor PDGF was sufficient to prevent cells transiently depleted of growth factors from delaying mitosis, even though it did not completely restore protein synthesis rates (Larsson et al. 1985). Rather, the ability of PDGF to prevent cell cycle exit was consistent with the prediction of restriction point theory that cell cycle-specific proteins must accumulate to traverse the R point (Rossow et al. 1979; Pardee et al. 1981). In agreement with this model, PDGF was able to prevent mitotic delay in cyclohexamide-treated cells, even though it failed to restore the overall rate of protein synthesis.

2.5
Genes and Cancer Physiology

Normal, proliferating cells stop multiplying in a cell cycle specific manner, arresting and/or exiting the cell cycle in G1. In contrast, transformed cells are often nonresponsive to suboptimal conditions and continue dividing rather than exiting the cell cycle. The mitogen independence associated with cancer cells often arises through the mutation or upregulation of proteins involved in the acquisition and processing of mitogenic signals (Cantley et al. 1991; Bishop 1991). These oncogenes mimic the growth-factor exposed state and hence lead to unregulated proliferation (Lasko et al. 1991). Cancer cells also exhibit a loss of response to negative growth signals such as TGFB, cell-to-cell contact, and loss of adhesion (Koff et al. 1993; Massague 1996; Zhu et al. 1996).

Genetic mutations associated with cancer cells point the way towards understanding the molecular mechanisms responsible for development of the tumorogenic phenotype. The realization that these genes modulate cell cycle progression emphasized the necessity of understanding their relationship to the cell cycle clock, which is the ultimate recipient of all signals influencing cell proliferation (Hatakeyama et al. 1994). Thus it is not hyperbole to suggest that cancer can be described as a disease of the cell cycle.

3
Cell Cycle Machinery

The ascension of molecular biology has provided the link between physiologic and genetic descriptions of proliferation with their underlying mechanisms. Many of the cell cycle components have been identified, and efforts are increasingly focused on understanding how their interrelationships control cell cycle progression.

3.1
Cyclins and CDKs – More of a Good Thing

The catalyst of the cell cycle engine is a protein kinase called cyclin dependent kinase (CDK). This enzyme, through the simple act of transferring a phosphate from ATP to protein substrates, is ultimately responsible for catalyzing the transitions demarcating the major divisions of the cell cycle (Morgan 1996). CDKs, however, are only the means to an end; by themselves they are incapable of deciding whether cell cycle progression is the desired fate for the cell. For this reason the nuts and bolts of the cell cycle machinery are concerned with regulating CDK activity so that it is responsive to the extracellular environment. While in yeast a single CDK is responsible for catalyzing major transitions through the cell cycle, in mammalian cells each phase of the cell cycle is characterized by a unique pattern of CDK activity (Myerson et al. 1992; Nasmyth 1993). Thus far, eight CDKs have been identified in mammalian cells, each with a specific profile of activity. For our purposes, CDKs 2, 3, 4 and 6 are required for progression through the G1 phase (Sherr 1994a; Van den Heuval and Harlow 1994). CDK7 plays a unique role in that it appears to stand outside the cell cycle machinery and regulate activation of CDKs (Solomon 1994).

There are two main methods of directly modulating CDK kinase activity – through binding of regulatory proteins or by enzymatic modification. As the name implies, the kinase monomer is inactive for two reasons: (1) the bound ATP cannot be oriented properly for nucleophilic attack by the substrate; and (2) a portion of CDK called the T-loop is blocking the substrate binding site (Morgan 1995). Association with a cyclin subunit is the first step in CDK activation, and the diversity of CDKs is matched by a corresponding variety in the types of cyclins. While the cyclin family of proteins display no catalytic activity themselves, they activate CDKs upon binding by re-orienting the amino acid residues coordinating ATP to facilitate transfer to a substrate (Jeffrey et al. 1995). Cyclin binding also contributes to T loop movement out of the substrate binding site. While cyclin binding has substantial influences on and is required for CDK activation, full activation also requires phosphorylation on CDK threonine 160 (Morgan 1995). This phosphorylation is itself catalyzed by a cyclin-CDK complex termed CAK (CDK activating complex). Interestingly, CAK activity does not appear to be cell cycle regulated or limiting for cell cycle progression (Nigg 1996).

3.2 G1 Cyclins D and E

Most cyclins are expressed in a cell cycle specific manner, and their appearance is often rate limiting for cell cycle progression (Sherr 1993). Indeed, the control of cyclin expression is a major mechanism employed by the cell to establish and maintain the periodicity of CDK activity. The periodic nature of CDK activity may reflect the requirement for phosphorylating different substrates specific to each stage of the cell cycle. Implicit in this hypothesis, yet still largely unsupported by experimental data, is that cyclins impart substrate specificity to CDKs. The best example in which this appears to be the case is the phosphorylation of the retinoblastoma protein (Rb) by cyclin D-CDK4 complexes (see below). Alternatively, substrate availability could be an important determinant in specific cell cycle transitions. Cyclins might be involved in other processes, such as interactions with other regulatory molecules or directing CDKs to the proper cellular locale (Nasmyth 1996).

The primary mammalian G1 cyclins are cyclins of the D type (1, 2, and 3), which associate with CDKs 4 and 6; and cyclin E, which associates with CDK2 and perhaps CDK3 (Sherr 1993). Cyclin A is expressed at the G1/S boundary and is thought to be involved in controlling DNA replication. Cyclin C is also classified as a G1 cyclin, although its precise role is unclear. It associates with CDK8 and may regulate transcriptional activity, and hence may be involved in coordinating growth control with cell cycle progression (Tassan et al. 1995). Cyclins D, E, and C were first identified in a screen for mammalian genes that could compliment *cln* defects in yeast (Lew et al. 1991; Koff et al. 1991; Xiong et al. 1991). At the same time, cyclin D was also identified as a mitogen responsive gene and as a potential oncogene upregulated in some cancers (Matsushime et al 1991; Motokura et al 1991). The three D type cyclins are expressed in specific cell types (Sherr et al 1994).

As quiescent cells are stimulated to reenter the cell cycle by exposure to mitogens, cyclin D expression is transcriptionally upregulated in early G1, and its levels are maintained as long as mitogens are present (Fig. 3; Sherr et al. 1994). Cyclin D1 transcription is regulated by the *ras* growth factor responsive signal transduction pathways and possibly *c-myc* (Winston and Pledger 1993; Roussel et al. 1995). While it is unclear how *c-myc* effects cyclin D transcription, *ras* appears to implement its effects through the MAP kinase cascade (Albanese et al. 1995). Cyclin D levels decline after mitogen removal, providing further indication of a linear connection between cyclin D expression and extracellular mitogenic signals (Matsushime et al. 1991).

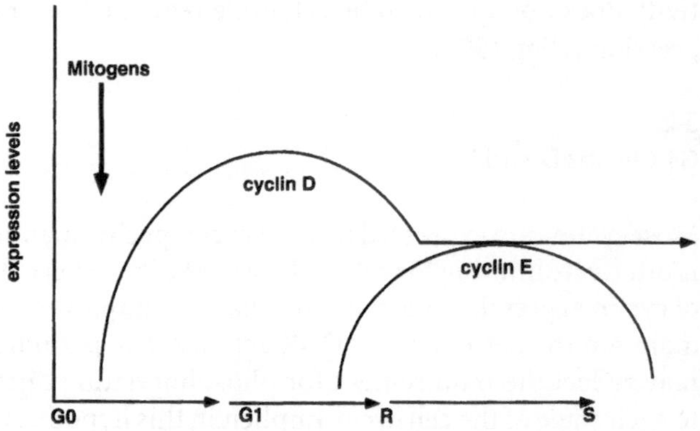

Fig. 3. Cyclin D and E expression from quiescence to the G1-S phase border. Depicted are the temporal order and relative levels of the G1 cyclins. Cyclin D peaks during mid G1 in response to mitogens and then declines to a steady state level

Cyclin D-CDK4 activity is also regulated posttranscriptionally. The assembly and activation of CDK-cyclin D complexes are regulated by mitogenic signals (Sherr et al. 1994). Importantly, the half-life of cyclin D is only ~15 min, which keeps the mitotic cycle under tight growth factor control by manipulation of cyclin D levels (Matsushime et al. 1992). Cyclin D1 turnover is mediated by phosphorylation of cyclin D1 on threonine-286 by an as yet unidentified kinase, leading to its ubiquitination and proteolytic degradation (Diehl et al. 1997). Once its role in passage through early G1 is complete, cyclin levels decline somewhat and do not fluctuate dramatically with the cell cycle (Sherr et al. 1994). This may reflect a specialized role in reentering the cell cycle from a quiescent state, but should not be taken to imply that cyclin D does not play an important role in cycling cells.

That cyclin D is a bona fide G1 cyclin is convincingly demonstrated by a variety of experimental data. Injection of monoclonal antibodies directed against cyclin D arrests cells in early G1 phase of the cell cycle (Baldin et al. 1993). Conversely, injection after the cell has progressed into late G1 has no effect, indicating that the function of cyclin D is completed in early G1. Overexpression of cyclin D shortens G1 and reduces the serum dependence of cell proliferation (Quelle et al. 1993). Thus, cyclin D is required for progression through early G1 and is at least partially rate limiting.

Cyclin D, while necessary, is not sufficient for G1 progression. Cyclin E is also required, and its expression pattern of RNA, protein, and kinase activity suggests that it functions after cyclin D (Dulic et al. 1992; Koff et al. 1992). The protein peaks at the G1/S phase border, after which it declines. As with cyclin D, antibodies injected against cyclin E arrest cells in G1 phase, while its overexpression shortens G1 (Ohtsubo et al. 1995). Cells overexpressing cyclin E are also smaller, perhaps reflecting an alteration in growth control during G1 (Ohtsubo and Roberts 1993). Unlike cyclin D, however, cyclin E periodicity is maintained in the cycling cell. This suggests that cyclin D may play a crucial role in establishing the mitotic program, which subsequently becomes self-activating through the activity of cyclin E. In support of this model, cyclin E is one of a large class of genes required for cell proliferation that is under control of the E2F transcriptional regulatory machinery, the implications of which are discussed below (Ohtani et al. 1995).

Cyclin E appears to be degraded by the ubiquitin-proteasome system after completing its job in late the G1 phase, although the mechanism by which this is achieved is somewhat different from that of cyclin D (Clurman et al. 1996; Won and Reed 1996). Cyclin E degradation is regulated by binding to CDK2 and the catalytic activity of the resulting complex. Cyclin E is phosphorylated when associated with CDK2, which may result in complex disassembly (by an unknown mechanism) and subsequent degradation of the free cyclin. One possibility is that linking cyclin E-CDK2 activity to destruction of cyclin E may help ensure periodicity and restrain cyclin E-CDK2 kinase activity to a specific point in G1.

4 Retinoblastoma Protein

Remarkably few cyclin CDK substrates have been unequivocally identified to date. One of the most important for G1 progression, however, is the product of the retinoblastoma gene (Rb). Rb is a 110-kDa protein which is the founding member of the so-called pocket protein family, whose other members include p130 and p107 (Weinberg 1995). As discussed below, Rb negatively regulates cell cycle progression through G1, and is the best characterized example of a tumor suppresser.

Befitting its central role in controlling cell proliferation, mutation or loss of Rb is one of the most common abnormalities associated with human cancers (Friend et al. 1987; Horowitz et al. 1990). Inappropriate inactivation of Rb by direct mutation or mutation of its regulators can lead to uncontrolled proliferation and ultimately cancer. Interestingly, the cell has attempted to prevent this calamity by linking loss of Rb to apoptosis, which requires the activity of the p53 tumor suppresser gene (Levine 1997). The apoptotic pathway therefore acts as a checkpoint ensuring that abnormal cell growth leads to cell death rather than tumor formation. Consistent with this model, many of the most aggressive cancers display loss of both Rb and p53 (Williams et al. 1994).

4.1 Holding the Reins on G1 Progression

Rb occupies a unique position in controlling progression through the R point during G1, and hence is a recipient of many diverse regulatory signals. Rb has four distinct protein binding domains: (1) the AB pocket binds LXCXE motif-containing proteins such as cyclin D and viral oncogenes; (2) the large A/B pocket binds the transcription factor E2F; (3) the C pocket binds Abl tyrosine kinase, and (4) the amino region interacts with a novel cell cycle-regulated kinase (Bartek et al. 1996).

Its main function, however, appears to be connecting the cell cycle clock with the transcriptional machinery necessary for initiating and completing events downstream of the R point. Active Rb holds the cell cycle in check by turning off this transcriptional program, and therefore must be inactivated for cells to progress through the R point. The connection between Rb and the cell cycle machinery was forged on the deceptively simple observation that the phosphorylation state of Rb varies depending on the point in the cell cycle. As cells progress through G1, Rb converts from the hypophosphorylated (underphosphorylated) to the hyperphosphorylated (multiple phosphorylations) form (Buchkovich et al. 1989; Chen et al. 1989). Rb stays hyperphosphorylated from late G1 to the exit from mitosis, whereupon it is dephosphorylated. The timing of phosphorylation correlates with the traversal of the R point, suggesting a simple model in which phosphorylation inactivates Rb and permits cell cycle progression.

This idea is substantiated by a variety of evidence: (1) tumor virus oncoproteins compromise Rb function by associating with the hypophosphorylated form (Ludlow et al. 1989; Imai et al. 1991); (2) mitogenic signals resulting in cell cycle progression do so in part by causing phosphorylation and inactivation of Rb (Sherr 1993); (3) forms of Rb that cannot be phosphorylated are dominantly growth suppressing (Hamel et al. 1992; Hinds et al. 1992); and (4) hypophosphorylated Rb interacts with regulatory molecules controlling other cell cycle events, while the hyperphosphorylated form does not (Taya 1997).

In the hypophosphorylated state, Rb prevents cell cycle progression by sequestering the transcription factor E2F, which inhibits its ability to drive gene expression (Nevins 1992; Weinberg 1995). In some cases Rb-E2F complexes bind to the promoter region of a target gene, but the presence of Rb prevents transcriptional activation (Krek et al. 1993). Phosphorylation of Rb leads to E2F dissociation and hence activation of transcription (Weinberg 1995). Cells arrested due to an inability to phosphorylate Rb can be rescued by E2F overexpression (DeGregori et al. 1995; Lukas et al. 1996). Ectopic expression of E2F also allows quiescent cells to transverse G1 and initiate S phase, and E2F can overcome an Rb-imposed cell cycle arrest (Lukas et al. 1996). These results imply that release of E2F from Rb is the rate determining factor for cell cycle progression through G1. Although the E2F interaction with Rb seems to predominate, at least in determining cell cycle progression, many other Rb binding proteins have been identified (Taya 1997).

E2Fs associate with Dp proteins to form DRTF1/E2F complexes which drive transcription (Bandara et al. 1993; Helin et al. 1993). Expression of a dominant negative Dp1 leads to cell cycle arrest, and it may also be regulated by phosphorylation in a cell cycle-dependent manner (Wu et al. 1996). The E2F family is composed of five related but distinct family members that are targeted to the consensus binding sequence TTCGCGC in the promoter region of genes (Lam and La Thangue 1994). E2F 1-3, but not 4 or 5, can induce S phase entry in quiescent fibroblasts (Lukas et al. 1996). Another Rb family member, p107, appears to interact with E2F 4 and 5 (Beijerbergen et al. 1994). It is unclear whether these multiple potential interactions regulate different or identical genes, or whether they function at discrete intervals of the cell cycle (Lam and Thangue 1994).

The E2F binding element is present in the promoter of genes important for growth control and cell cycle progression, such as *c-myc*, *B-myb*, *cdc2*, thymidine kinase, and dihydrofolate reductase (Lam and La Thangue 1994). Activation of the many E2F responsive genes is complex and proceeds in a temporal fashion that is not well understood. Furthermore, in some cases the E2F binding sites actually function as cis-acting repressing sequences, since mutations preventing E2F binding result in constitutive transcription of some genes (Singh et al. 1995).

4.2 Rb Null Mice

The generation of mice lacking Rb has been invaluable for elucidating the role of this tumor suppressor in regulating the cell cycle. Rb null mice are normal until day 12-13 of gestation, whereupon they die of defects in erthropoiesis and neuronal development (Clarke et al. 1992; Jacks et al. 1992; Lee et al. 1992). Thus much of the early developmental program has been implemented in the absence of Rb, which may reflect basic differences between early embryonic and somatic cell cycles (as described earlier). Mice heterozygotic for Rb survive longer, but eventually succumb to cancer after they lose the other Rb allele. In contrast, p130 or p107 mice are viable, fertile, and show no obvious abnormalities (Lee et al. 1996). However, the double p107/p130 minus mice are not fully compensated for by Rb as there are complications during limb development (Cobrinik et al. 1996).

Fibroblasts from Rb knockout mice have lost R point criteria (Herrera et al. 1996). These cells do not arrest in G1 in response to the protein synthesis inhibitor cyclohexamide, and are prone to failure of the DNA damage checkpoint operating in G1. Rb minus cells also display a tendency to escape the G1 arrest associated with limiting growth factors or nutrients, although eventually most of these cells arrest in G0 in response to low serum (Lukas et al. 1995a; Herrera et al. 1996). This suggests the presence of other Rb-independent factors contributing to R point control, with the cyclin-dependent kinase inhibitors as potential candidates (see below). Consistent with the role of Rb in restraining E2F activity, Rb minus cells show increased levels of transcription of E2F responsive genes, leading to a shorter G1 phase and smaller cells which grow faster than wild type (Almasan et al. 1995). Importantly, cyclin E expression occurs 5-6 h prematurely in Rb minus fibroblasts, suggesting Rb suppresses cyclin E expression (Almasan et al. 1995; Herrera et al. 1996). This is not surprising given the presence of E2F promoter elements in the cyclin E gene, and likely has important consequences for the timing of G1 progression (see below).

4.3 Linking Cell Growth with Cell Cycle Progression

As discussed in the first sections, the link between growth control and cell cycle progression has remained enigmatic. Recent results indicate that Rb may play a crucial role in this process by suppressing global protein synthesis in response to nutrient deprivation. In addition to regulating the specific transcription of E2F responsive genes by RNA polymerase II, Rb also regulates general transcription by RNA polymerases I and III (Cavanaugh et al. 1995; White et al. 1996). Polymerase I synthesizes large ribosomal RNA (rRNA), while polymerase III synthesizes small stable RNAs like 5S rRNA and tRNA.

The ability of Rb to simultaneously suppress cell growth and proliferation may explain how these two processes are balanced. An intriguing hypothesis is that Rb places polymerases I and III under cell cycle control, although direct evidence is lacking. However, preliminary results indicate that TFIIIB is repressed in the early part of G1 and increases as cells enter S phase (White et al. 1996). Consistent with this idea, Zetterberg and Larsson(1991) found that cells restrained before the R point fail to grow; after passing the R point (and hence phosphorylating Rb), however, they begin to increase in size. It will be interesting to re-analyze early experiments showing growth can be uncoupled from proliferation in light of this new information.

5
Cyclin-CDK Control of G1 Progression

5.1
Catalyzing G1 Progression

The diverse processes negatively regulated by Rb are linked to the cell cycle machinery by the cyclin dependent kinases, which are responsible for phosphorylating and inactivating Rb (Weinberg 1995). There are more than 12 potential CDK phosphorylation sites on Rb (S/T-P motifs), which likely reflects the complexity of Rb regulation in response to a wide array of mitogenic signals. Indeed, experiments with Rb mutants lacking certain phosphorylation sites show that different residues are required for inhibiting distinct Rb activities (Knudsen and Wang 1996). Thus, different cyclin-CDK complexes may preferentially phosphorylate specific sites to elicit the required response. Although definitive evidence is lacking, it is tempting to speculate that hyperphosphorylated Rb may require the sequential phosphorylation of specific sites by multiple kinases, thereby providing an attractive way to temporally order cell cycle progression.

While several cyclin-CDKs will utilize Rb as a substrate in vitro, such experiments only suggest possible relationships which may or may not exist in the cell. The challenge, therefore, has been to assimilate the wealth of data on the relation between G1 cyclin-CDKs and Rb into a coherent picture of progression through G1. One approach has been to determine the phosphorylation state of Rb in cells and correlate this with the expression and activity of G1 cyclin-CDK complexes. Starting with a synchronized population of cells, these experiments indicate that phosphorylation of Rb begins in G1 well before transition into the S phase (Buchkovich et al. 1989; Chen et al. 1989).

Cyclin D is synthesized and activated during this period in G1, and is most clearly implicated in phosphorylation of Rb on the basis of diverse data (Sherr et al. 1994). Anti-mitogenic signals which interfere with cyclin D-associated kinase activity prevent Rb phosphorylation in vivo (Lukas et al. 1995b, Medema et al. 1995). Similarly, inhibition of ras in cycling cells causes a G1 arrest by downregulating cyclin D levels and causing accumulation of hypophosphorylated Rb (Mulcahy et al. 1985; Dobrowolski et al. 1994; Peeper et al. 1997). When these experiments were repeated in Rb minus cells, however, inactivation of ras had no effect on cell cycle progression. Thus, the growth-factor activated, ras-dependent signal transduction pathway mediates its proliferative effects on cell cycle progression by inactivating Rb. Interestingly, ras is required for exit from quiescence whether or not Rb is present, indicating that components other than Rb are required and controlled by the ras signaling pathway during the G0 to G1 transition (Peeper et al. 1997).

Direct inactivation of cyclin D by microinjection of cyclin D antibodies or overexpression of p16 also results in the accumulation of hypophosphorylated Rb and cell cycle arrest in G1 (Lukas et al. 1995a; Medema et al. 1995). To determine if inactivation of Rb was the only function of cyclin D-CDK, these experiments were repeated in Rb minus cells. Surprisingly, cells proceeded uninterrupted through the cell cycle in the absence of cyclin D-CDK kinase activity (Lukas et al. 1995). Conversely, overexpression of cyclin D accelerates progression through G1 only if the cells are Rb positive (Quelle et al. 1993). These results suggest that phosphorylation of Rb is the only function of cyclin D required for cell cycle progression (Lukas et al. 1995).

Cyclin E is also a potential candidate which has been implicated in Rb phosphorylation, although this point is still controversial. Purified cyclin E-CDK2 complexes phosphorylate Rb in vitro, and this effects binding of Rb to E2F (Dynlacht et al. 1994). Furthermore, the levels of cyclin E mRNA and protein rise during midG1 when Rb is phosphorylated. Rb is also phosphorylated in a normal manner in DNA tumor virus infected cells, even though the viral oncoproteins block cyclin D binding to Rb (DeCaprio 1989). Like cyclin D, cyclin E is partially rate limiting for progression through G1, and in some instances ectopic expression of cyclin E correlates with increased Rb phosphorylation in cells (Ohtsubo and Roberts 1993; Ohtsubo et al. 1995). However, similar experiments using a tetracycline repressible system indicated that induction of cyclin D, but not cyclin E, triggered the rapid phosphorylation of Rb (Resnitzky and Reed 1995).

A cyclin E mutant has been identified which when overexpressed only accelerates G1 in Rb minus cells (Kelly and Roberts, unpublished). In vitro analysis indicates that this mutant is deficient in Rb phosphorylation, but phosphorylates histone H1(a ubiquitous CDK substrate) as well as wild type cyclin E-CDK2. These results suggest that cyclin E can influence G1 progression in at least two distinct ways – by phosphorylating Rb and some as yet unidentified substrate. Consistent with this view, Resnitzky and Reed(1995) have suggested that the effects of cyclin D and cyclin E overexpression on G1 acceleration are additive. Thus, despite intense efforts, it is still unclear whether cyclin E normally phosphorylates Rb in the cell. Experiments using overexpressed cyclins are always subject to the caveat of potential overlap in substrate selectivity between cyclin-CDK complexes.

The resolution of this controversy has important implications for understanding progression through G1, because there are E2F responsive elements in the cyclin E and A genes (Ohtani et al. 1995; Schulze et al. 1995). Thus, the induction of cyclin E by E2F may establish a positive feedback loop regulating cyclin E expression (Fig. 4). In this model cyclin E phosphorylates Rb and releases E2F, which in turn up-regulates E expression. The subsequent activation of cyclin E-CDK2 makes the loop autonomous, perhaps explaining how Rb phosphorylation becomes independent of cyclin D and mitogenic stimulation. The activation of cyclin D-CDK complexes in early G1 could serve to activate this loop by releasing a small amount of E2F, thereby establishing the temporal order of CDK activity. The positive feedback model has been proposed as a mechanism for progression through the R point and commitment to cell cycle progression (Hatakeyama et al. 1994).

In contrast to its requirement during G1, cyclin E may also play a role in preparation for subsequent mitotic events in G2/M. At least in the case of *Drosophila* embryogenesis, cyclin E appears to assist in preparations for downstream mitotic events in the G2/M phase (Knoblich et al. 1994). Driving cells through the S phase by ectopic cyclin E expression also results in the accumulation of mitotic factors normally not present in G1(A, B, string). This causes the cells to initiate M phase inappropriately, and implies that expression of mitotic proteins may depend in some instances on cyclin E expression. Cyclin E stimulates accumulation of mitotic factors at the protein level rather than by upregulating their transcription, indicating that translation or stability of mitotic factors is affected in G1. One possibility, with precedence in yeast, is that

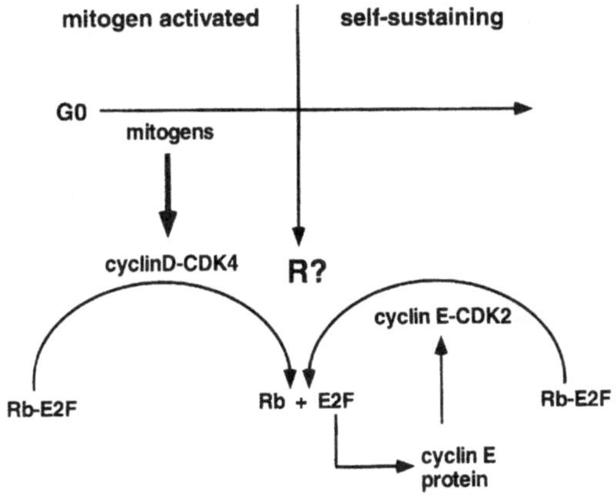

Fig. 4. Model of positive feedback loop regulating cyclin E levels. On the *left* is mitogen activation of cyclin D-CDK4 resulting in Rb phosphorylation and release of E2F. E2F transcriptionally activates cyclin E-CDK2, which in turn phosphorylates more Rb to generate the self-sustaining loop (*right*)

degradation of mitotic components is normally active in G1 and is inactivated by cyclin E at the G1/S phase boundary in preparation for mitosis (Nasmyth 1996).

5.2 Inhibiting G1 Progression

Physiologic signals preventing growth and proliferation often do so by transducing information that prevents Rb phosphorylation. Exposure of cells to TGFB, cAMP, and contact inhibition all culminate in the expression of a CDK inhibitor (CKI) which inhibits CDK activity and hence Rb phosphorylation (Peter and Herskowitz 1994). Given the large array of cyclins and CDKs employed by the cell to catalyze transitions through the cycle, it is perhaps not surprising that an equally diverse assortment of inhibitors are employed to counteract them. There are two main classes of CKIs employed for this purpose which are distinguished by their targets: (1) the CDK4 inhibitor proteins (INK4; p15, p16, p18, and p19), which specifically target CDK4/6 kinases; and 2) the Cip/Kip family of inhibitors (p21, p27, and p57), which are thought to be more general CDK inhibitors (Sherr and Roberts 1995). Although these molecules target different CDKs and exhibit different mechanisms of inhibition, in almost all cases their overexpression results in a G1 cell cycle arrest.

The INK4 family of CKIs contains four ankyrin repeats, i.e, a structural element involved in binding interactions with other proteins (Serrano et al. 1993). INK4 proteins can inhibit kinase activation by either binding monomeric CDK4/6 and preventing association with the D type cyclins, or by directly binding and inhibiting cyclin D-CDK complexes. In contrast to the Cip/Kip family, the INK4 CKIs are strongly implicated as candidate tumor suppresser genes (Kamb 1994). This likely reflects their pivotal role in regulating the phosphorylation state of Rb by controlling cyclin D kinase activity. Consistent with this idea, p16 overexpression fails to arrest cells in the absence of Rb, just as the absence of Rb renders cyclin D superfluous for cell cycle progression (Lukas et al. 1995a; Medema et al. 1995).

The Cip/Kip family members bind cyclin-CDK complexes better than either component alone, and inhibit kinase activity in part by occupying the ATP binding site (Sherr and Roberts 1995). Intriguingly, these CKIs utilize cyclin binding motifs similar to those employed by the pocket proteins (Zhu et al. 1995). Furthermore, a p21 or p27-induced cell cycle arrest can be alleviated by co-expression of viral oncogenes (Mal et al. 1996). This raises the interesting possibility that the Cip/Kip family may share functional as well as structural similarity to pocket proteins. In fact, members of the Cip/Kip family also appear to be regulated by cyclin-CDK phosphorylation, as described below. Surprisingly, despite heroic efforts, Cip/Kip family members have yet to be unequivocally identified as tumor suppresser genes. This implies, albeit indirectly, that

Cip/Kip family members may play a minimal role in controlling the phosphorylation state of Rb.

p21 is thought to be involved in both the checkpoint response to DNA damage and the withdrawal from the cell cycle required for terminal differentiation (El-Deiry et al. 1993; Parker et al. 1995). Paradoxically, however, p21 levels also rise as cells exit quiescence and enter the cell cycle, in apparent contradiction to its role as a CDK inhibitor (Yo Li et al. 1994; Firpo et al. 1994). Thus p21 also appears to be important in cycling cells, perhaps as an assembly factor for cyclin-CDK complexes or in substrate selectivity. Consistent with these ideas, in some cases p21 appears to associate with active cyclin-CDK complexes, implying that p21 binding per se is not sufficient for CDK inhibition and may require the formation of higher order structures containing multiple p21 molecules (Zhang et al. 1994; Harper et al. 1995). Despite its intimate involvement in cell cycle control, mice lacking p21 are viable and show little physiologic effects, although cell lines established from these animals do show a defective response to DNA damage (Deng et al. 1995).

p27 is a mitogen responsive gene expressed in most cell types, while its closely related family member p57 is more restricted to terminally differentiated tissues (Harper and Elledge 1996). Levels of p27 are high in quiescent cells and those inhibited by cell-to-cell contact (Roberts et al. 1994). Agents which inhibit cell proliferation, such as cAMP in macrophages and rapamycin in T-lymphocytes, do so in part by preventing p27 downregulation (Nourse et al. 1994; Kato et al. 1994). The fall in p27 levels as cells reenter the cell cycle correlates with increased insensitivity to the inhibitory effects of TGFB, suggesting that upregulating p27 may be part of a generalized mechanism responsible for withdrawal from the cell cycle (Polyak et al. 1994a; Slingerland et al. 1994). Consistent with this idea, decreasing p27 levels using antisense led to decreased serum dependence of cells growing in culture; i.e., cells were more likely to remain in the mitotic cycle in the absence of serum (Coats et al. 1996). Thus, p27 is implicated in the inhibition of cyclin-CDK complexes that appears to be required for cell cycle exit into G0 (but see below).

The importance of p27 in cell cycle progression is underscored by the p27 null mice, which grow twice as fast as wild types and have increased numbers of cells in all lineages (Fero et al. 1996; Kiyokawa et al. 1996; Nakayama et al. 1996). These mice also develop tumors in the intermediary lobe of the pituitary, similar to the Rb minus mice, suggesting a functional connection between Rb and p27. Surprisingly, however, primary cells lacking p27 withdraw from the cell cycle with the same kinetics as wild type cells, indicating that p27 is not absolutely required for exit into G0 (Fero et al. 1996). Importantly, cyclin E-CDK2 complexes are still inhibited upon serum withdrawal, even in the complete absence of p27 (Coats and Roberts, unpublished). This result suggests that cell cycle exit requires components in addition to p27, any one of which is sufficient to cause cell cycle arrest. Furthermore, it points out an important difference between primary and immortalized cells: immortalized fibroblasts growing in culture possess multiple genetic defects, which may involve loss of negative regulatory controls.

Members of the INK4 and Cip/Kip family often work in concert in response to antimitogenic signals such as TGFB (Reynisdottir et al. 1995). Although the TGFB family of cytokines mediates different effects on proliferation in different cell types, it is a useful paradigm for transduction of anti-proliferative signals. TGFB forms a disulfide-linked dimer, which upon binding to its cell surface receptor causes receptor dimerization and subsequent activation of cytoplasmic kinases (Massague 1996). This initiates a signal transduction cascade culminating in the expression of genes required to implement antiproliferative effects in G1. Exposure to TGFB ultimately blocks the activation of cyclin D-CDK4 complexes by inducing expression of the CKI p15 as well as downregulating CDK4 expression (Hannon and Beach 1994; Ewen et al. 1995). One consequence of p15 expression is displacement of p21/p27 from cyclin D-CDK complexes to cyclin E-CDK2 (Reynisdottir et al. 1995). Thus, TGFB induces multiple alterations in the cell cycle machinery which ensures cell cycle progression halts in G1.

Diverse evidence has led to the hypothesis that CKIs establish an inhibitory threshold which must be surpassed in order for the cell cycle to proceed (Sherr and Roberts 1995). In this view, stimulating quiescent cells to reenter the cell cycle first requires that cyclin D-CDK complexes exceed a threshold maintained by CKIs before phosphorylating Rb. Cyclin D-CDK complexes therefore titrate CKIs and allow activation of cyclin E-CDK2 needed for downstream events. The inhibitory threshold therefore provides a way to control the timing of G1 and prevent inappropriate cell proliferation.

5.3
CKI Control of G1 Progression

The role of CKIs in the cycling cell is less clear, in part due to a lack of information about their regulation during the cell cycle. Initial results suggested that perhaps the levels of p27 do not vary during the cell cycle (Toyoshima and Hunter 1994). However, deregulation of p27 negative signaling is potentially one way that cells could adapt to culture. Later work in different cell lines suggested that p27 protein levels may be regulated by degradation as well as translational mechanisms in a cell cycle-dependent manner (Pagano et al. 1995; Hengst and Reed 1996). In some cases p27 levels appear to be highest in early G1 phase, decreasing as cells progress towards S phase and accumulate cyclin E-CDK2 activity (Hengst et al 1994). Although the signals responsible for controlling p27 levels are unclear, the downregulation of p27 during G1 suggests it may play an as yet unidentified role in the G1 to S phase transition in cycling cells.

While CKIs are generally thought of as negative regulators of CDK activity, the potential for CDK regulation of CKI activity also exists. All members of the Cip/Kip family contain CDK consensus phosphorylation sites which are conserved among species (Polyak et al. 1994b; Lee et al. 1995). Furthermore, these sites are phosphorylated by purified CDKs in vitro, albeit with profoundly different kinetics than the standard CDK substrate histone H1 (Zhang et al. 1994; Sheaff et al. 1997). A case in point is p27, which has been depicted simply as a tight-binding general inhibitor of cyclin-CDK complexes. Surprisingly, our observations suggest that the opposite relationship also exists – cyclin E-CDK2 regulates p27 (Sheaff et al. 1997). In vivo, co-expression of cyclin E-CDK2 kinase reverses the inhibitory effect that p27 has on cell cycle progression and initiates a pathway leading to elimination of p27 from the cell. p27 is a substrate of cyclin E-CDK2 in vitro, and downregulation of p27 in vivo is prevented both by mutation of its single major CDK2 phosphorylation site and by mutations that decrease the catalytic activity of cyclin E-CDK2 complexes. Together, these data suggest that cyclin E-CDK2 may regulate p27 directly via phosphorylation of T187.

A kinetic analysis shows how p27 can be both an inhibitor and a substrate of cyclin E-CDK2. p27 initially engages cyclin E-CDK2 in a transient, loosely bound state (Sheaff et al. 1997). At this point the enzyme-p27 complex can slowly convert to a catalytically inactive form which is likely to closely resemble the recently solved crystal structure of p27-cyclin A-CDK2 (Russo et al. 1996). If instead CDK2 binds ATP, the enzyme becomes committed to p27 phosphorylation and release. Thus, the equilibrium between the loosely and tightly bound states can be modulated by the ambient ATP concentration (and possibly other extrinsic factors), thereby determining whether p27 is mainly a CDK inhibitor or a CDK substrate. Since physiological ATP concentrations favor p27 phosphorylation over CDK inhibition in vitro, a similar relationship may exist in the cell.

In one view, p27 and cyclin E-CDK2 reciprocally inhibit each other, with cell cycle progression hanging in the balance (Fig. 5). In a quiescent cell CDK2 is inhibited by p27, while in a proliferating cell CDK2 is active and p27 is a CDK2 substrate. This model embodies the general idea that a switch between two physiological states can be governed by two mutually inhibitory molecules, the classic example of which is the control of bacteriophage lambda lysogeny by the interaction between repressor and Cro (Alberts et al. 1989). In the case of lambda the balance between repressor and Cro is modulated by extrinsic factors that link phage reproduction to the physiological state of the host cell. The same could apply to the interaction between cyclin E-CDK2 and p27. Growth factors or other mitogenic cues may tip the balance one way or the other and switch the cell from a nonproliferating state in which CDK2 is inactive to a proliferating state in which CDK2 is active. The position of this switch should be ultrasensitive to small changes in the amount of active cyclin E-CDK2. Any stimulus

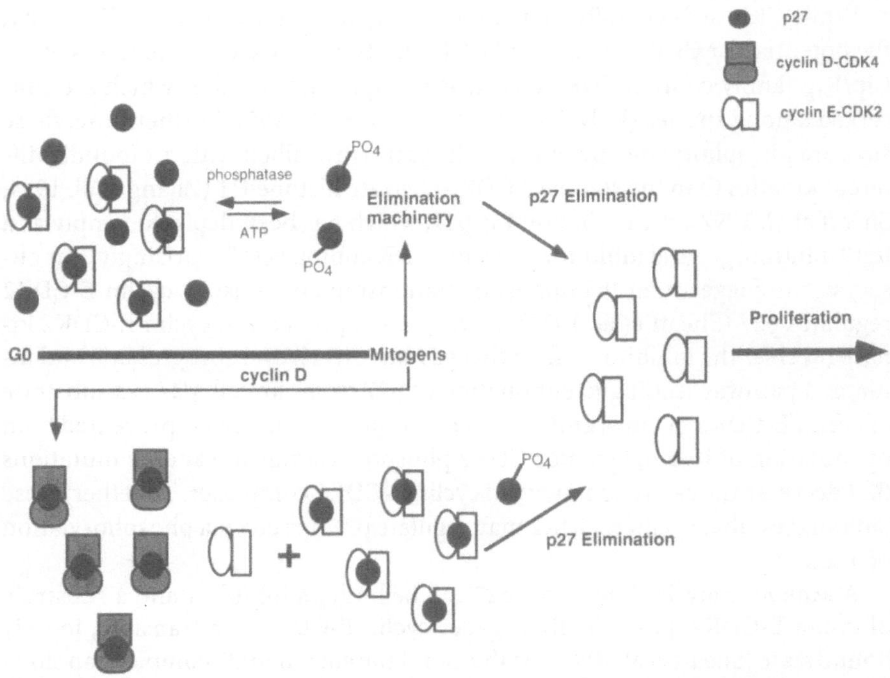

Fig. 5. Control of proliferation by p27 and its CDK-catalyzed elimination. In the *upper panel* cyclin E-CDK2 actively phosphorylates p27, but a pool of free p27 is maintained by the counterbalancing effect of phosphatases. Mitogens tip the balance in favor of cell cycle progression by activating elimination machinery so that phosphorylated p27 is eliminated from the cell. The *lower panel* depicts an alternative model in which cyclin E-CDK2 complexes are inhibited in G0 by p27. Mitogenic activation of cyclin D-CDK4 results in the redistribution of p27 to these complexes, and the subsequent activation of low levels of active cyclin E-CDK2. These in turn catalyze p27 elimination by phosphorylation, thereby activating more cylin E-CDK2 and driving G1 progression

which caused accumulation of active cyclin E-CDK2 should trigger a feedback loop in which cyclin E-CDK2 downregulates p27 and activates more cyclin E-CDK2.

One possibility is that the pathway which recognizes phosphorylated p27 and promotes its degradation may itself be growth factor-dependent (Fig. 5). Alternatively and as previously suggested, mitogen-dependent synthesis of cyclin D may tip the balance, perhaps by sequestering some p27 into cyclin D-CDK4 complexes (Fig. 5; Polyak et al. 1994b). This could result in the accumulation of a small amount of cyclin E-CDK2 in excess of a p27 inhibitory threshold, thereby shifting the equilibrium toward p27 elimination and promoting cell cycle progression.

A more radical viewpoint is that in vivo p27 may be solely a substrate of cyclin E-CDK2 and an inhibitor of other cyclin-CDK complexes (Fig. 6). In this model, mitogen-stimulated progression from the G0 to S phase would activate cyclin E-CDK2, which would then downregulate p27 and thereby allow activation of downstream CDKs. There are parallels between this model for the regulation of p27 in mammalian cells and the biology of the CKI p40Sic1 in *S. cerevisiae*. p40Sic1 is an inhibitor of the clb-cdc28 kinase activity required for entry into the S phase, but a substrate of G1 cln-cdc28 complexes (Mendenhall 1993, Schwob et al. 1994). Cln-dependent phosphorylation of Sic1 promotes its degradation through the ubiquitin mediated proteolytic pathway, thereby freeing clb-cdc28 complexes from inhibition by p40Sic1 and promoting progression into the S phase (Schneider et al. 1996; Fig. 6).

Analogously, mammalian G1 cyclins may be resistant to inhibition by p27 in cells. For example, cyclin D-CDK4/6 complexes are active early in G1 when p27 levels are high, and cyclin D-CDK complexes containing p27 have demonstrable Rb kinase activity (Soos et al. 1996). Thus like Sic1, p27 could control the timing of S phase initiation by inhibiting a downstream target such as cyclin A-CDK2. Consistent with this model, p27 is regulated in part by the ubiquitin proteasome pathway. In this view cyclin E-CDK2 plays a crucial role in establishing the threshold level of p27 during G1 progression. One intriguing possibility is that p27 is the critical substrate which must be phosphorylated by cyclin E-CDK2 in Rb minus cells. While the ultimate significance of these early results is still unclear, it is safe to say that the role of p27 is likely to be more complicated than previously thought.

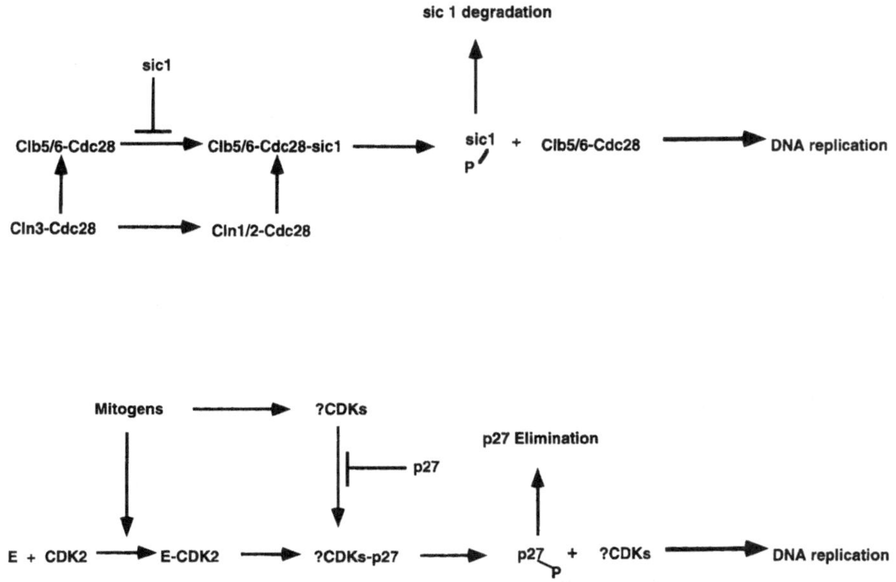

Fig. 6. Controlling the timing of DNA replication by modulation of p27 levels. The *top panel* shows how yeast control the timing of DNA replication by phosphorylation and elimination of the inhibitor Sic1 (see text for details). The *lower panel* hypothesizes a similar method of control in higher eukaryotes. *CDKs* represents cyclin-CDK complexes required for initiation of DNA replication and inhibited by p27

5.4
DNA Damage Checkpoint

As exemplified by the control of Rb, cell duplication in multicell organisms is tightly regulated so that a new cell is produced only when needed (Raff 1996). In this way the organism avoids the unpleasant consequences of deregulated proliferation. Current evidence strongly supports the idea that multiple mutations are required to develop the tumorogenic phenotype and that the cell cycle machinery is affected in most, if not all, cancers (Knudson 1986; Renan 1993). Given the central position of Rb in the cell cycle network, it is perhaps not surprising that mutations in Rb and its regulators are most often culprits in the development of cancer. Nevertheless, the generation of tumorogenic cells likely requires more than simply mutations in cell cycle regulatory genes, leading to the paradox of how multiple mutations can occur in the lifetime of the ordinary person. The answer likely lies in the observation that a hallmark of tumor cells may be genetic instability, which in turn increases the mutation rate to sufficiently high levels to allow cancer development (Tlsty et al. 1994).

In support of this notion, the cell has evolved highly rigorous quality control methods to maintain the integrity of the genome. Replicative enzymes possess inherent error rates, and environmental carcinogens can increase the frequency of these errors significantly. Therefore, the cell continually monitors the process of DNA replication and halts cell cycle progression if genomic stability or fidelity is threatened. Cell cycle progression is reinitiated only when the damage is repaired, thereby ensuring that initiation of downstream events (such as mitosis) is dependent on the completion of upstream events (such as DNA replication). Inactivation of these checkpoints is thought to be a major factor in development of the cancer cell, since it allows accumulation of genetic errors and deregulated proliferation (Hartwell et al. 1994).

Damaged DNA induces the p53 tumor suppressor protein, which upregulates transcription of the CKI p21 (El-Deiry et al. 1993). p21 subsequently inactivates CDKs, leading to cell cycle arrest in G1, and may also target proliferating cell nuclear antigen (PCNA), a polymerase delta processivity factor (Chen et al. 1995). Inhibition of CDKs and PCNA ensures that DNA replication is not initiated until the DNA damage has been repaired. Importantly, p21 does not inhibit the repair of DNA damage which must occur during this time (R. Li et al. 1994). It is unclear how cells sense DNA damage, although a p53 inducible component that is always present at low levels in cycling cells is one possibility.

6
Establishing and Maintaining the G1 Clock

Much progress has been made in identifying and elucidating the underlying mechanisms responsible for G1 progression, but there are still many confounding issues. Physiologically, it is clear that the R point represents a critical period when the cell decides whether or not to continue the mitotic cycle. Mechanistically, Rb phosphorylation by cyclin D-CDK4 is strategically situated to represent this decision. Does Rb then represent a single switch within a linear series of events, or are there complex feedback loops in place that create a multistep branched pathway involving downstream components? Given the wide range of processes requiring Rb regulation, the more likely model suggests a multicomponent biochemical pathway rather than a single master regulator (Bartek et al. 1996).

Based on the original definition of the R protein proposed by Pardee, negative regulators such as Rb or CKIs cannot be the R protein. The D type cyclins appear more likely candidates, since they are: (1) positive, labile proteins regulating mid to late G1; (2) growth factor sensors, i.e., they are induced; and (3) linked to anchorage dependence and cytoskeletal integrity. Furthermore, cyclin D overexpression is implicated in tumorogenesis, as is ablation of p16 (Motokura et al. 1991; Serrano et al. 1993). Mutational alteration of CDK4 so that it is resistant to inhibition by p16 also contributes to uncontrolled proliferation (Wolfel et al. 1995). Given the complex regulation of cyclin D-CDK activity by many other proteins, it seems unlikely that the designation of R protein can be assigned to any one molecule (Bartek et al. 1996).

Thus we are left to sort out circuitry composed of many different components, all of which contribute to traversal of G1. G1 progression can be thought of as two interconnected cycles – the establishment of cyclin D-CDK activity and subsequent Rb phosphorylation, both of which are under mitogenic control, and the establishment of cyclin E-CDK2 activity, which appears to be less mitogen-dependent and more self-regulatory (Fig. 7). The challenge is to understand how these two cycles are connected such that the one leads to the other in a timely and ordered fashion.

In the first cycle, preparation for Rb inactivation requires the accumulation of cyclin D, its association with CDK4/6, and phosphorylation of this complex by CAK (Fig. 7). Eventually, cyclin D levels surpass the threshold established by CKIs. Manipulation of CKI levels in response to extracellular signals or internal regulatory mechanisms can therefore influence the timing of progression through early G1. These initial events in G1 culminate in cyclin D-CDK inactivation of Rb and release of E2F, which in turn upregulates cell cycle genes required for the next stages of cell cycle progression, as well as genes required for downstream events such as DNA replication. Rb may also be involved in the inhibition of general RNA transcription that is likely in operation during this time, so that its inactivation permits the cell growth required before division. Once its job is complete, cyclin D is downregulated by proteolysis and perhaps upregulation of p16 (Tam et al. 1994).

A critical function of E2F appears to be activating transcription of cell cycle genes necessary for constructing the circuitry needed for controlling progression through the latter stages of G1. In particular, E2F upregulates cyclin E,

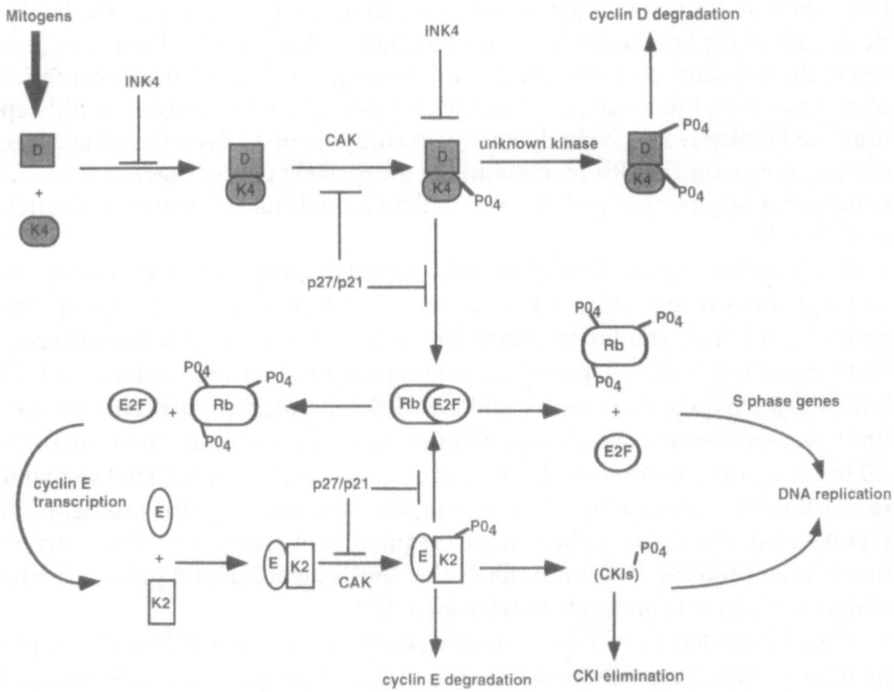

Fig. 7. Model of cell cycle events mediating progression through G1. The activation of cyclin D-CDK4 by mitogenic stimulation results in phosphorylation of Rb and release of E2F. E2F activates cyclin E transcription, as well as downstream components needed for initiation of DNA replication. Cyclin E has the potential to form self-regulating circuitry which is responsible for the periodicity of cyclin E-CDK2 kinase activity. The conversion from a cyclin D-activating cycle to a cyclin E self-sustaining cycle may represent the R point switch from mitogen dependence to mitogen independence. Cyclin E may control the timing of DNA replication in two ways: by activating components required for initiation and/or by inactivating the negative regulator p27. *CAK* CDK-activating kinase. (After Sherr and Roberts 1995)

which becomes responsible for controlling further advancement through G1. Thus the control of E2F activity by cyclin D, combined with the ability of E2F to up-regulate cyclin E, may represent the mechanistic link between early and mid to late G1.

The fact that cyclin D destruction is mediated by another kinase while cyclin E self-destructs most likely reflects the different roles of these cyclins in G1 progression. If cyclin D functions before the R point, the cell may require that its regulation be responsive to further mitogenic signaling. Once Rb is phosphorylated and the restriction point is passed, however, the cell is committed to the mitotic cycle. If the main role of cyclin E is phosphorylation of substrates required to initiate DNA replication, then its degradation need not be responsive to mitogenic signaling and instead is organized to give the necessary spike of cylin E-CDK2 kinase activity. The potential for cyclin E to control its own activity, both through a positive feedback loop involving E2F and through self-mediated destruction, may represent the switch from mitogen dependence to mitogen independence demarcated by the R point.

After commitment to completion of the mitotic cycle, emphasis shifts from establishing the cell cycle machinery to controlling its timing and monitoring its progress. Checkpoints monitoring the fidelity and completion of downstream events must be maintained, and many are activated in the latter stages of G1. Still unanswered is whether somatic cells operate a checkpoint ensuring a minimum size is reached before proceeding with the cycle. Given the proposal that CKIs act as an inhibitory threshold, it is tempting to speculate that they may be important for regulating progression through G1 and by extension perhaps link proliferation to size control.

Much less is known about the normal mechanisms controlling the timing of cyclin E catalyzed events. It is likely that a crucial role of cyclin E-CDK2 is phosphorylating substrates required for the initiation of DNA replication, although these have yet to be identified. Nevertheless, as investigations in yeast have revealed, much of the G1 cell cycle machinery appears to be involved in controlling the timing of DNA replication (Nasmyth 1996). In somatic cells, evidence is beginning to emerge suggesting CKIs such as p27 or p21 may play important roles in G1 progression by controlling cyclin-CDK2 activity required for initiation of DNA replication. The additional possibility that cyclin-CDKs control CKIs may provide just the self-regulating system of checks and balances the cell needs for timing progression through the G1 phase of the cell cycle (Fig. 7).

Acknowledgments. RJS was supported by the Damon Runyon-Walter Winchell Cancer Research fund 1313.

References

Albanese C, Johnson J, Watanabe G, Eklund N, Vu D, Arnold A, Pestell R (1995) Transforming p21ras mutants and c-Ets-2 activate the cyclin D promoter through distinguishable regions. J Biol Chem 270:23589–23597

Alberts B, Bray D, Lewis J, Raff M, Roberts K, Watson J (1989) Molecular biology of the cell. Garland, New York

Almasan A, Yin Y, Kelly RE, Lee EY-HP, Bradley A, Li W, Bertino JR, Walh GM (1995) Deficiency of retinoblastoma protein leads to inappropriate S-phase entry, activation of E2F-responsive genes, and apoptosis. Proc Natl Acad Sci USA 92:5436–5440

Auer G, Zetterberg A, Foley GE (1970) The relationship of DNA synthesis to protein accumulation in the cell nucleus. J Cell Physiol 76:357–363

Baldin V, Lukas J, Marcote MJ, Pagano M, Draetta G (1993) Cyclin D 1 is a nuclear protein required for cell cycle progression in G1. Genes Dev 7:812–821

Bandara LR, Buck VM, Zmanian M, Johnston LH, La Thangue NB (1993) Functional synergy between DP-1 and E2F-1 in the cell cycle-regulating transcription factor DRTF1/E2F. EMBO J 13:4317–4324

Bartek J, Bartkova J, Lukas J (1996) The retinoblastoma protein pathway and the restriction point. Curr Opin Cell Biol 8:805–814
Baserga R (1984) Growth in cell size and cell DNA-replication. Exp Cell Res 151:1–5
Baserga R (1985) The biology of cell reproduction. Harvard University Press, Cambridge
Beijerbergen RL, Kerkhoven RM, Zhu L, Carlee L, Voorhoeve FM, Bernards R (1994) E2F-4, a new member of the E2F gene family, has oncogenic activity and associates with p107 in vivo. Genes Dev 8:2680–2690
Bishop JM (1991) Molecular themes in oncogenesis. Cell 64:235–248
Brooks RF (1977) Continuous protein synthesis is required to maintain the probability of entry into S phase. Cell 12:311–317
Buchkovich K, Duffy LA, Harlow E (1989) The retinoblastoma protein is phosphorylated during specific phases of the cell cycle. Cell 58:1097–1105
Cantley L, Auger K, Carpenter C, Duckworth B, Graziani A, Kapeller R, Soltoff S (1991) Oncogenes and signal transduction. Cell 64:281–302
Cavanaugh AH, Hempel WM, Taylor LJ, Rogalsky V, Todorov G, Rothblum LI (1995) Activity of RNA polymerase I transcription factor UBF blocked by Rb gene product. Nature 374:177–180
Chen J, Jackson PK, Kirschner MW, Dutta A (1995) Separate domains of p21 involved in the inhibition of cdk kinase and PCNA. Nature 374:386–388
Chen PL, Scully P, Shew JY, Wang JYJ, Lee WH (1989) Phosphorylation of the retinoblastoma gene product is modulated during the cell cycle and cellular differentiation. Cell 58:1193–1198
Clarke AR, Maandag ER, van Roon M, van der Lugt NM, van der Valk M, Hooper ML, Berns A, te Riele H (1992) Requirement for a functional Rb-1 gene in murine development. Nature 359:328–330
Clurman BE, Sheaff RJ, Thress K, Groudine M, Roberts JM (1996) Turnover of cyclin E by the ubiquitin-proteasome pathway is regulated by CDK2 binding and cyclin phosphorylation. Genes Dev 10:1979–1990
Coats S, Flanagan WM, Nourse J, Roberts JM (1996) Requirement of p27(kip1) for restriction point control of the fibroblast cell cycle. Science 272:877–880
Cobrinik D, Lee MH, Hannon G, Mulligan G, Bronson RT, Dyson N, Harlow E, Beach D, Weinberg RA, Jacks T (1996) Shared role of the pRb-related p130 and p107 proteins in limb development. Genes Dev 10:1633–1644
Das HR, Lavin M, Sicuso A, Young DV (1983) The uncoupling of macromolecular synthesis from cell division in SV-3T3 cells by glucocorticoids. J Cell Physiol 117:241–248
DeCaprio JA, Ludlow JW, Lynch D, Furukawa Y, Griffin J, Piwnica-Worms H, Huang CM, Livingston DM (1989) The product of the retinoblastoma susceptibility gene has properties of a cell cycle regulatory element. Cell 58:1085–1095
DeGregori J, Leone G, Ohtani K, Miron A, Nevins JR (1995) E2F-1 accumulation bypasses a G1 arrest resulting from the inhibition of G1 cyclin-dependent kinase activity. Genes Dev 9:2873–2887
Deng C, Zhang P, Harper JW, Elledge SJ, Leder P (1995) Mice lacking p21cip1/waf1 undergo normal development, but are defective in G1 checkpoint control. Cell 82:675–684
Diehl JA, Zindy F, Sherr CJ (1997) Inhibition of cyclin D1 phosphorylation on threonine-286 prevents its rapid degradation via the ubiquitin-proteasome pathway. Genes Dev 11:957–972
Dobrowolski S, Harter M, Stacey DW (1994) Cellular ras activity is required for passage through multiple points of the G0/G1 phase in Balb/c3T3 cells. Mol Cell Biol 14:441–5449
Dulic V, Lees E, Reed SI (1992) Association of human cyclin E with a periodic G1-S phase protein kinase. Science 257:1958–1961
Dynlacht BD, Flores O, Lees JA, Harlow E (1994) Differential regulation of E2F transactivation by cylin/cdk2 complexes. Genes Dev 8:1772–1786
El-Deiry WS, Tokino T, Velculescu VE, Levy DB, Parsons R, Trent JM, Lin D, Mercer E, Kinzler KW, Volgelstein B (1993) WAF1, a potential mediator of p53 tumor suppression. Cell 75:817–825
Ewen ME, Oliver CJ, Sluss HK, Miller SJ, Peeper DS (1995) p53 dependent repression of CDK4 translation in TGF beta-induced G1 cell cycle arrest. Genes Dev 9:204–217

Fero ML, Rivkin M, Tasch M, Porter P, Carow CE, Firpo E, Polyak K, Tsai LH, Broudy V, Perlmutter RM, Kaushansky K, Robert JM (1996) A syndrome of multiorgan hyperplasia with features of gigantism, tumorigenesis, and female sterility in p27kip1-deficient mice. Cell 85:733–744
Firpo EJ, Koff A, Solomon MJ, Roberts JM (1994) Inactivation of a cdk2 inhibitor during interleukin 2-induced proliferation of human T lymphocytes. Mol Cell Biol 14:4889-4901
Friend SH, Horowitz JM, Gerber MR, Wang XF, Bogenmann E, Li FP, Weinberg RA (1987) Deletion of a DNA sequence in retinoblastomas and mesenchymal tumors: organization of the sequence and its encoded protein. Proc Natl Acad Sci USA 84:9059–9063
Gerhart J, Wu M, Kirschner MW (1984) Cell cycle dynamics of an M-phase-specific cytoplasmic factor in Xenopus laevis oocytes and eggs. J Cell Biol 98:1247–1255
Hamel PA, Gill RM, Phillips RA, Gallie BL (1992) Regions controlling hyperphosphorylation and confirmation of the retinoblastoma gene product are independent of domains required for transcriptional repression. Oncogene 7:693–701
Hannon GJ, Beach D (1994) p15INK4b is a potential effector of cell cycle arrest mediated by TGF beta. Nature 371:257–261
Harper JW, Elledge SJ (1996) Cdk inhibitors in development and cancer. Curr Opin Genet Dev 6:55–64
Harper JW, Elledge SJ, Keyomarsi K, Dynlacht B, Tsai LH, Zhang P, Dobrowolski S, Bai C, Connell-Crowley L, Swindell E, Fox MP, Wei N (1995) Inhibition of cyclin dependent kinases by p21. Mol Biol Cell 6:387–400
Hartwell L (1991) Twenty-five years of cell cycle genetics. Genetics 129:975–980
Hartwell LH, Unger MW (1977) Unequal division in Saccharomyces cerevisiae and its implications for the control of cell division. J Cell Biol 75:422–425
Hartwell LH, Weinert TA (1989) Checkpoints: controls that ensure the order of cell cycle events. Science 246:629–634
Hartwell L, Mortimer K, Culotti J, Culotti M (1973) Genetic control of the cell division cycle in yeast. V. Genetic analysis of cdc2 mutants. Genetics 74:267–286
Hartwell L, Weinert T, Kadyk L, Gavrik B (1994) Cell cycle checkpoint, genomic integrity, and cancer. Cold Spring Harb Symp Quant Biol 59:259–263
Hatakeyama M, Herrera RA, Makela T, Dowdy SF, Jacks T, Weinberg RA (1994) The cancer cell and the cell cycle clock. Cold Spring Harb Symp Quant Biol 59:1–10
Helin K, Wu CL, Fattaey AR, Lees JA, Dynlacht BD, Ngwu C, Harlow E (1993) Heterodimerization of the transcription factors E2F-1 and DP-1 leads to cooperative trans-activation. Genes Dev 7:1850–1861
Hengst L, Reed SI (1996) Translational control of p27kip1 accumulation during the cell cycle. Science 271:1861–1864
Hengst L, Dulic V, Slingerland JM, Lees E, Reed SI (1994) A cell cycle-regulated inhibitor of cyclin-dependent kinases. Proc Natl Acad Sci USA 91:5291–5295
Herrera RE, Sah VP, Williams BO, Makela TP, Weinberg RA, Jacks T (1996) Altered cell cycle kinetics, gene expression, and G1 restriction point regulation in Rb-deficient fibroblasts. Mol Cell Biol 16:2402–2407
Hinds PW, Mittnacht S, Dulic V, Arnold A, Reed SI, Weinberg RA (1992) Regulation of retinoblastoma protein functions by ectopic expression of human cyclins. Cell 70:993–1006
Horowitz JM, Park S, Bogenmann E, Cheng J, Yandell DW, Kaye FJ, Minna JD, Dryja TP, Weinberg RA (1990) Frequent inactivation of the retinoblastoma anti-oncogene is restricted to a subset of human tumor cells. Proc Natl Acad Sci USA 87:2775–2779
Imai Y, Matsushima Y, Sugimura T, Terada M (1991) Purification and characterization of human papillomavirus type 16 E7 protein with preferential binding capacity to the underphosphorylated form of the retinoblastoma gene product. J Virol 65:4966–4972
Jacks T, Fazeli A, Schmitt EM, Bronson RT, Goodell MA, Weinberg RA (1992) Effects of an Rb mutation in the mouse. Nature 359:295–300
Jeffrey PD, Russo AA, Polyak K, Gibbs E, Hurwitz J, Massague J, Paveltich NP (1995) Crystal structure of a cyclin A-cdk2 complex at 2.3A: mechanism of cdk activation by cyclins. Nature 376:313–320

Kamb A (1994) Role of a cell cycle regulator in hereditary and sporadic cancer. Cold Spring Harb Symp Quant Biol 49:39–47
Kato JM, Matsuoka M, Polyak K, Massague J, Sherr CJ (1994) Cyclic AMP-induced G1 phase arrest mediated by an inhibitor (p27kip1) of cyclin-dependent kinase-4 activation. Cell 79:487–496
Killander D, Zetterberg A (1965a) Quantitative cytochemical studies on interphase growth. I. Determination of DNA, RNA, and mass content of age determined mouse fibroblasts in vitro and of intercellular variation in generation time. Exp Cell Res 38:272–284
Killander D, Zetterberg A (1965b) A quantitative cytochemical investigation of the relationship between cell mass and initiation of DNA synthesis in mouse fibroblasts in vitro. Exp Cell Res 40:12–20
Kiyokawa H, Kineman RD, Manova-Tododova KO, Soares VC, Hoffman ES, Ono M, Khanam D, Hayday AC, Frohman LA, Koff A (1996) Enhanced growth of mice lacking the cyclin-dependent kinase inhibitor function of p27kip1. Cell 85:721–732
Knoblich JA, Sauer K, Jones L, Richardson H, Saint R, Lehner CF (1994) Cyclin E controls S phase progression and its down-regulation during Drosophila embryogenesis is required for the arrest of cell proliferation. Cell 77:1–20
Knudsen ES, Wang JYJ (1996) Differential regulation of retinoblastoma protein function by specific Cdk phosphorylation sites. J Biol Chem 271:8313–8320
Knudson AG Jr (1986) Genetics of human cancer. Annu Rev Genet 20: 231–251
Koff A, Cross F, Fisher A, Schumacher J, Leguelle K, Philippe M, Roberts JM (1991) Human cyclin E, a new cyclin that interacts with two members of the CDC2 gene family. Cell 66:1217–1228
Koff A, Giordano A, Desai D, Yamashita K, Harper JW, Elledge S, Nishimoto T, Morgan DO, Franza R, Roberts JM (1992) Formation and activation of a cyclin E/CDK2 complex during the G1 phase of the human cell cycle. Science 257:1689–1693
Koff A, Ohtsuki M, Polyak K, Roberts JM, Massague J (1993) Negative regulation of G1 in mammalian cells: inhibition of cyclin E dependent kinase by TGF beta. Science 260:536–539
Krek W, Livingston DM, Shirodkar S (1993) Binding to DNA and the retinoblastoma gene product promoted by complex formation of different E2F family members. Science 262:1557–1560
Laiho M, DeCaprio JA, Ludlow JW, Livingston DM, Massague J (1990) Growth inhibition of TGF beta linked to suppression of retinoblastoma protein. Cell 62:175–185
Lam EWF, La Thangue NB (1994) DP and E2F proteins: coordinating transcription with cell cycle progression. Curr Opin Cell Biol 6:859–866
Larsson O, Zetterberg A, Engstrom W (1985) Cell-cycle-specific induction of quiescence achieved by limited inhibition of protein synthesis: counteractive effect of addition of purified growth factors. J Cell Sci 73:375–387
Larsson O, Dafgard E, Engstrom W, Zetterberg A (1986) Immediate effects of serum depletion on dissociation between growth in size and cell division in 3T3 cells. J Cell Physiol 127:267–273
Lasko D, Cavenee W, Nordenskjold M (1991) Loss of constitutional heterozygozity in human cancer. Annu Rev Genet 25:281–314
Lee EY-HP, Chang CY, Hu N, Wang YC, Lai CC, Herrup K, Lee WH, Bradley A (1992) Mice deficient for Rb are nonviable and show defects in neurogenesis and haematopoiesis. Nature 359:288–294
Lee MH, Reynisdottir I, Massague J (1995) Cloning of p57Kip2, a cyclin-dependent kinase inhibitor with unique domain structure and tissue distribution. Genes Dev 9:639–649
Lee MH, Williams BO, Mulligan G, Mukai S, Bronson RT, Dyson N, Harlow E, Jacks T (1996) Targeted disruption of p107: functional overlap between p107 and Rb. Genes Dev 10:1621–1632
Levine AJ (1997) p53, the cellular gatekeeper for growth and division. Cell 88:323–331
Lew DJ, Dulic V, Reed SI (1991) Isolation of three novel human cyclins by rescue of G1 cyclin (cln) function in yeast. Cell 66:1197–1206
Li R, Waga S, Hannon GJ, Beach D, Stillman B (1994) Differential effects by the p21 CDK inhibitor on PCNA dependent DNA replication and repair. Nature 371:534–537
Li Y, Jenkins CW, Nichols MA, Xiong Y (1994) Cell cycle expression and p53 regulation of the cyclin-dependent kinase inhibitor p21. Oncogene 9:2261–2268

Ludlow JW, DeCaprio JA, Huang CM, Lee WH, Paucha E, Livingston DM (1989) SV40 large T antigen binds preferentially to an underphosphorylated member of the retinoblastoma susceptibility gene product family. Cell 56:57–65
Lukas J, Bartkova J, Rhode M, Strauss M, Bartek J (1995a) Cyclin D1 is dispensable for G1 control in retinoblastoma gene-deficient cells independently of cdk4 activity. Mol Cell Biol 15:2600–2611
Lukas J, Parry D, Aagaard L, Mann DJ, Bartkova J, Strauss M, Peters G, Bartek J (1995b) Retinoblastoma-protein-dependent cell cycle inhibition by the tumor suppresser p16. Nature 375:503–506
Lukas J, Peterson BO, Holm K, Bartek J, Helin K (1996) Deregulated expression of E2F family members induces S-phase entry and overcomes p16^{INK4a}-mediated growth suppression. Mol Cell Biol 16:1047–1057
Mal A, Poon RYC, Howe PH, Toyoshima H, Hunter T, Harter ML (1996) Inactivation of p27(kip1) by the viral E1A oncoprotein in TGF beta-treated cells. Nature 380:262–265
Massague J (1996) TGF beta signaling: receptors, transducers, and mad proteins. Cell 85:947–950
Matsushime H, Roussel M, Ashmun R, Sherr CJ (1991) Colony-stimulating Factor 1 regulates novel cyclins during the G1 phase of the cell cycle. Cell 65:701–713
Matsushime H, Ewen ME, Strom DK, Kato J, Hanks SK, Roussel MF, Sherr CJ (1992) Identification and properties of an atypical catalytic subunit (p34PSKJ3/CDK4) for mammalian D-type cyclins. Cell 71:323–334
Medema RH, Herrera RE, Lam F, Weinberg RA (1995) Growth suppression by p16INK4 requires functional retinoblastoma protein. Proc Natl Acad Sci USA 92:6289–6293
Mendenhall M (1993) An inhibitor of p34 CDC28 protein kinase activity from Saccharomyces cerevisiae. Science 259:216–219
Mercer HE, Avignolo C, Galanti N, Ruse KM, Hyland JK, Jacob ST, Baserga R (1984) Cellular DNA replication is dependent on the synthesis and the accumulation of ribosomal RNA. Exp Cell Res 150:118–130
Mitchison JM (1971) The biology of the cell cycle. Cambridge University Press, London
Morgan D (1995) Principles of CDK regulation. Nature 374:131–134
Morgan D (1996) The dynamics of cyclin dependent kinase structure. Curr Opin Cell Biol 8:767–772
Motokura T, Bloom T, Kim HG, Juppnre H, Ruderman JV, Kronenberg HM, Arnold A (1991) A BCL1-linked candidate oncogene which is rearranged in parathyroid tumors encodes a novel cyclin. Nature 350:512–525
Mulcahy LS, Smith MR, Stacey DW (1985) Requirement for ras proto-oncogene function during serum-stimulated growth of NIH 3T3 cells. Nature 313:241–243
Murray A, Hunt T (1993) The cell cycle: an introduction. Freeman, New York
Myerson M, Enders GH, Wu C, Su L, Gorka C, Nelson C, Harlow E, Tsai L (1992) The human cdc2 kinase family. EMBO J 11:2909–2017
Nakayama K, Ishida N, Shirane M, Inomata A, Inoue T, Shishido N, Horii I, Loh DY, Nakayama KI (1996) Mice lacking p27kip1 display increased body size, multiple organ hyperplasia, retinal dysplasia, and pituitary tumors. Cell 85:707–720
Nasmyth K (1993) Control of the yeast cell cycle by the Cdc28 protein kinase. Curr Opin Cell Biol 5:166–179
Nasmyth K (1996) At the heart of the budding yeast cell cycle. Trends Genet 12:405–412
Nevins JR (1992) E2F: a link between the Rb tumor suppresser protein and viral oncoproteins. Science 258:424–429
Newport JW, Kirschner MW (1982) A major developmental transition in early Xenopus embryos. II. Control of the onset of transcription. Cell 30:687–696
Newport JW, Kirschner MW (1984) Regulation of the cell cycle during early Xenopus development. Cell 37:731–742
Nigg E (1996) Cyclin-dependent kinase 7: at the crossroads of transcription, DNA repair, and cell cycle control. Curr Opin Cell Biol 8: 312–317

Nourse J, Firpo EJ, Flanagan M, Meyerson M, Polyak K, Lee MH, Massague J, Crabtree G, Roberts JM (1994) Il-2 mediated elimination of the p27kip1 cyclin-Cdk kinase inhibitor prevented by rapamycin. Nature 372:570–573

Ohtani K, DeGregori J, Nevins JR (1995) Regulation of the cyclin E gene by transcription factor E2F. Proc Natl Acad Sci USA 92:12146–12150

Ohtsubo M, Roberts JM (1993) Cyclin-dependent regulation of G1 in mammalian fibroblasts. Science 259:1908–1012

Ohtsubo M, Theodoras AM, Schumacher J, Roberts JM, Pagano M (1995) Human cyclin E: a nuclear protein essential for the G1 to S phase transition. Mol Cell Biol 15:2612–2624

Pagano M, Tam SW, Theodoras AM, Beer-Romero P, Del Sal G, Chau V, Yew PR, Draetta GF, Rolfe M (1995) Role of the ubiquitin-proteasome pathway in regulating amounts of the cyclin-dependent kinase inhibitor p27. Science 269:682–685

Pardee AB (1974) A restriction point for control of normal animal cell proliferation. Proc Natl Acad Sci USA 71:1286–1290

Pardee AB (1989) G1 events and regulation of cell proliferation. Science 246:603–608

Pardee AB, Medrano EE, Rossow PW (1981) The biology of normal human growth. Raven Press, New York

Parker SB, Eichele G, Zhang P, Rawls A, Sands AT, Bradley A, Olson EN, Harper JW, Elledge SJ (1995) p53 independent expression of $p21^{CIP1}$ in muscle and other terminally differentiating cells. Science 267:1024–1027

Peeper DS, Upton TM, Ladha MH, Neuman E, Zalvide J, Bernards R, DeCaprio JA, Ewen ME (1997) Ras signaling linked to the cell-cycle machinery by the retinoblastoma protein. Nature 386:177–181

Peter M, Herskowitz I (1994) Joining the complex: cyclin dependent kinase inhibitory proteins and the cell cycle. Cell 79:181–184

Prescott DM (1976) Reproduction of eukaryotic cells. Academic Press, New York

Polyak K, Kato J, Solomon MJ, Sherr CJ, Massague J, Roberts JM, Koff A (1994a) p27kip1, a cyclin-cdk inhibitor, links transforming growth factor beta and contact inhibition to cell cycle arrest. Genes Dev 8:9–22

Polyak K, Lee M, Erdjument-Bromage H, Koff A, Roberts JM, Tempst P, Massague J (1994b) Cloning of p27kip1, a cyclin-dependent kinase inhibitor and a potential mediator of extracellular antimitogenic signals. Cell 78:59–66

Quelle DE, Ashmun RA, Shurtleff SA, Kato J, Barsagi D, Roussell MF, Sherr CJ (1993) Overexpression of mouse D-type cyclins accelerates G1 phase in rodent fibroblasts. Genes Dev 7:1559–1571

Raff M (1996) Size control: the regulation of cell numbers in animal development. Cell 86:173–175

Renan MJ (1993) How many mutations are required for tumorigenesis? Implications from human cancer data. Mol Carcinog 7:139–146

Resnitzky D, Reed SI (1995) Different roles for cyclins D1 and E in regulation of the G1-to-S phase. Mol Cell Biol 15:3463–3469

Reynisdottir I, Polyak K, Iavarone A, Massague J (1995) Kip/cip and INK4 Cdk inhibitors cooperate to induce cell cycle arrest in response to TGF beta. Genes Dev 9:1831–1845

Roberts JM, Koff A, Polyak K, Firpo E, Collins S, Ohtsubo M, Massague J (1994) Cyclins, Cdks, and cyclin kinase inhibitors. Cold Spring Harb Symp Quant Biol 59:31–38

Ronning OW, Pettersen EO (1984) Doubling of cell mass is not necessary in order to achieve cell division in cultured human cells. Exp Cell Res 155:267–272

Rossow PW, Riddle VG, Pardee AB (1979) Synthesis of labile serum-dependent protein in early G1 controls animal cell growth. Proc Natl Acad Sci USA 76:4446–4450

Roussel MF, Theodoras AM, Pagano M, Sherr CJ (1995) Rescue of defective mitogenic signaling by D-type cyclins. Proc Natl Acad Sci USA 92:6837–6841

Russo AA, Jeffrey PD, Patten A, Massague J, Pavletich NP (1996) Crystal structure of the p27kip1 cyclin-dependent kinase inhibitor bound to the cyclin A-CDK2 complex. Nature 382:325–331

Schneider BL, Yang QH, Futcher AB (1996) Linkage of replication to Start by the Cdk inhibitor Sic1. Science 272:560–562

Schulze A, Zerfass K, Spitkovsky D, Middendorp S, Berges J, Helin K, Jansen-Durr P, Henglein B (1995) Cell cycle regulation of the cyclin A gene promoter is mediated by a variant of the E2F site. Proc Natl Acad Sci USA 92:11264–11268
Schwob E, Bohm T, Mendenhall MD, Nasmyth K (1994) The B-type cyclin kinase inhibitor p40sic1 controls the G1 to S phase transition in S. cerevisiae. Cell 79:233–244
Serrano M, Hannon GJ, Beach D (1993) A new regulatory motif in cell cycle control causing specific inhibition of cyclin D/cdk4. Nature 366:704–707
Sheaff RJ, Groudine M, Gordon M, Roberts JM, Clurman BE (1997) Cyclin E-CDK2 is a regulator of p27kip1. Genes Dev 11:1464–1478
Sherr C (1993) Mammalian G1 cyclins. Cell 73:10591–1065
Sherr C (1994a) G1 phase progression: cycling on cue. Cell 79:551–555
Sherr CJ (1994b) The ins and outs of Rb: coupling gene expression to the cell cycle clock. Trends Cell Biol 4:15–19
Sherr CJ, Roberts JM (1995) Inhibitors of mammalian G1 cyclin-dependent kinases. Genes Dev 9:1149–1163
Sherr C, Kato J, Quell D, Matsuoka M, Roussel M (1994) D-type cyclins and their cyclin-dependent kinases: G1 phase integrators of the mitogenic response. Cold Spring Harb Symp Quant Biol 49:11–19
Singh P, Coe J, Hong W (1995) A role for retinoblastoma protein in potentiating transcriptional activation by the glucocorticoid receptor. Nature 374:562–565
Slingerland JM, Hengst L, Pan CH, Alexander D, Stampfer MR, Reed SI (1994) A cell-cycle regulated inhibitor of cyclin-dependent kinases. Proc Natl Acad Sci USA 91:5291–5295
Solomon MJ (1994) The function(s) of CAK, the p34cdc2 activating kinase. Trends Biochem Sci 19:496–500
Soos TJ, Kiyokawa H, Shi Yan J, Rubin MS, Giordano A, DeBlasio A, Bottega S, Wong B, Mendelsohn J, Koff A (1996) Formation of p27-CDK complexes during the human mitotic cell cycle. Cell Growth Diff 7:135–146
Tam SW, Shay JW, Pagano M (1994) Differential expression and cell cycle regulation of the cyclin-dependent kinase 4 inhibitor p16INK4. Cancer Res 54:5816–5820
Tassan JP, Jaquenoud M, Leopold P, Schultz SJ, Nigg EA (1995) Identification of human cyclin-dependent kinase 8, a putative protein kinase partner for cyclin C. Proc Natl Acad Sci USA 92:8871–8875
Taya Y (1997) Rb kinases and Rb binding proteins: new points of view. Trends Biol Sci 22:14–17
Tlsty T, White A, Livanos E, Sage M, Roelofs H, Briot A, Poulose B (1994) Genomic integrity and the genetics of cancer. Cold Spring Harb Symp Quant Biol 59:265–275
Toyoshima H, Hunter T (1994) p27, a novel inhibitor of cyclin/cdk kinase activity, is related to p21. Cell 78:67–74
Van den Heuval S, Harlow E (1994) Distinct roles for cyclin-dependent kinases in cell cycle control. Science 262:2050–2054
Weinberg RA (1995) The retinoblastoma protein and cell cycle control. Cell 81:323–330
White RJ, Gottlieb TM, Downes CS, Jackson SP (1995) Cell cycle regulation of RNA polymerase III transcription. Mol Cell Biol 15:6653–6662
White RJ, Trouche D, Martin K, Jackson SP, Kouzarides T (1996) Repression of RNA polymerase III transcription by the retinoblastoma protein. Nature 382:88–90
Winston JT, Pledger WJ (1993) Growth regulation of cyclin D1 mRNA expression through protein synthesis-dependent and -independent mechanisms. Mol Biol Cell 4:1133–1144
Williams BO, Remington L, Albert DM, Mukai S, Bronson RT, Jacks T (1994) Cooperative tumorigenic effects of germline mutations in Rb and p53. Nature [Genet] 10:480–484
Wolfel T, Hauer M, Schneider J, Serrano M, Wolfel C, Klehmann-Hieb E, De Plaen E, Hankeln T, Meyer zum Buschenfelde KH, Beach D (1995) A p16INK4a-sensitive cdk4 mutant targeted by cytolytic T lymphocytes in a human melanoma. Science 269:1281–1284
Won KA, Reed SI (1996) Activation of cyclin E/CDK2 is coupled to site-specific autophosphorylation and ubiquitin-dependent degradation of cyclin E. EMBO J 15:4182–4193
Wu CL, Classon M, Dyson N, Harlow E (1996) Expression of dominant-negative mutant DP-1 blocks cell cycle progression in G1. Mol Cell Biol 16:3698–3706

Xiong Y, Connolly T, Futcher B, Beach D (1991) Human D-type cyclin. Cell 65:691–699
Zetterberg A, Larsson O (1985) Kinetic analysis of regulatory events in G1 leading to proliferation of quiescent Swiss 3T3 cells. Proc Natl Acad Sci USA 82:5365–5369
Zetterberg A, Larsson O (1991) Coordination between cell growth and cell cycle transit in animal cells. Cold Spring Harb Symp Quant Biol 56:137–147
Zetterberg A, Engstrom W, Larsson O (1982) Growth activation of resting cells. Ann N Y Acad Sci 397:130–147
Zetterberg A, Engstrom W, Dafgard E (1984) The relative effects of different types of growth factors on DNA-replication, mitosis and cellular enlargement. Cytometry 5:368–375
Zhang H, Hannon GJ, Beach D (1994) p21-containing cyclin kinases exist in both active and inactive states. Genes Dev 8:1750–1758
Zhu L, Harlow E, Dynlacht BD (1995) p107 uses a p21(Cip1)-related domain to bind cyclin/cdk2 and regulate interactions with E2F. Genes Dev 9:1740–1752
Zhu X, Ohtsubo M, Bohmer RM, Roberts JM, Assoian R (1996) Adhesion-dependent cell cycle progression linked to the expression of cyclin D1, activation of cyclin E-Cdk2, and phosphorylation of the retinoblastoma protein. J Cell Biol 133:391–403

Regulation of S Phase

A. Dutta[1]

1 Introduction

The regulation of S phase can be divided into factors and activities that promote DNA replication at the G1-S boundary and those that prevent DNA replication once a segment of DNA has replicated. Considerable regulation is achieved by the synthesis of new proteins (and mRNA coding for these proteins) essential for DNA replication. In this chapter, however, we restrict our discussion to regulation of the DNA replication apparatus alone. The process of DNA replication is itself divisible into the initiation of bidirectional replication, elongation of the new DNA strands, and the termination of DNA replication. Since much of the recent work has illuminated the process of replication initiation, we have opted to focus our discussion to this process alone.

The molecular analysis of the initiation of DNA replication in eukaryotes has entered a new era with the recent discovery of initiator protein complexes central to the process. In prokaryotes and in eukaryotic viruses replication usually initiates from origins of DNA replication adjacent to DNA signal sequences which are essential to the replication process. These signal sequences, referred to as replicator elements, function by promoting the assembly of initiator protein complexes at or near them (Stillman 1993). The latter facilitates the unwinding of the double-stranded DNA and the recruitment of DNA polymerases and accessory proteins involved in initiating DNA synthesis on the unwound DNA. The actual sites at which replication initiates may be separate from the replicator element and may even be diffusely distributed over a segment of DNA. A similar distribution of functional elements is probably conserved for chromosomal replication in eukaryotes, although the details may differ.

[1] Dept. of Pathology, Brigham and Women's Hospital, Harvard Medical School, 75 Francis Street, Boston, Massachusetts 02115, USA, E-mail: adutta@bustoff.bwh.harvard.edu

2 Origins of DNA Replication

2.1 Replicator Sequences

In *Saccharomyces cerevisiae*, origins of replication were discovered as sequences that permit extrachromosomal DNA sequences to replicate autonomously (autonomous replication sequences or ARS; Newlon 1993). Comparison of several ARS revealed an 11 base ARS consensus sequence (ACS) which was genetically essential for the activity of the ARS. Closer examination indicated that, besides ACS (or A element), there are secondary sequences (B elements) which have significant homology to the A element (9/11 matches) and are required for optimal replication activity.

In *Schizosaccharomyces pombe*, origins of replication are now being defined in equivalent detail. Linker scanning mutations across an ARS (ars3001) indicate that sequences of about 30–55 base-pairs are required for optimal origin function, and similar sequences are present on other ARS (Dubey et al. 1996). This small consensus sequence could be the essential replicator sequence in *S. pombe*.

In *Xenopus laevis* egg extracts, almost any DNA sequence can be replicated, raising doubts as to whether there is a specific sequence element necessary for initiating DNA replication (Hyrien and Mechali 1992; Mahbubani et al. 1992). One explanation could be that the egg extracts have stockpiles of replication factors (to accommodate the rapid cell division after fertilization), thereby allowing productive interactions between initiation factors and suboptimal replicator sequences. This leads to a relaxation of the sequences necessary for initiation of DNA replication. However, when intact eukaryotic nuclei were added to the egg extracts, replication initiated from defined regions of the DNA, indicating that structural features of an organized nucleus may play a role in selecting sites of initiation of DNA replication (Gilbert et al. 1995; Wu and Gilbert 1996).

In mammalian cells, physical methods of mapping sites of initiation of DNA replication indicate that initiation occurs at multiple sites spread over long stretches of DNA (Brewer 1994; Hamlin et al. 1994). One has to nevertheless keep in mind that sites of initiation of replication are likely to be different from replicator sequences (genetic elements essential for replication, probably because they bind to initiation factors). When deletions of specific regions of DNA near the beta-globin locus were correlated with the direction from which a replication fork progressed through the region, it became evident that certain regions of DNA were necessary for the generation of replication forks coming from that direction (Kitsberg et al. 1993). When those regions were deleted, the DNA was still replicated, but by a fork coming from the opposite direction and presumably initiating from a different initiation site. Thus genetic elements are required for the generation of DNA replication forks in mammalian chromosomes. What they are, how degenerate they are, and how distributed they are over a length of DNA are questions remaining to be answered.

2.2 Initiator Protein Complexes

Three proteins (or protein complexes) appear central to the process of initiation of DNA replication in eukaryotes. Much of what is known is derived from experiments in the yeasts *S. cerevisiae* and *S. pombe* and from extracts of *Xenopus* eggs. Several features of these initiation protein complexes are conserved across species, but details of their regulation are likely to be different between species. However, constituents of all three protein complexes have been identified in all species ranging from yeasts to humans. These complexes, the origin recognition complex (ORC) proteins, CDC6/Cdc18, and the mini-chromosome-maintenance (MCM) proteins, will now be briefly reviewed. For an exhaustive list of references please see Dutta and Bell 1997.

The physical state of these protein complexes, particularly properties that determine whether they can assemble at origins of DNA replication, are affected by the position of the cell in the cell cycle (Fig. 1). These changes are suspected to link the regulation of the initiation of DNA replication with the cellcycle machinery.

3 Initiation Factors

3.1 ORC

A six-subunit protein complex was identified in *S. cerevisiae* which bound to the A (or ACS) element of ARS sequences (Bell and Stillman 1992). Ranging in size from 104 to 50 kDa, the subunits were named in descending order of size as Orc1 to Orc6. The genes for all six subunits are essential for cell survival, and conditional mutations in some of the genes show them to be essential for the initiation of DNA replication. Point mutations in the A element which interfere with the binding of ORC concordantly diminish the activity of the A element as a replicator sequence (Bell and Stillman 1992; Rowley et al. 1995). Of the six yeast subunits, Orc1 and Orc5 have sequence motifs characteristic of proteins that bind nucleotide triphosphates (NTP), and indeed the ORC protein requires to bind to an NTP to associate with the A element (Klemm et al. 1997). The NTP binding motif of Orc1 is essential for viability. In addition, Orc1 possesses a weak ATPase activity inhibited by the binding of ORC to the A element. These findings suggest that ATP-bound ORC can remain associated with the A element until the ATPase activity is stimulated by unidentified cellular factors.

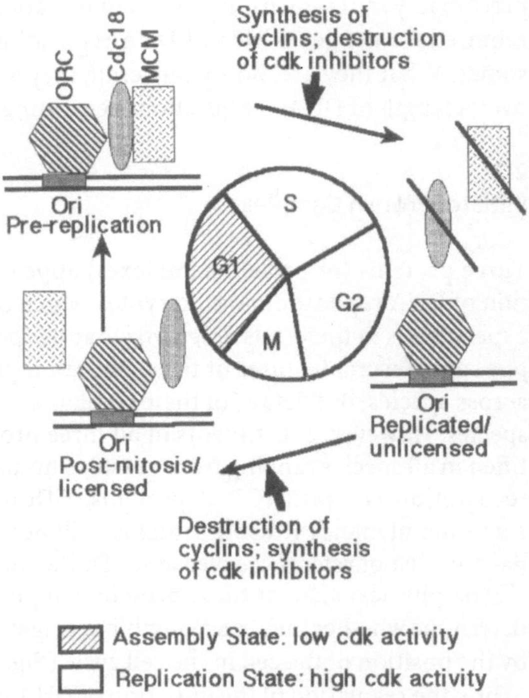

Fig. 1. Proposed changes in the replication initiation complex assembled at origins of DNA replication throughout the cell-cycle based on studies in yeast, *Xenopus* and mammals. ORC, CDC6/Cdc18 and MCM proteins are described in the text. In yeast CDC6/Cdc18 destruction in S phase makes re-replication impossible until the next mitosis and new Cdc18 synthesis. In *Xenopus*, CDC6/Cdc18 is not destroyed, but displaced from the chromatin during S. An unknown factor is required to reassemble the initiation complex after mitosis. In humans, Cdc18 persists throughout the cell cycle but may be displaced from the nucleus to the cytoplasm at the onset of S, thereby limiting the availability of this crucial adapter protein until the next G1. MCM proteins are also displaced from the chromatin during S phase and reacquired on the chromatin in G1. In addition, the cell in G1 is in a state that promotes the assembly of the initiator protein complex at origins of replication (the assembly state) which correlates with low CDK activity. Once replication has initiated, the factors are in a state that does not favor assembly of the entire initiator protein complex (the replication state), and this correlates with elevated CDK activity. Thus initiation of DNA replication is impossible in G2 until the cell enters a new assembly state, which can happen only after the cell has completed mitosis and has destroyed the mitotic cyclins so as to diminish CDK activity.

Homologues of ORC subunits have been identified in *S. pombe, D. melanogaster, X. laevis*, *M. musculus* and *H. sapiens* (Gavin et al. 1995; Gossen et al. 1995; Muzi and Kelly 1995; Carpenter et al. 1996; Leatherwood et al. 1996; Rowles et al. 1996; Quintana et al., submitted). In *Drosophila* and *Xenopus* a multiprotein complex has been isolated which contains some of the known ORC subunits (Gossen et al. 1995; Romanowski et al. 1996). The *Xenopus* protein complex restores DNA replication activity to *X. laevis* egg extracts selectively depleted of ORC by antibodies to one of the subunits. The depletion of ORC from these extracts inhibited the association of CDC6 and MCM proteins (see below), with chromatin indicating that ORC association with DNA is of primary importance for the recruitment of other members of the replication initiator complex. Not all ORC subunits have yet been molecularly cloned in any species other than *S. cerevisiae.*

Very little is known about the regulation of ORC activity through the cell cycle. The footprint of protein complexes bound to *S. cerevisiae* A elements in vivo suggests that ORC may remain bound to the A element through all phases of the cell cycle (Brown et al. 1991; Diffley and Cocker 1992; Diffley et al. 1994). Immunofluorescence studies in *Xenopus* egg extracts, however, indicate that one of the ORC subunits is bound to chromatin through all of interphase (including G2, when the origins are not competent to initiate DNA replication) but is released to the cytoplasm during mitosis (Romanowski et al. 1996). Since origins of DNA replication are competent to initiate only in G1 and early S, both lines of evidence suggest that ORC is associated with the origins of replication in the G2 phase of the cell cycle and therefore ORC binding is not the rate-limiting event that prevents DNA replication in G2 phase in *S. cerevisiae* and *X. laevis*. Regulation of ORC function during the mammalian cell cycle could be more complicated. Of the three subunits of human ORC identified to date, HsOrc1, HsOrc2 and HsOrc4, the former two have been shown to associate with each other only with overexpressed epitope-tagged recombinant proteins (Gavin et al. 1995; Quintana et al., unpubl.). Antibodies to HsOrc2 or HsOrc4, however, co-immunoprecipitate the other protein (implying they are in a complex) but not HsOrc1. Such a result could be explained by the antibodies disrupting the ORC, by only small proportions of HsOrc1, HsOrc2 and HsOrc4 being associated with each other, or by the association of all ORC subunits into one complex being restricted to a short phase in the cell cycle when the origins are competent to initiate DNA replication.

3.2
CDC6/Cdc18

CDC6 (in *S. cerevisiae*; Zhou et al. 1989; Hogan and Koshland 1992) or Cdc18 (in *S. pombe;* Kelly et al. 1993) is the second putative component of the replication initiator complex. Identified in the yeasts, in *Xenopus* and in humans (Coleman et al. 1996; Williams et al. 1997; Saha et al., submitted), it is a single polypeptide of about 66 kDa with homology to the largest subunit of ORC (Orc1). The homology is most marked around a central region that includes a putative nucleotide triphosphate binding motif. Conditional mutants in the yeasts (Piatti et al. 1995) and immunodepletion of the protein in *Xenopus* egg extracts indicate it to be essential for the initiation of DNA replication. In yeasts, an extended footprint from bound proteins is noted at A elements in G1 (Diffley et al. 1994). The extended footprint shrinks to a smaller footprint after the initiation of DNA replication which is similar to that expected from ORC binding alone. This extended footprint at origins of DNA replication in G1 probably represents the binding of the replication initiator complex to the origins (technically called the pre-replication complex). Formation of the pre-replication complex requires active CDC6. In *X. laevis*, immunodepletion of CDC6 did not affect the binding of ORC to the chromatin but interfered with the association of MCM proteins (see below) with chromatin (Coleman et al. 1996). CDC6/Cdc18 interacts with one or two ORC subunits both genetically and biochemically in yeast, *Xenopus* and mammals. Thus CDC6/Cdc18 could be an adaptor protein which in G1 recognizes ORC bound to DNA and facilitates the recruitment of other members of the replication initiator complex (Fig. 1).

CDC6/Cdc18 appears to be central to the mechanism that ensures that initiation of DNA replication occurs once per cell cycle (Fig. 1). In both *S. cerevisiae* and *S. pombe*, CDC6/Cdc18 is an unstable protein expressed transiently in G1 from an mRNA that is itself expressed selectively at this phase of the cell cycle. Thus the degradation of the protein after the origins have fired may prevent re-replication until after the next mitosis when new CDC6/Cdc18 protein will again be synthesized. In support of this hypothesis overexpression of *S. pombe* Cdc18 in G2 results in re-initiation of DNA replication without going through mitosis (Nishitani and Nurse 1995). The situation is slightly different in *Xenopus* eggs and in mammalian cells, where the steady state level of CDC6/Cdc18 protein appears unchanged as cells proceed from G1 through S, G2 and M (Coleman et al. 1996; Saha et al., unpubl.). Therefore destruction of CDC6/Cdc18 protein may not be the mechanism by which animal cells restrict the frequncy of DNA replication to once per cell cycle. In both these instances, however, CDC6/Cdc18 protein is removed from the nucleus to the cytoplasm during S phase. Selective translocation of CDC6/Cdc18 out of the nucleus into the cytoplasm after onset of S (and presumably a selective re-entry into the nucleus after M) could be an alternative mechanism by which animal cells restrict the frequency of the initiation of DNA replication to once per cell cycle.

"Replication licensing" is a convenient term invoked to explain the observation that *Xenopus* factors essential for DNA replication disappear from the nucleus during S phase (loss of license) and must re-enter the nucleus at mitosis (or through the artificial permeabilization of nuclei in G2) to make the nucleus competent (licensed) for another round of DNA replication (Blow and Laskey 1988; Fig. 1). The exclusion of CDC6 from the nucleus during S phase suggests that CDC6 may be part of this licensing system. However, CDC6 has not yet been purified as one of the proteins responsible for the licensing activity.

3.3 MCM Proteins

The mini-chromosome maintenance (MCM) proteins are a complex of six related proteins essential for DNA replication, all of which contain NTP binding motifs (Chong et al. 1996). Originally identified in *S. cerevisiae* as genes essential for the maintenance of mini-chromosomes (MCM) (Yan et al. 1991) or as cell division cycle (CDC) mutants (Hennessy et al. 1990), some of the MCM proteins were independently identified because they co-purify with DNA polymerase alpha from mammalian cells (Thommes et al. 1992). Conditional mutants of MCM genes in yeast and immunodepletion of MCM protein from *Xenopus* egg extracts indicate that these proteins are essential for DNA replication. ORC and CDC6 have to be present for the loading of MCM proteins on to chromatin in *Xenopus* egg extracts, implying that at least some of the MCM proteins are likely to be at origins of DNA replication as part of the initiator protein complex (Coleman et al. 1996; Romanowski et al. 1996; Rowles et al. 1996). Synthetic lethality between certain alleles of ORC subunits and of MCM subunits in yeast also implies that the MCM proteins function at origins of replication.

MCM proteins play an important role in the regulation of DNA replication during the cell cycle. While MCM is associated with the DNA during G1, there is a dramatic decrease in the affinity of the MCM proteins for some nuclear anchor during DNA replication (Fig. 1). In *S. cerevisiae*, the protein is excluded from the nucleus to the cytoplasm during S phase to reenter the nucleus during the next G1 in time to participate in the next round of initiation of DNA replication (Hennessy et al. 1990). In *Xenopus* the protein also disappears from the chromatin during DNA replication and has been purified as part of a complex of proteins essential for licensing the nucleus to replicate after mitosis (Chong et al. 1995; Kubota et al. 1995; Madine et al. 1995). In mammalian cells, although the protein remains in the nucleus throughout the cell cycle, its affinity for a subnuclear anchor changes during S phase such that it is easily extracted in high-salt buffers (Todorov et al. 1994). The increased extractability of MCM2 has been correlated with increased phosphorylation of the protein as the cell cycle progresses through S phase. Thus the MCM proteins could be part of the replication initiator complex which is displaced (perhaps by phosphorylation) as origins are fired. The MCMs are prevented from reloading on the chromatin in S, G2 and M (perhaps by the phosphorylation of MCM and/or the exclusion of CDC6/Cdc18 from the nucleus). This stops the nucleus from re-replicating until the switch is re-set in G1 such that the MCM can again load at origins of DNA replication.

3.4
CDC45, CDC14

Additional CDC genes have been implicated in origin function by genetic studies in *S. cerevisiae*. Mutations in *CDC45* result in yeast unable to replicate their DNA with defects in origin firing (Hopwood and Dalton 1996; Zou et al. 1997). Genetic and biochemical interactions between *CDC45* and *MCM* also implicate the CDC45 protein in origin function. The level and nuclear localization of CDC45 protein in *S. cerevisiae* is unchanged through the cell cycle. Homologues have been identified in higher eukaryotes (Saha and Dutta, unpublished).

Defects in the *CDC14* gene result in yeast arresting with a 2n DNA content (in mitosis). However, yeast with these mutations show a defect in plasmid maintenance which is rescued by adding multiple copies of an ARS to a given plasmid (Hogan and Koshland 1992). A similar phenotype has been described for proteins suspected to be present in pre-replication complexes CDC6, ORC2, ORC3, ORC5 (see above) and DBF4 (see below). Increasing the number of origins on a plasmid increases the likelihood of at least one origin firing on a plasmid, compensating for the decreased activity of these putative initiator proteins. By this criterion, *CDC14* (and *CDC5*, see below) could be involved in pre-replication complex formation. However, a more direct biochemical demonstration is necessary of the involvement of the CDC14 protein in origin function. The biochemical activity of the CDC14 protein is unclear and no homologues have been described in other species.

4 Protein Kinases at Origins of Replication

4.1 Cyclin-Dependent Kinases

Cyclin-dependent kinases (CDKs) active in the S phase of the cell cycle have been implicated in the initiation of DNA replication in animal cells and extracts (Fig. 1). Cell fusion experiments by Rao and Johnson (1970) showed that S phase nuclei have an activator which hastens the onset of replication of G1 nuclei. This activator, the S phase promoting factor (SPF), could be partly composed of CDKs. DNA replication from the SV40 origin of replication and unwinding of the origin DNA were less efficient in extracts from G1 cells compared to extracts from S phase cells (Roberts and D'Urso 1988). Some of this difference could be attributed to the activity of cyclin-CDK, because addition of recombinant cyclin A or purified CDK to G1 extracts stimulated replication (D'Urso et al. 1990; Dutta and Stillman 1992). A similar effect was obtained in *Xenopus* egg extracts, where depletion of CDK (Blow and Nurse 1990) or inhibition of the kinase by specific inhibitors like p21/CIP1/Waf1 dramatically inhibits DNA replication (Strausfeld et al. 1994; Chen et al. 1995; Jackson et al. 1995). The addition of G1 HeLa cell nuclei to S phase extracts also stimulated DNA replication, and the stimulatory action of S phase extracts was mimicked by the addition of recombinant cyclin-CDK (Krude et al. 1997).

Consistent with the result described above, microinjection of antibodies to cyclins and CDKs into mammalian cells inhibits the onset of S phase, indicating the requirement for cyclin-CDKs for the G1-S transition (Pagano et al. 1993; Ohtsubo et al. 1995). Similarly, overexpression of inhibitors of cyclin-CDKs, p21 or p27, results in a block of cells at the G1-S transition (Harper et al. 1993). Mutations in the overexpressed p21 or p27 molecules demonstrated that they had to inhibit cyclin-dependent kinases to produce this G1-S block (Chen et al. 1995; Luo et al. 1995).

CDKs are required for S phase in *S. cerevisiae*. CDC28 (cdk1) in *S. cerevisiae* codes for the catalytic component of yeast cyclin-CDKs. Conditional mutations in this gene demonstrated a requirement for active kinase for entry into S phase (Pines 1994a, b). The regulatory cyclin subunit required is best inferred from the time of appearance of the relevant subunit relative to S phase. CLB5 and CLB6 are the cyclins in *S. cerevisiae* which appear at the onset of S phase (Epstein and Cross 1992; Schwob and Nasmyth 1993). Mutation of these two genes results in a prolongation of S phase. Absolute block to S phase is not seen because the other B-type cyclins, particularly CLB3 and CLB4, can functionally substitute for CLB5 and 6. Several lines of evidence indicate that the G1 cyclins (CLNs) are indirectly essential for S phase because they induce the expression of the S phase cyclins (CLB5 and CLB6) and remove an inhibitor of the S phase CDK. Not only is CLB5 expression activated by overexpression of the G1 cyclin CLN3, but CLB5 overexpression also bypasses the requirement of all the G1 cyclins (CLNs). In addition, deletion of the *SIC1* gene, an inhibitor of the S phase cyclins CLB5 and 6, also makes the three CLN genes dispensable for viability (Schneider et al. 1996). The G1 cyclin-CDK1 complexes phosphorylate SIC1 and target it for degradation by the CDC34/CDC53/CDC4/Skpp1 proteolytic machinery (King et al. 1996). Therefore, the CLNs inactivate SIC1 and in parallel induce the synthesis of the S phase cyclins (CLB5 and 6), thereby triggering DNA replication

Cdks are essential for DNA replication in *S. pombe*. Here, Cdc2 (CDK1) is the catalytic subunit and Cig1 and Cig2/Cyc17 the inferred S phase cyclins. Wild-type Cdc2 is essential for S phase. The Cig1 and Cig2 cyclins are expressed at the G1-S transition, although the mitotic cyclin, Cdc13, can compensate for their absence (Fisher and Nurse 1996; Mondesert et al. 1996). Evidence for the requirement of Cig2 in S phase comes from a strain mutated in *Cdc13*. Having lost the mitotic cyclin, these cells undergo repeated rounds of S phase without undergoing mitosis (see later). However, the repeated firing of origins of replication is prevented by a simultaneous mutation in *Cig2*, implicating this cyclin in the onset of DNA replication. Repression of cyclin-CDK kinase activity is probably utilized physiologically to block the onset of S phase because Rum1, an inhibitor of cyclin-CDK, is required for *S. pombe* cells to arrest in G1 following nitrogen starvation (Moreno and Nurse 1994).

Of course, several of these experiments described above indicate only that cyclin-CDKs are essential for S phase, not that the kinase directly activates the replication apparatus. The kinase could be required to activate the transcription of genes necessary for S phase (which they are), or to remodel the nucleus or the chromatin to facilitate DNA replication. DNA replication in *Xenopus* egg extracts, however, does not require the synthesis of new proteins and yet requires active cyclin-CDK kinase as evidenced by depletion of cyclin E or addition of p21. Therefore CDK is necessary for directly activating the DNA replication apparatus at least in *Xenopus* egg extracts.

The co-localization of cyclin-CDKs with other replication proteins and with sites of new DNA synthesis also suggests an involvement of these kinases in replication initiation and elongation (Cardoso et al. 1993). In mammalian cells (and *Xenopus* egg extracts) cyclin-CDKs are localized at replication foci, although how they are recruited to these foci is unclear. Cyclin-CDKs associate physically with CDC6/Cdc18, a component of the initiator protein complex, so that the kinase could be recruited to pre-replication foci through such associations (Elsasser et al. 1996; Piatti et al. 1996; Saha et al., unpubl.). In *S. pombe*, GST-Orc2, a component of the origin recognition complex, co-precipitates with CDK1, providing added evidence that cyclin-CDKs associate with the pre-replication complex (Leatherwood et al. 1996).

Requirement of a certain type of cyclin-CDK for the onset of S phase imposes a strong link between the position of the cell in the cell cycle and the firing of the origins of DNA replication. In most organisms there are two types of cyclin-CDKs, some with cyclins present and active only in G1-S and others with cyclins present and active in G2-M. The prevailing view is that the former is required for the initiation of DNA replication. As we shall discuss later, the G2-M CDKs are not only poor substitutes at G1-S but may actively inhibit the assembly of pre-replication complexes essential for the origins to become competent to initiate DNA replication. The alternating appearance and activity of G1-S cyclins (activator of S phase) and G2-M cyclins (inhibitor of pre-replication complex formation) in the cell cycle could ensure that each S phase alternates with an M phase. How the various cyclin-CDKs promote DNA replication or prevent pre-replication complex assembly are important directions of future research.

4.2 DBF4-CDC7

DBF4-CDC7 is yet another protein kinase whose activity is required in the yeasts for the onset of DNA replication (Johnston and Thomas 1982; Patterson et al. 1986; Hollingsworth and Sclafani 1990; Yoon and Campbell 1991; Kitada et al. 1992). Mutations in *S. cerevisiae CDC7* result in defects in DNA replication, repair and recombination. CDC7 kinase requires another protein subunit, DBF4, for it to function as a protein kinase (Jackson et al. 1993). Mutations in *DBF4* also produce yeast that fail to enter S phase. A homologue of the *CDC7* gene has been discovered in *S. pombe* (*Hsk1*) and in *Xenopus* and humans (Masai et al. 1995). Hskl is essential for viability in *S. pombe*, with the newly germinated spores carrying the null mutation blocked at the G1-S transition. Although not active as a kinase by itself, it appears to have kinase activity when isolated from the cognate host, implying that as in *S. cerevisiae*, a regulatory subunit and/or posttranslational modifications are necessary for this kinase to be active.

Although it is not yet certain that CDC7 kinase activates the DNA replication apparatus directly, the evidence progressively points in that direction. DBF4 not only associates with CDC7, but also associates with proteins bound to yeast origins of DNA replication (Dowell et al. 1994). *CDC7* and *ORC2* interact with each other genetically and the two proteins associate (perhaps indirectly) in a two-hybrid assay (Hardy 1996). Therefore the DBF4-CDC7 kinase is most likely localized to the replication initiation complexes. A mutation in the *BOB1* gene produces yeast strains that are no longer dependent on *CDC7* for entry into S phase. *BOB1* has been recently identified as *MCM5* (*CDC46*) which, as we have discussed above, is a component of the initiator complex (Hardy et al. 1997). Thus specific mutations in the MCM protein allow the pre-replication complex to fire without the activity of CDC7, suggesting that in wild-type strains CDC7 kinase is required to overcome an inhibitory effect of MCM5 on the initiation of DNA replication.

It is interesting that a factor essential for DNA replication (MCM5) and formation of pre-replication complexes could simultaneously be an inhibitor for the firing of the origins that must be overcome by the action of DBF4-CDC7. If true, there is a converse similarity between the MCM5 protein and CLB-CDC28 kinase in this respect. The latter is an activator of origin firing which is simultaneously an inhibitor of pre-replication complex formation (Fig. 1). Conversely, the MCM5 protein is an activator of pre-replication complex formation (Fig. 1) which may simultaneously be an inhibitor of origin firing. Perhaps other factors involved in the initiation of DNA replication will fall into these two converse groups: those that stimulate pre-replication complex assembly but inhibit origin firing (like MCM5) and those that stimulate origin firing but inhibit pre-replication complex assembly (like CDKs). The cell cycle can be divided into two mutually exclusive states: a replication state (S, G2 and M), during which competent origins can fire, and an assembly state (G1), during which origins are made competent by the assembly of the initiator protein complex (Fig. 1). The alternation between these two states has been proposed to be governed by the activation (replication state) or inactivation (assembly state) of CDKs. If factors involved in the assembly state (like MCM5) are simultaneously inhibitors of the replication state which must be inactivated by CDK activity before origin firing, we might have a superficial explanation of why the onset of DNA replication requires CDK activity.

Because CLB-CDC28 phosphorylates and activates CDC7 kinase (Yoon et al. 1993), the question arises: are CLB-CDC28 and DBF4-CDC7 kinases in the same or different pathways for origin firing? The two kinases may have different substrates , e.g., Cdc18 or ORC2 for CLB-CDC28, and MCM5 or a protein that interacts with MCM5 for DBF4-CDC7, the phosphorylation of all of which is essential for origin firing, so that the *bob1* mutation while bypassing CDC7 may still require CLB-CDC28 kinase activity for DNA replication. Alternatively, the *bob1* mutation could simultaneously bypass the requirement for both DBF4-

CDC7 and CLB-CDC28 for origin firing, suggesting that the two kinases are in the same pathway required to de-repress MCM and permit origin firing.

4.3 DBF4-CDC5

A third protein kinase has been recently implicated in the G1-S transition. Overexpression of *S. cerevisiae* CDC5 rescued a temperature-sensitive *dbf4* allele (Kitada et al. 1993). DBF4 and CDC5 proteins associate with each other physically (Hardy and Pautz 1996). *cdc5-1* and *orc2-1* are synthetically lethal, suggesting an interaction between CDC5 and ORC. The high frequency of plasmid loss seen in *cdc5-1* mutants is partially rescued by plasmids containing multiple origins, which, as we have discussed earlier could indicate that mutations in *CDC5* result in defects in origin firing. These results indicate that CDC5 interacts with ORC2 and with the origin of replication, components of the pre-replication complex, implicating the protein in G1-S transition. However, cells deficient in CDC5 arrest at anaphase/telophase with a postreplication complex at the origin, implicating the protein in mitosis.

Homologues of CDC5 discovered in other species also support a role of the protein in both G2-M and G1-S. The *Drosophila* homologue, Polo (Fenton and Glover 1993) and the *S. pombe* homologue, Plo (Ohkura et al. 1995) are both required at mitosis. A homologue in *Xenopus*, Plx, phosphorylates and activates the mitotic Cdc25C phosphatase (Kumagai and Dunphy 1996). Several homologues have been identified in mice and humans and based on their time of expression could be divided into G1-S kinases (Prk, Snk, Fnk) or G2-M kinases (Plk; Fode et al. 1994; Golsteyn et al. 1994; Donohue et al. 1995; Lee et al. 1995; Li et al. 1996). Prk kinase, for example, is induced within 2 h of addition of serum or cytokines to quiescent cells. Fnk kinase was discovered by a differential display screen for immediate early genes induced after addition of FGF1. In contrast, Plk is a murine kinase which is induced 17 h after serum addition during G2 and M, and the protein is intimately associated with the mitotic spindle. Plk therefore resembles *S. cerevisiae* CDC5, *S. pombe* Plo or *Drosophila* Polo in its activity profile relative to the cell cycle.

Is requirement of CDC5-DBF4 at G1-S independent of its activity at G2-M? The genetic interaction of CDC5 with components of the pre-replication complex could be explained by an indirect action of CDC5 on pre-replication complex assembly. Since *cdc5* cells are defective in mitosis, the cells arrest with highly active mitotic CDK, which indirectly inhibits the establishment of pre-replication complexes. Any additional defect in pre-replication complex components (as in *orc2-1* strains) could lead to synthetic lethality with *cdc5*. However, the existence in mammalian cells of a group of CDC5-like kinases which are induced when cells move from G0 to S tends to support an independent role of CDC5-like kinases in G1-S and G2-M.

5
Stimulation of DNA Replication by Protein Kinase at G1-S

Despite the list of protein kinases implicated in the stimulation of S phase, almost nothing is known about how they stimulate replication. The substrates of some of these kinases could be components of the pre-replication complex or factors involved in elongation of replication products.

One clue to the possible role of cyclin-dependent kinases comes from a determination of which step of DNA replication in cell-free extracts is blocked upon inhibition of these kinases. In *Xenopus* egg extracts the inhibition is specific to DNA replication on double stranded DNA, while DNA synthesis on single-stranded DNA or on already initiated replication forks is unimpeded. In the SV40 in vitro DNA replication reaction, the origin unwinding step was specifically diminished in extracts from G1 cells which lacked cyclin-CDK kinase activity. Therefore, opening of double-stranded DNA at origins of replication could require cyclin-CDK kinase activity. However, we cannot rule out that the origins are slightly unwound or even that small primers are already laid down at the origins (small enough to be undetected by current techniques) before the onset of S phase. In such a scenario, activation of S phase by cyclin-CDKs may involve the removal of some barrier to extension of these primers (and accompanying DNA unwinding). There is no experimental evidence to address this issue.

Could the cyclin-CDKs phosphorylate components of the pre-replication complex and activate DNA replication? The sequence of yeast ORC subunits shows putative CDK phosphorylation sites but their functional importance is unknown. CDC6/Cdc18 has multiple putative substrate sites for cyclin-CDK kinases, and human Cdc18 is a robust substrate for these kinases in vitro (Saha et al., unpubl.). In *S. pombe*, overexpression of an inhibitor of CDKs, Rum1, or of *S. cerevisiae* Sic1 suppresses a conditional mutation in the *Cdc18* gene implying that CDKs and Cdc18 are in opposition (Jallepalli and Kelly 1996). However, no evidence has been published to date that indicates whether the ORC or CDC6/Cdc18 proteins are phosphorylated in vivo. In contrast, the MCM proteins have been shown to be phosphorylated in vivo beginning from the G1-S transition to M, and at least in one case a CDK was implicated as the kinase that phosphorylates the protein in vivo (Hendrickson et al. 1996).

Even if the CDKs phosphorylate initiator proteins at the G1-S transition, the biochemical mechanism by which this phosphorylation translates into the initiation of DNA replication is not clear. The CDKs that promote origin firing are simultaneously responsible for preventing the assembly of pre-replication complexes, so that a minimal hypothesis could be that disassembly of the pre-replication complexes is all that is required for the initiation of DNA replication. For example, ectopic overexpression of CLB2 in G1 inhibits pre-replication complex formation (Piatti et al. 1996). However, this inhibition is overcome by the presence of CDC6 protein mutated to eliminate all potential sites of phosphorylation by CLB-CDK1 (J. Li, pers. comm.). Therefore CDKs could promote origin firing by initiating the disassembly of the pre-replication complex through the phosphorylation of CDC6 protein. Several MCM proteins are also phosphorylated at the onset of S phase (Todorov et al. 1995; Hendrickson et al. 1996). The phosphorylation of MCM2 is correlated with increased extractability from chromatin. A mutation in the *S. cerevisiae MCM5* allows the yeast to bypass the requirement of CDC7-DBF4 for the onset of S phase. Therefore the CDK and CDC7 protein kinases could promote the dissociation of MCM proteins from the pre-replication complex, consistent with the minimal hypothesis cited above. It still remains possible, however, that the disassembly of the pre-replication complex is an entirely independent event that is coincident with but not the cause of the initiation of DNA replication, in which case we are still left searching for the biochemical mechanism by which the CDKs stimulate the DNA replication apparatus.

CDKs phosphorylate several DNA replication factors involved in elongation, notably the 34-kDa subunit of RPA and the 180-kDa subunit of DNA polymerase alpha (Nasheuer et al. 1991; Dutta and Stillman 1992). Both these proteins are phosphorylated from the onset of S phase, and remain phosphorylated through G2 up to the end of mitosis. Although the synchronous onset of phosphorylation with the onset of DNA synthesis was interpreted to suggest that the phosphorylation activated these replication factors, there is no direct biochemical evidence that such is the case. The persistence of the phosphorylated state up to mitosis could also be a mechanism by which the DNA replication proteins are marked as S phase proceeds so that they cannot support DNA replication until the next M phase. Alternatively, the phosphorylation could earmark the proteins for replication related activities such as DNA repair or recombination.

6 Inhibition of Pre-replication Complex Assembly in G2-M

Besides the activation of DNA replication at the G1-S transition, the cell cycle also inhibits the replication apparatus such that replication is not initiated during G2-M. Once the origins of replication have been fired, the cell prevents the newly replicated origins from firing until after the next mitosis. In this manner the chromosomal complement is not needlessly duplicated. In essence, at each origin the switch is flipped to an off position until it is reset at the next mitosis. The mechanism by which this is achieved has recently become clearer (Fig. 1).

6.1
Limiting CDC6/Cdc18.

An important mechanism in the yeasts is to degrade the CDC6/Cdc18 protein such that pre-replication complex cannot be assembled until new CDC6/Cdc18 is synthesized in the next G1 phase. In mammalian cells and in *Xenopus* egg extracts, the CDC6/Cdc18 is not degraded but the protein may be excluded out of the nucleus during DNA replication, achieving the same end result of limiting the availability of this critical protein until after the next mitosis.

6.2
Active Mitotic Cyclin Dependent Kinases.

In addition, active CDKs appear essential for preventing re-replication until after mitosis. Mutations in *S. pombe Cdc2* or *Cdc13* gene (the mitotic cyclin B) inactivated the G2-M cyclin and permitted re-replication without mitosis (Broek et al. 1991; Hayles et al. 1994). Similarly transient overexpression of Rum1, which is the inhibitor of mitotic CDK, also allowed the cell to re-replicate its DNA without going through a mitosis (Moreno and Nurse 1994; Correa-Bordes and Nurse 1995). Transient expression of dominant-negative mutants of *cdc2* which inactivate the mitotic kinase also inactivate the off switch. In *S. cerevisiae*, transient overexpression of SIC1, the inhibitor of CLB-CDK1, resets the off switch in G2 such that replication is allowed without passing through mitosis (Dahmann et al. 1995). Thus the same CDK whose activity is required to fire the origin at the G1-S transition is also responsible for preventing the origin from refiring.

The active mitotic CDK prevents the assembly of pre-replication complexes. Artificial induction of CDC6 in *S. cerevisiae* did not promote DNA replication or produce a footprint at origins of DNA replication corresponding to the pre-replication complex when the induction was done in the presence of active CLB-CDK. Induction of CDC6 in the presence of inactive CLB-CDK, however, promoted pre-replication complex assembly.

The two mechanisms for preventing initiator protein assembly, namely, limitation of CDC6/Cdc18 and increased CDK activity, may be interdependent. Overexpression of Rum1 in *S. pombe* complements a temperature-sensitive mutation of *Cdc18* and increases the abundance of the protein (Jallepalli and Kelly 1996). Therefore, increased activity of CDK somehow destabilizes the Cdc18 protein, perhaps by keeping the proteolysis mechanism active and preventing the accumulation of the protein. Since Cdc18 is phosphorylated in vitro by cyclin-CDKs, these kinases may destabilize the Cdc18 protein by phosphorylating it and targeting it for proteolysis. Conversely, since very high overexpression of Cdc18 in *S. pombe* induces re-replication, it has been surmised that at least in *S. pombe* limiting quantities of Cdc18 is an important mechanism that prevents re-replication in G2. However, Cdc18 has multiple substrate sites for cyclin-CDKs and associates stably with cyclin-CDKs so that overexpressed Cdc18 could titrate away CDKs from their relevant substrates in G2, in effect acting as an inhibitor of G2-M kinases. Consistent with the latter possibility, overexpression of a fragment of Orc1 which contains putative substrate sites for CDK resulted in re-replication in *S. pombe* (Wolf et al. 1996).

6.3
Other factors

Other mutations that allow re-replication in G2 include an intriguing mutation in *SEC72* encoding a protein required for translocation of proteins into the endoplasmic reticulum of *S. cerevisiae* (Dahmann et al. 1995). How this mutation causes re-replication is unclear, although a decrease in mitotic CDK activity was noted. In *S. cerevisiae*, mutations in *CDC16* and *CDC27*, components of the anaphase promoting complex, i.e., the proteolytic machinery that becomes active in G2-M, also lead to re-replication of DNA despite high mitotic CDK activity (Heichmanand Roberts 1996). This implies that there are yeast factors essential for formation of the initiator protein complex and/or for firing of origins of DNA replication which are normally degraded by the G2-M proteolytic machinery. Limitation of these factors could be important for preventing re-replication in G2-M. The factor(s) stabilized in the *CDC16* and *CDC27* mutants cannot be restricted to CDC6/Cdc18 because overexpression of CDC6 alone in *S. cerevisiae* does not stimulate DNA replication in the presence of active mitotic CDK (Piatti et al. 1996). Thus destruction of unknown factors by the G2-M proteolytic machinery contributes to the prevention of re-replication until after mitosis.

7
Conclusion

It is now clear that the cell cycle machinery is intimately involved in the regulation of the initiation of DNA replication. The activity of cyclin-dependent kinases and the state of two different ubiquitin-mediated proteolytic pathways change as the cell progresses through the cell cycle, and these changes are translated into whether the DNA replication factors are in a state which promotes assembly of initiator proteins at origins or the movement of replication forks away from the origins. The assembly and replication states alternate with each other through each round of the cell cycle, and it is this alternation that drives the cell through alternate bouts of DNA replication and mitosis.

Much remains to be learned, however, about the regulation of DNA replication in eukaryotes. Although several initiation factors have been discovered, it is still unclear how they establish the DNA replication apparatus at the origins and how the transition to actively moving bidirectional replication forks is achieved. In animal cells even the DNA sequences responsible for acting as origins of DNA replication are unclear. The regulated localization of DNA replication factors in the cytosol, the nucleus or in different subnuclear compartments through different phases of the cell cycle is an understudied mechanism by which S phase may be regulated in animal cells. The role of chromatin assembly in the initiation of DNA replication is unclear, as is the mechanism by which the DNA repair apparatus is tightly linked to the replication process so that errors in DNA replication are quickly corrected before being fixed in the progeny. We can only look forward to an ever-increasing comprehension of these intricate processes that form the very foundation of life.

References

Bell SP, Stillman B (1992) ATP-dependent recognition of eukaryotic origins of DNA replication by a multiprotein complex. Nature 357: 128–134

Blow JJ, Laskey RA (1988) A role for the nuclear envelope in controlling DNA replication within the cell cycle. Nature 332: 546–548

Blow JJ, Nurse P (1990) A cdc2-like protein is involved in the initiation of DNA replication in *Xenopus* egg extracts. Cell 62: 855–862

Brewer BJ (1994) Intergenic DNA and the sequence requirements for replication initiation in eukaryotes. Curr Opin Genet Dev 4: 196–202

Broek D, Bartlett R, Crawford K, Nurse P (1991) Involvement of p34cdc2 in establishing the dependency of S phase on mitosis. Nature 349: 388–393

Brown JA, Holmes SG, Smith MM (1991) The chromatin structure of *Saccharomyces cerevisiae* autonomously replicating sequences changes during the cell division cycle. Mol Cell Biol 11: 5301–5311

Cardoso MC, Leonhardt H, Nadal GB (1993) Reversal of terminal differentiation and control of DNA replication: cyclin A and CDK2 specifically localize at subnuclear sites of DNA replication. Cell 74: 979–992

Carpenter PB, Mueller PR, Dunphy WG (1996) Role for a *Xenopus* Orc2-related protein in controlling DNA replication. Nature 379: 357–360

Chen J, Jackson PK, Kirschner MW, Dutta A (1995) Separate domains of p21 involved in the inhibition of CDK kinase and PCNA. Nature 374: 386–388

Chong JP, Mahbubani HM, Khoo CY, Blow JJ (1995) Purification of an MCM-containing complex as a component of the DNA replication licensing system. Nature 3755: 418–421

Chong JP, Thommes P, Blow JJ (1996) The role of MCM/P1 proteins in the licensing of DNA replication. Trends Biochem Sci 21: 102–106

Coleman TR, Carpenter PB, Dunphy WG (1996) The *Xenopus* Cdc6 protein is essential for the initiation of a single round of DNA replication in cell-free extracts. Cell 87: 53–63

Correa-Bordes J, Nurse P (1995) p25rum1 orders S phase and mitosis by acting as an inhibitor of the p34cdc2 mitotic kinase. Cell 83: 1001–1009

D'Urso G, Marraccino RL, Marshak DR, Roberts JM (1990) Cell cycle control of DNA replication by a homologue from human cells of the p34cdc2 protein kinase. Science 250: 786–791

Dahmann C, Diffley JF, Nasmyth KA (1995) S-phase-promoting cyclin-dependent kinases prevent re-replication by inhibiting the transition of replication origins to a pre-replicative state. Curr Biol 5: 1257–1269

Diffley JF ,Cocker JH (1992) Protein-DNA interactions at a yeast replication origin. Nature 357: 169–172

Diffley JF, Cocker JH, Dowell SJ, Rowley A (1994) Two steps in the assembly of complexes at yeast replication origins in vivo. Cell 78: 303–316

Donohue PJ, Alberts GF, Guo Y, Winkles JA (1995) Identification by targeted differential display of an immediate early gene encoding a putative serine/threonine kinase. J Bio Chem 270: 10351–10357
Dowell SJ, Romanowski P, Diffley JF (1994) Interaction of Dbf4, the Cdc7 protein kinase regulatory subunit, with yeast replication origins in vivo. Science 265: 1243–1246
Dubey DD, Kim SM, Todorov IT, Huberman JA (1996) Large, complex modular structure of a fission yeast DNA replication origin. Curr Biol 6: 467–473
Dutta A, Bell SP (1997) Initiation of DNA replication in eukaryotic cells. Annu Rev Cell Dev Biol 13: 293–332
Dutta A, Stillman B (1992) cdc2 family kinases phosphorylate a human cell DNA replication factor, RPA, and activate DNA replication. EMBO J 11: 2189–2199
Elsasser S, Lou F, Wang B, Campbell JL, Jong A (1996) Interaction between yeast Cdc6 protein and B-type cyclin/Cdc28 kinases. Mol Biol Cell 7: 1723–1735
Epstein CB, Cross FR (1992) CLB5: a novel B cyclin from budding yeast with a role in S phase. Genes Dev 6: 1695–1706
Fenton B, Glover DM (1993) A conserved mitotic kinase active at late anaphase-telophase in syncytial *Drosophila* embryos. Nature 363: 637–640
Fisher DL, Nurse P (1996) A single fission yeast mitotic cyclin B p34cdc2 kinase promotes both S-phase and mitosis in the absence of G1 cyclins. EMBO J 15: 850–860
Fode C, Motro B, Yousefi S, Heffernan M, Dennis JW (1994) Sak, a murine protein-serine/threonine kinase that is related to the Drosophila polo kinase and involved in cell proliferation. Proc Natl Acad of Sci USA 91: 6388–63992
Gavin KA, Hidaka M, Stillman B (1995) Conserved initiator proteins in eukaryotes. Science 270: 1667–1671
Gilbert D, Miyazawa H, DePamphilis M (1995) Site-specific initiation of DNA replication in *Xenopus* egg extract requires nuclear structure. Mol Cell Biol 15: 2942–2954
Golsteyn RM, Schultz SJ, Bartek J, Ziemiecki A, Ried T, Nigg EA (1994) Cell cycle analysis and chromosomal localization of human Plk1, a putative homologue of the mitotic kinases *Drosophila* polo and *Saccharomyces cerevisiae* Cdc5. J Cell Sci 107: 1509–1517
Gossen M, Pak DT, Hansen SK, Acharya JK, Botchan MR (1995) A *Drosophila* homolog of the yeast origin recognition complex. Science 270: 1674–1677
Hamlin JL, Mosca PJ, Levenson VV (1994) Defining origins of replication in mammalian cells. Biochim Biophys Acta 1198: 85–111
Hardy CF (1996) Characterization of an essential Orc2p-associated factor that plays a role in DNA replication. Mol Cell Biol 16: 1832–1841
Hardy CF, Pautz A (1996) A novel role for Cdc5p in DNA replication. Mol Cell Biol 16: 6775–6782
Hardy CFJ, Dryga O, Pahl PMP, Sclafani RA (1997) *mcm5/cdc46-bob1* bypasses the requirement for the S phase activator Cdc7p. Proc Natl Acad Sci USA 94: 3151–3155
Harper JW, Adami GR, Wei N, Keyomarsi K, Elledge SJ (1993) The p21 CDK-interacting protein Cip1 is a potent inhibitor of G1 cyclin-dependent kinases. Cell 75: 805–816
Hayles J, Fisher D, Woollard A, Nurse P (1994) Temporal order of S phase and mitosis in fission yeast is determined by the state of the p34cdc2-mitotic B cyclin complex. Cell 78: 813–822
Heichman KA, Roberts JM (1996) The yeast CDC16 and CDC27 genes restrict DNA replication to once per cell cycle. Cell 85: 39–48
Hendrickson M, Madine M, Dalton S, Gautier J (1996) Phosphorylation of MCM4 by cdc2 protein kinase inhibits the activity of the minichromosome maintenance complex. Proc Natl Acad Sci USA 93: 12223–12228
Hennessy KM, Clark CD, Botstein D (1990) Subcellular localization of yeast CDC46 varies with the cell cycle. Genes Dev 4: 2252–2263
Hogan E, Koshland D (1992) Addition of extra origins of replication to a minichromosome suppresses its mitotic loss in cdc6 and cdc14 mutants of *Saccharomyces cerevisiae*. Proc Natl Acad Sci USA 89: 3098–31102
Hollingsworth RJ, Sclafani RA (1990) DNA metabolism gene CDC7 from yeast encodes a serine (threonine) protein kinase. Proc Natl Acad Sci USA 87: 6272–6276

Hopwood B, Dalton S (1996) Cdc45p assembles into a complex with Cdc46p/Mcm5p, is required for minichromosome maintenance, and is essential for chromosomal DNA replication. Proc Natl Acad Sci USA 93: 12309–12314

Hyrien O, Mechali M (1992) Plasmid replication in *Xenopus* eggs and egg extracts: a 2D gel electrophoretic analysis. Nucleic Acids Res 20: 1463–1469

Jackson AL, Pahl PM, Harrison K, Rosamond J, Sclafani RA (1993) Cell cycle regulation of the yeast Cdc7 protein kinase by association with the Dbf4 protein. Mol Cell Biol 13: 2899–2908

Jackson PK, Chevalier S, Philippe M, Kirschner MW (1995) Early events in DNA replication require cyclin E and are blocked by p21CIP1. J Cell Biol 130: 755–769

Jallepalli PV, Kelly TJ (1996) Rum1 and Cdc18 link inhibition of cyclin-dependent kinase to the initiation of DNA replication in *Schizosaccharomyces pombe*. Genes Dev 10: 541–552

Johnston LH, Thomas AP (1982) A further two mutants defective in initiation of the S phase in the yeast *Saccharomyces cerevisiae*. Mol Gen Genet 186: 445–448

Kelly TJ, Martin GS, Forsburg SL, Stephen RJ, Russo A, Nurse P (1993) The fission yeast cdc18+ gene product couples S phase to START and mitosis [see comments]. Cell 74: 371–382

King RW, Deshaies RJ, Peters JM, Kirschner MW (1996) How proteolysis drives the cell cycle. Science 274: 1652–1659

Kitada K, Johnston LH, Sugino T, Sugino A (1992) Temperature-sensitive cdc7 mutations of *Saccharomyces cerevisiae* are suppressed by the DBF4 gene, which is required for the G1/S cell cycle transition. Genetics 131: 21–29

Kitada K, Johnson AL, Johnston LH, Sugino A (1993) A multicopy suppressor gene of the *Saccharomyces cerevisiae* G1 cell cycle mutant gene dbf4 encodes a protein kinase and is identified as CDC5. Mol Cell Biol 13: 4445–4457

Kitsberg D, Selig S, Keshet I, Cedar H (1993) Replication structure of the human beta-globin domain. Nature 366: 588–590

Klemm RD, Austin RJ, Bell SP (1997) Coordinate binding of ATP and origin DNA regulates the ATPase activity of the origin recognition complex. Cell 88: 493–502

Krude T, Jackman M, Pines J, Laskey RA (1997) Cyclin/CDK-dependent initiation of DNA replication in a human cell-free system. Cell 88: 109–119

Kubota Y, Mimura S, Nishimoto S, Takisawa H, Nojima H (1995) Identification of the yeast MCM3-related protein as a component of *Xenopus* DNA replication licensing factor. Cell 81: 601–609

Kumagai A, Dunphy WG (1996) Purification and molecular cloning of Plx1, a Cdc25-regulatory kinase from *Xenopus* egg extracts. Science 273: 1377–1380

Leatherwood J, Lopez GA, Russell P (1996) Interaction of Cdc2 and Cdc18 with a fission yeast ORC2-like protein. Nature 379: 360–363

Lee KS, Yuan YL, Kuriyama R, Erikson RL (1995) Plk is an M-phase-specific protein kinase and interacts with a kinesin-like protein, CHO1/MKLP-1. Mol Cell Biol 15: 7143–7151

Li B, Ouyang B, Pan H, Reissmann PT, Slamon DJ, Arceci R, Lu L, Dai W (1996) Prk, a cytokine-inducible human protein serine/threonine kinase whose expression appears to be down-regulated in lung carcinomas. J Biol Chem 271: 19402–19408

Luo Y, Hurwitz J, Massague J (1995) Cell-cycle inhibition by independent CDK and PCNA binding domains in p21Cip1. Nature 375: 159–161

Madine MA, Khoo CY, Mills AD, Laskey RA (1995) MCM3 complex required for cell cycle regulation of DNA replication in vertebrate cells. Nature 375: 421–424

Mahbubani HM, Paull T, Elder JK, Blow JJ (1992) DNA replication initiates at multiple sites on plasmid DNA in *Xenopus* egg extracts. Nucleic Acids Res 20: 1457–1462

Masai H, Miyake T, Arai K (1995) hsk1+, a *Schizosaccharomyces pombe* gene related to *Saccharomyces cerevisiae* CDC7, is required for chromosomal replication. EMBO J 14: 3094–3104

Mondesert O, McGowan CH, Russell P (1996) Cig2, a B-type cyclin, promotes the onset of S in *Schizosaccharomyces pombe*. Mol Cell Biol 16: 1527–1533

Moreno S, Nurse P (1994) Regulation of progression through the G1 phase of the cell cycle by the rum1+ gene. Nature 367: 236-242

Muzi FM, Kelly TJ (1995) Orp1, a member of the Cdc18/Cdc6 family of S-phase regulators, is homologous to a component of the origin recognition complex. Proc Natl Acad Sci USA 92: 12475–12479

Nasheuer HP, Moore A, Wahl AF, Wang TS (1991) Cell cycle-dependent phosphorylation of human DNA polymerase alpha. J Biol Chem 266: 7893–7903

Newlon C (1993) The structure and function of yeast ARS elements. Curr Opin Genet Dev. 3: 752–758

Nishitani H, Nurse P (1995) p65cdc18 plays a major role controlling the initiation of DNA replication in fission yeast. Cell 83: 397–405

Ohkura H, Hagan IM, Glover DM (1995) The conserved *Schizosaccharomyces pombe* kinase plo1, required to form a bipolar spindle, the actin ring, and septum, can drive septum formation in G1 and G2 cells. Genes Dev 9: 1059–1073

Ohtsubo M, Theodoras AM, Schumacher J, Roberts JM, Pagano M (1995) Human cyclin E, a nuclear protein essential for the G1-to-S phase transition. Mol Cell Biol 15: 2612–2624

Pagano M, Pepperkok R, Lukas J, Baldin V, Ansorge W, Bartek J, Draetta G (1993) Regulation of the cell cycle by the CDK2 protein kinase in cultured human fibroblasts. J Cell Biol 121: 101–111

Patterson M, Sclafani RA, Fangman WL, Rosamond J (1986) Molecular characterization of cell cycle gene CDC7 from *Saccharomyces cerevisiae*. Mol Cell Biol 6: 1590–1598

Piatti S, Lengauer C, Nasmyth K (1995) Cdc6 is an unstable protein whose de novo synthesis in G1 is important for the onset of S phase and for preventing a 'reductional' anaphase in the budding yeast *Saccharomyces cerevisiae*. EMBO J 14: 3788–3799

Piatti S, Bohm T, Cocker JH, Diffley JF, Nasmyth K (1996) Activation of S-phase-promoting CDKs in late G1 defines a "point of no return" after which Cdc6 synthesis cannot promote DNA replication in yeast. Genes Dev 10: 1516–1531

Pines J (1994a) The cell cycle kinases. Semin Cancer Biol 5: 305–313

Pines J (1994b) Protein kinases and cell cycle control. Semin Cell Biol 5: 399–408

Rao PN, Johnson RT (1970) Mammalian cell fusion: studies on the regulation of DNA synthesis and mitosis. Nature 225: 159–164

Roberts JM, D'Urso G (1988) An origin unwinding activity regulates initiation of DNA replication during mammalian cell cycle. Science 241: 1486–1489

Romanowski P, Madine MA, Rowles A, Blow JJ, Laskey RA (1996) The *Xenopus* origin recognition complex is essential for DNA replication and MCM binding to chromatin. Curr Biol 6: 1416–1425

Rowles A, Chong JPJ, Brown L, Howell M, Evan GI, Blow JJ (1996) Interaction between the origin recognition complex and the replication licensing system in *Xenopus*. Cell 87: 287–296

Rowley A, Cocker JH, Harwood J, Diffley JFX (1995) Initiation complex assembly at budding yeast replication origins begins with the recognition of a bipartite sequence by limiting amounts of the initiator, ORC. EMBO J 14: 2631–2641

Schneider BL, Yang QH, Futcher AB (1996) Linkage of replication to start by the CDK inhibitor Sic1. Science 272: 560–562

Schwob E, Nasmyth K (1993) CLB5 and CLB6, a new pair of B cyclins involved in DNA replication in *Saccharomyces cerevisiae*. Genes Dev 7: 1160–1175

Stillman B (1993) DNA replication. Replicator renaissance. Nature 366: 506–507

Strausfeld UP, Howell M, Rempel R, Maller JL, Hunt T, Blow JJ (1994) Cip1 blocks the initiation of DNA replication in *Xenopus* extracts by inhibition of cyclin-dependent kinases. Current Biol 4: 876–883

Thommes P, Fett R, Schray B, Burkhart R, Barnes M, Kennedy C, Brown NC, Knippers R (1992) Properties of the nuclear P1 protein, a mammalian homologue of the yeast Mcm3 replication protein. Nucleic Acids Res 20: 1069–1074

Todorov IT, Pepperkok R, Philipova RN, Kearsey SE, Ansorge W, Werner D (1994) A human nuclear protein with sequence homology to a family of early S phase proteins is required for entry into S phase and for cell division. J Cell Sci 107: 253–265.

Todorov IT, Attaran A ,Kearsey SE (1995) BM28, a human member of the MCM2-3-5 family, is displaced from chromatin during DNA replication. J Cell Biol 129: 1433–1445

Williams RS, Shohet RV, Stillman, B (1997) A human protein related to yeast Cdc6p. Proc Natl Acad Sci USA 94: 142–147

Wolf DA, Wu D, McKeon F (1996) Disruption of re-replication control by overexpression of human ORC1 in fission yeast. J Biol Chem 271: 32503-32506

Wu JR, Gilbert DM (1996) A distinct G1 step required to specify the Chinese hamster DHFR replication origin. Science 271: 1270–1272

Yan H, Gibson S, Tye BK (1991) Mcm2 and Mcm3, two proteins important for ARS activity, are related in structure and function. Genes Dev 5: 944–957

Yoon HJ, Campbell JL (1991) The CDC7 protein of *Saccharomyces cerevisiae* is a phosphoprotein that contains protein kinase activity. Proc Natl Acad Sci U S A 88: 3574–3578

Yoon HJ, Loo S, Campbell JL (1993) Regulation of *Saccharomyces cerevisiae* CDC7 function during the cell cycle. Mol Biol Cell 4: 195–208

Zhou C, Huang SH, Jong AY (1989) Molecular cloning of *Saccharomyces cerevisiae* CDC6 gene. Isolation, identification, and sequence analysis. J Biol Chem 264: 9022–9029

Zou L, Mitchell J, Stillman B (1997) CDC45, a novel yeast gene that functions with the origin recognition complex and Mcm proteins in initiation of DNA replication. Mol Cell Biol 17: 553–563

Regulation of the G2 to M Transition

J. Pines[1]

1
Introduction

A cell is faced with a number of problems to solve at each division:

1. A cell should only divide when the genome has been completely replicated.
2. A cell should only divide when the genome is undamaged.
3. The daughter cells should inherit identical genomes.
4. The daughter cells should each have sufficent cytoplasmic organelles to be viable.

The cell achieves these objectives by a complex process of division called mitosis, in which almost the entire architecture of the cell is rearranged in a highly coordinated fashion that is regulated mostly by phosphorylation and dephosphorylation. In this review I will outline the basic mechanism which controls the entry into mitosis, and how this is made responsive to the integrity of the genome and to the components required to form the mitotic apparatus (the spindle). The key players in this processs are the cyclin-dependent kinases (CDKs) in complexes with their partner mitotic cyclins. Thus many of the controls on the entry into mitosis act through regulating the cyclin-CDK complexes. However, I will only briefly outline the mechanisms by which cyclin-CDK complexes are regulated – through phosphorylation, cyclin synthesis and cyclin destruction – because they are dealt with in greater detail in subsequent chapters.

1.1
Mitosis Is the Dominant State of the Cell Cycle

It is essential that the cell regulates the entry into mitosis because experiments performed almost 30 years ago demonstrated that mitosis is the dominant state in the cell cycle. When a tissue culture cell in mitosis was fused with an interphase cell the contents of the mitotic cell caused the other cell to condense its chromosomes and attempt to enter mitosis, regardless of the replication state of its DNA (Johnson and Rao 1970). The first clues to the nature of the compo-

[1] Wellcome/CRC Institute, Tennis Court Road, Cambridge, CB2 1QR, UK,
E-mail: jp103@mole.bio.cam.ac.uk

nents in a mitotic cell able to drive a cell into mitosis came from studies on the initiation of meiosis in frog oocytes. Masui and Markert (1971) found that the contents of a cell in M phase (an egg) would cause a cell in G2 phase (an oocyte) to enter M phase. They called the factors responsible for initiating meiosis 'maturation promoting factor' or MPF. Subsequent work showed that MPF activity was also present in mitotic cells and hence MPF has come to stand for M phase-promoting factor.

1.2
Cyclin B-CDK1: The Key to G2 to M

Masui's observation eventually formed the basis for a cell free assay by which Lohka and colleagues (1988) purified MPF from *Xenopus* egg extracts, and its essential component was found to be a protein kinase composed of a B-type cyclin and CDK1, the homologue of fission yeast cdc2/budding yeast CDC28p (Arion et al. 1988; Dunphy et al. 1988; Labbé et al. 1989; Gautier et al. 1990). Thus at heart, the proper regulation of entry into mitosis is the proper regulation of cyclin B-CDK1 activity, and it is no surprise to find that it is this process that is responsive to the replication state and integrity of the genome.

However, at this point I should stress that cyclin B-CDK1 is not sufficent to drive an interphase somatic cell into mitosis. When injected into cells, active cyclin B-CDK1 causes the cell to initiate some of the cytoskeletal changes that characterise the beginning of mitosis – cells begin to detach from a substrate and round up – but the nuclear envelope does not breakdown and the cell eventually reverses these changes and flattens out (Lamb et al. 1990). Therefore there must be other, as yet unidentified, components of MPF, and biochemical fractionation experiments have suggested that some of these may also be protein kinases (Kuang and Ashorn 1993). Recently, a number of protein kinases have been identified that have important roles in mitosis, e.g. the polo-like kinases, the NIMA kinase, Mps1 and MAP kinase, some of which might also be components of MPF. Nevertheless, both biochemical and genetic evidence have shown that cyclin B-CDK1 is the primary activity required for a cell to enter mitosis in all eukaryotic cells.

2
Regulating Cyclin B-CDK1 Complexes

2.1
Cyclin B Levels

The cyclin B-CDK1 complex is regulated at a number of levels to ensure that it is not activated until DNA replication is complete. In human cells cyclin B is not detectable by immunoblot until the end of S phase (Pines and Hunter 1989;

Sherwood et al. 1994). In part this is due to cell cycle regulated transcription, although the cell cycle regulated transcription factors responsible for this have not been unambiguously identified (Piaggio et al. 1995). Moreover, this is not a highly conserved level of regulation because in rodent cell lines there is detectable cyclin B at the beginning of S phase (Yamashita et al. 1991). The synthesis of cyclin B can also be regulated at the level of mRNA stability in human cells. Cyclin B1 mRNA may be more stable in G2 phase than in G1 phase (Pines and Hunter 1989), and it becomes more unstable in cells that are damaged by radiation treatment (Maity et al. 1995).

A more evolutionarily conserved means of regulation is at the level of protein stability. Mitotic B-type cyclins become unstable after the cell enters anaphase. They remain unstable through G1 phase until either the cell has committed itself to another round of DNA replication (e.g. in budding yeast cells; Amon et al. 1994), or until DNA replication begins (e.g. in mouse cells; Brandeis and Hunt 1996). Cyclin degradation is dealt with in more detail in chapter 6 this Vol.

2.2
CDK Phosphorylation

Once cyclin B protein appears, there is a large pre-existing pool of CDK1 to which it could bind. However, although the level of CDK1 remains approximately constant through the cell cycle, CDK1 is turned over more rapidly after S phase begins (McGowan et al. 1990; Welch and Wang 1992). This is compensated for by an increase in the rate of CDK1 mRNA transcription, and it is to newly synthesised CDK1 that the B-type cyclins appear preferentially to bind. After the cyclin B-CDK1 complex is formed it is stabilised by phosphorylation on the 'T-loop' threonine by CDK-activating kinase (CAK; reviewed in Morgan 1995, see also Chapter 4, this Vol.). Although this could be a regulated event, thus far CAK activity appears to be constitutive through the cell cycle.

The combination of cyclin binding and phosphorylation of the T-loop threonine converts CDK1 from its completely inert, monomeric form to an active kinase. The complex is then held in check by phosphorylation on one (tyrosine 15), or in animal cells two (threonine 14 and tyrosine 15) conserved residues in the ATP binding region of CDK1. This allows a pool of inactive cyclin B-CDK1 complexes to accumulate throughout G2 phase that can be rapidly activated by dephosphorylation when the cell is ready to begin mitosis.

3
Entry into Mitosis Is Regulated by Checkpoints

In the rapid early embryonic cell cycles of *Xenopus* and *Drosophila*, the cell relies on timing to coordinate DNA replication with mitosis (reviewed in Murray and Kirschner 1989; Murray and Hunt 1993) The stockpiles of enzymes and components required for DNA replication, and the absence of transcription, ensure that each round of DNA replication can be rapidly completed in a fairly constant time interval. In the *Xenopus* embryo the replication state of the DNA is unable to influence when mitosis begins until the amount of DNA reaches a threshold level (Dasso and Newport 1990). Instead the rate of mitotic cyclin synthesis determines when mitosis begins. However, once zygotic transcription and cell differentiation begin, cell cycles become asynchronous and their duration becomes cell type specific. Therefore it is crucial that a cell should be able to prevent mitosis until its genome is completely replicated, or if the DNA is damaged, until it is repaired. This is achieved through 'checkpoints' (Hartwell and Weinert 1989) in S phase and G2 phase – mechanisms which arrest the cell cycle and are activated by unreplicated DNA or DNA damage. Checkpoints are normally silent, but are activated when the cell detects an error committed in the preceding phase of the cell cycle. Because the regulation of the phosphorylation state of CDK1 Y15 and T14 is the critical control on the entry into mitosis, it is the primary mechanism underlying the G2 DNA damage checkpoint.

4 Tyrosine 15 and Threonine 14 Phosphorylation

Cyclin B-CDK1 complexes are kept inactive by the Wee1/Mik1/Myt1 family of protein kinases that phosphorylate CDK1 in the ATP binding region. Tyrosine 15 phosphorylation is conserved through evolution (Gould and Nurse 1989; Krek and Nigg 1991); it is carried out by the wee1 and mik1 gene products in fission yeast (Russell and Nurse 1987b; Featherstone and Russell 1991; Lundgren et al. 1991; Parker et al. 1991), Swe1p in budding yeast (Booher et al. 1993), and the WEE1/MIK1 gene product in animal cells (Parker and Piwnica-Worms 1992; Heald et al. 1993; McGowan and Russell 1993). In fission yeast this pathway responds to the state of the DNA, indeed the ability of unreplicated or damaged DNA to prevent mitosis is primarily dependent on the phosphorylation of tyrosine 15 of cdc2 (Enoch and Nurse 1991; Enoch et al. 1992; Rowley et al. 1992; Rhind et al. 1997). CDK1 is also maintained in the tyrosine-phosphorylated state in response to DNA damage in animal cells. In contrast, phosphorylation of the tyrosine in the ATP binding site of budding yeast (Y19) by Swe1p is not required to regulate mitosis properly in the presence of unreplicated or damaged DNA (Amon et al. 1992; Sorger and Murray 1992). Instead, phosphorylation is important in a 'morphogenesis checkpoint' that monitors whether cells have budded correctly earlier in G1 phase (Lew and Reed 1993, 1995).

In animal cells, cyclin B-CDK1 is regulated by phosphorylation on both Y15, which interferes with the transfer of the γ-phosphate from ATP to substrate, and the adjacent T14 residue (Krek and Nigg 1991; Norbury et al. 1991), which interferes with ATP binding (Atherton Fessler et al. 1993). It is unclear why an additional level of control has been introduced in animal cells. T14 and Y15 are phosphorylated by the dual-specificity MYT1 kinase (Mueller et al. 1995b; Liu et al. 1997), whose sequence places it in the same subfamily as Wee1 and Mik1. The most striking difference in sequence between MYT1 and wee1 or mik1 is that MYT1 has a transmembrane domain. WEE1 has been shown to be a nuclear protein in human cells (Heald et al. 1993; Baldin and Ducommun 1995), whereas biochemical fractionation showed that MYT1 is membrane-associated in *Xenopus* extracts (Kornbluth et al. 1994; Mueller et al. 1995b), and immunofluorescence studies demonstrated that human MYT1 colocalises with the ER and Golgi apparatus (Liu et al. 1997). This raises the possibility that cyclin B-CDK1 complexes may be regulated differently according to their location in the cell. Moreover, there are at least two different types of B-cyclin in human cells, and perhaps as many as five in *Xenopus* (T Hunt, pers. comm.), and one of the main differences between human cyclin B1 and human cyclin B2 is that B1 is apparently associated with the cytoskeleton, whereas B2 is associated with the membrane compartment, especially the Golgi apparatus (Jackman et al. 1995). Thus it is possible that phosphorylation on T14 and Y15 is especially important for the membrane-associated cyclin B2-CDK1 complexes.

The aspect of spatial control in the regulation of CDK1 tyrosine phosphorylation is also apparent in fission yeast. The primary mitotic kinase is composed of the B-type cyclin, cdc13, which is bound to the CDK, cdc2. From immunofluorescence studies this complex is nuclear (Booher et al. 1989; Alfa et al. 1990), as is its negative regulatory kinase, Wee1 (Wu et al. 1996). However, one of the negative regulators of Wee1, the protein kinase Nim1, is a cytoplasmic protein (Wu et al. 1996). Nim1 directly inactivates Wee1 by phosphorylation (Russell and Nurse 1987a; Coleman et al. 1993; Wu and Russell 1993), therefore there may be a cell cycle-dependent relocalisation of either Nim1 or Wee1 at mitosis to accomplish this. Relocalisation may also be important in the activation of cyclin B by the Cdc25 phosphatases (see below).

5 Chk1: A Link Between Damaged DNA and Cyclin B-CDK1?

Unreplicated or damaged DNA is able to prevent mitosis by keeping the cyclin B-CDK1 complexes in their inactive, Y15 phosphorylated state. This could be accomplished by enhancing the activity of the Wee1/Myt1 kinases and/or by inhibiting the Cdc25 phosphatases that act on phosphorylated Y15 and T14. The Wee1 protein kinase in fission yeast is hyperphosphorylated and thereby more active in fission yeast cells arrested by DNA damage. The chk1 protein kinase may be the link between Wee1 and the gene products that sense damaged DNA, because recent data show that chk1 is able to phosphorylate Wee1, although this has not yet been demonstrated to stimulate Wee1 kinase activity (O'Connell et al. 1997). *chk1* was originally isolated as a suppressor of a cdc2 mutant (Walworth et al. 1993) and is required for fission yeast to arrest in G2 phase in response to DNA damage (Walworth et al. 1993; Carr et al. 1995). chk1 is activated in response to DNA damage and acts as an effector for a number of *rad* gene pathways, including the *S. pombe rad3* gene that is a member of the ATM and DNA-PK family of protein kinases (Walworth et al. 1993; Walworth and Bernards 1996). *chk1* is also required to prevent G1 cells that have yet to begin DNA synthesis from entering mitosis (Carr et al. 1995).

The *chk1* gene may be a conserved element linking DNA integrity to the cell cycle machinery. Recently a chk1 homologue has been identified in *Drosophila* as the product of the *grapes* gene (Fogarty et al. 1994; Sibon et al. 1997). This gene is related in sequence to *chk1* and has also been implicated as a link between the replication state of the DNA and entry into mitosis. Flies with mutations in the *grapes* gene are unable to switch from the rapid embryonic cell cycle – characterised by successive rounds of replication and mitosis with no G1 or G2 delay – to the zygotic cell cycle which has a G2 phase at the mid-blastula transition. This appears to be because they are insensitive to an inhibitory signal preventing mitosis in the presence of unreplicated DNA (Sibon et al. 1997).

Although *chk1* is required for non-replicated, or fully replicated but damaged DNA to inhibit mitosis, in fission yeast cells it is not required to prevent cells arrested in S phase from entering mitosis, in contrast to the role implied for the grapes kinase in *Drosophila*. Instead this checkpoint acts through the Cdc25 phosphatases.

6 Unreplicated DNA Prevents Mitosis by Inhibiting Cdc25 Phosphatases

Cell fusion experiments had first indicated that nuclei actively replicating their DNA generated a signal that prevented G2 nuclei from initiating mitosis (Rao and Johnson 1970). It is now known that components of the replication machinery itself – such as DNA polymerases α and ε, and the budding yeast CDC6 and fission yeast cdc18 proteins – are required to generate this signal (Bueno and Russell 1992; Kelly et al. 1993; Navas et al. 1995). The signal appears to prevent the activation of the Cdc25 phosphatases that remove the phosphates from Y15 (and T14) in CDK1, because in mutant fission yeast cells in which Cdc25 activity is deregulated, cells are able to enter mitosis regardless of the replication state of their DNA (Enoch and Nurse 1990). Cdc25 proteins are activated by phosphorylation as part of a feedback loop with cyclin B-CDK1 (Galaktionov and Beach 1991; Hoffmann et al. 1993) and are kept inactive by a PP2A containing phosphatase (Clarke et al. 1993). Okadaic acid is able to drive cells prematurely into mitosis at concentrations sufficient only to inhibit PP2A, and this appears to be through generating the phosphorylated, active form of Cdc25 (Yamashita et al. 1991). A PP2A phosphatase activity has also been implicated in maintaining Wee1 in its active, hypophosphorylated state.

Only one Cdc25 has been identified in fission yeast, although a second phosphatase, Pyp3 (Millar et al. 1992), can also contribute to the induction of mitosis. In contrast there are three isoforms of Cdc25 in human cells: Cdc25A, B and C (Galaktionov and Beach 1991; Honda et al. 1993). One of these, Cdc25A is probably involved in regulating the G1/S transition (Hoffmann et al. 1994; Jinno et al. 1994; Galaktionov et al. 1996), but both Cdc25B and C appear to be involved in regulating mitosis and are able to dephosphorylate cyclin B-CDK1 (Galaktionov and Beach 1991; Kumagai and Dunphy 1991, 1992; Kakizuka et al. 1992; Hoffmann et al. 1993; Honda et al. 1993; Sebastian et al. 1993; Gabrielli et al. 1996). Whether these two types of Cdc25 have distinct roles in mitosis is unclear, but again there are data to suggest that they differ in their subcellular localisation (see below).

7 Rapid Cyclin B-CDK1 Activation Is Achieved via a Positive Feedback Loop

The Cdc25 proteins are activated by phosphorylation on the amino terminus. One consequence of this phosphorylation is that they are recognised by members of the 14-3-3 family (Conklin et al. 1995), which may be important for their activation. Moreover, recent evidence shows that phosphorylated Cdc25 can be bound by the peptidyl-prolyl isomerase Pin1 (Ranganathan et al. 1997; S. Kornbluth, pers. comm.). Peptidyl-prolyl isomerases convert prolines between the *cis* and *trans* conformations, and are important in protein folding. Pin1 was first isolated as a protein that interacts with the NIMA kinase (see below) and is required for mitosis in yeast and human cells (Lu et al. 1996). Pin1 appears to recognise prolines that are C-terminal to a negatively charged residue, and in particular to phosphorylated serine/threonine. It is probably not by coincidence that a serine/threonine with a C-terminal proline is the basic element of a consensus CDK phosphorylation site.

The kinases which activate Cdc25 at mitosis probably include the mitotic CDKs themselves, which would set up a positive feedback loop through the mutual activation of Cdc25 and cyclin B-CDK1 (Galaktionov and Beach 1991; Kumagai and Dunphy 1992; Hoffmann et al. 1993). However, the question remains as to which kinase(s) effectively initiate(s) mitosis by activating Cdc25. The kinase is probably not another CDK because Cdc25 can be phosphorylated and activated in interphase *Xenopus* extracts devoid of CDK1 and CDK2 activity (Izumi and Maller 1995). Recently a member of the polo kinase family, Plx1, has been shown to phosphorylate and activate Cdc25 in *Xenopus* extracts (Kumagai and Dunphy 1996), but it has not yet been demonstrated that this kinase is activated at the correct time to initiate mitosis.

The rapid activation of cyclin B1-CDK1 at mitosis is further enhanced by the downregulation of Wee1 activity, again achieved by phosphorylation. In fission yeast, Wee1 is inhibited by the Nim1/Cdr1 kinase which phosphorylates Wee1 directly in the kinase domain (Russell and Nurse 1987a; Coleman et al. 1993; Parker et al. 1993; Tang et al. 1993; Wu and Russell 1993). Nim 1 homologues have not yet been found in vertebrates, and an alternative Wee1 inhibitory kinase activity has been observed in *Xenopus* egg extracts (Tang et al. 1993; Mueller et al. 1995a). This inhibits Wee1 kinase activity to a greater extent than Nim1 by hyperphosphorylating the N-terminus of Wee1. Human WEE1/MIK1 is phosphorylated in mitosis (McGowan and Russell 1995; Watanabe et al. 1995), but the resulting shift in mobility of WEE1 is much smaller than that of *Xenopus* Wee1, and its significance for downregulating WEE1 activity is not clear. Human WEE1 kinase activity is also decreased in mitosis by a reduction in WEE1 protein levels (Watanabe et al. 1995). MYT1 activity also decreases at mitosis (Mueller et al. 1995b), but it is not known whether this is through hyperphosphorylation, and, if so, whether this is by a Nim1/Cdr kinase.

8 Entry into Mitosis Is Also Regulated by a Spindle Assembly Checkpoint

Ensuring that DNA is fully replicated and undamaged is only one consideration for a cell before it initiates mitosis. The cell must also have duplicated its centrosomes (in animal cells) or spindle pole bodies (in yeast), which set up the bipolar spindle that will ensure that the chromosomes are properly distributed between daughter cells. In *Drosophila* and human cells a member of the polo family of serine/threonine kinases is localised to the centrosomes in interphase (Golsteyn et al. 1994). *Drosophila* with a defective polo kinase exhibit a number of abnormalities in mitosis, including monopolar spindles (Llamazares et al. 1991). Microinjecting anti-human polo (PLK1) antibodies into normal human diploid fibroblasts prevents centrosomes from growing – perhaps by inhibiting the recruitment of γ-tubulin – and from separating (Lane and Nigg 1996). This arrests normal cells in G2 phase, and suggests that PLK1 may link the centrosome cycle to the chromosome cycle. Moreover, some transformed cells microinjected with anti-PLK1 antibodies go on to enter mitosis but are unable to form a proper spindle, indicating that transformed cells may lack the centrosomal checkpoint.

Mps1 is another conserved protein kinase that has also been implicated in the link between the centrosome cycle and the chromosome cycle. In budding yeast-Mps1p is required for the spindle pole bodies to duplicate (Lauze et al. 1995; Weiss and Winey 1996). But Mps1 also has a role in mitosis itself, i.e. in the checkpoint mechanism that ensures that chromosome separation (anaphase) does not start until the sister chromosomes are properly aligned on the metaphase plate. A human homologue of Mps1 has been isolated and shown to be essential for this anaphase checkpoint. The anaphase checkpoint is intricately associated with the regulation of ubiquitin-dependent degradation that is dealt with in Chapter 6. (this Vol.).

9 Further Regulation of Cyclin B-CDK1 Activity

9.1 Inhibitors

Cyclin B-CDK1 complexes can be held in check by mechanisms other than through the phosphorylation of T14 and Y15 in CDK1. A mutant form of CDK1 that cannot be downregulated by phosphorylation at T14 and Y15 is still inhibited when added to *Xenopus* extracts (Kumagai and Dunphy 1995; Lee and Kirschner 1996), and this appears to be due to a membrane-bound inhibitor activity. This activity may be also be present in human tissue culture cells because the same mutant form of CDK1 is only able to cause a minor percentage of transformed cells, and even fewer normal diploid fibroblasts, to enter mitosis when expressed at the DNA damage checkpoint (Jin et al. 1996), although an alternative explanation could be that the cyclin B-CDK complex is still regulated by its subcellular localisation (see below). Budding yeast are also able to arrest before mitosis with the analogous mutation in their CDK. Indeed, as mentioned above, tyrosine phosphorylation is only required for the morphogenesis checkpoint (Lew and Reed 1993, 1995). A cyclin B-CDK inhibitor has been identified in budding yeast, the Sic1 protein (Schwob et al. 1994; Tyers 1996). However, Sic1p primarily inhibits the Clb5/6-Cdc28 complexes that initiate DNA replication, although it may also have a role in exit from mitosis (Donovan et al. 1994). In fission yeast, the Rum1 protein (Moreno and Nurse 1994) inhibits the S phase cig2-cdc2 and the mitotic cdc13-cdc2 complexes (Correa Bordes and Nurse 1995; Martin Castellanos et al. 1996), but it is only normally present in G1 cells where it is required to prevent cells that have not yet begun DNA synthesis from prematurely entering mitosis (Moreno and Nurse 1994; Jallepalli and Kelly 1996).

9.2 Subcellular Localisation

In many cases, different members of the multigene families of cell cycle regulators – such as the mitotic cyclins, the Cdc25 phosphatases and the Wee1/Mik1/Myt1 kinases – are localised to different compartments within the cell. This has obvious implications for substrate specificity, and potentially for the mechanics of cell cycle regulation because some cell cycle regulators also translocate between compartments in a cell cycle-dependent manner. In human cells Cdc25C is a nuclear protein (Millar et al. 1991; Girard et al. 1992), whereas its putative substrate, cyclin B-CDK1, is cytoplasmic (Pines and Hunter 1991; Gallant and Nigg 1992; Ookata et al. 1992). The major type of cyclin B-CDK1 complex translocates into the nucleus just before prophase (Pines and Hunter 1991; Gallant and Nigg 1992; Ookata et al. 1992), raising the possibility that this may be another way in which the cell prevents cyclin B-CDK1 from being prematurely activated. However, this does not appear to be an evolutionarily conserved mechanism because in rodent cells Cdc25C is cytoplasmic (Seki et al. 1992) and translocates into the nucleus in the same manner as cyclin B-CDK1 (Heald et al. 1993). A further complication is that a second member of the Cdc25 family, Cdc25B, may also be required for mitosis (Honda et al. 1993; Sebastian et al. 1993). Cdc25B localised to the cytoplasm (Gabrielli et al. 1996) – although when overexpressed the majority appeared nuclear – and when a dominant negative Cdc25B mutant was overexpressed, it caused cells to accumulate in G2 phase, with microtubules nucleating from centrosomes in a manner indicative of an interphase rather than a prophase state (Gabrielli et al. 1996).

It is not clear which isoform of Cdc25 first activates cyclin B-CDK1, or whether both Cdc25B and C activate cyclin B1-CDK1 simultaneously. This will clearly be influenced by when exactly cyclin B-CDK1 is activated – before, during or after translocation. In starfish meiotic maturation, CDK1 is activated before cyclin B-CDK1 relocates to the nucleus (Ookata et al. 1992), but it is not known where in the somatic cell cyclin B1/-CDK1 is first activated to levels which will initiate mitosis.

9.2.1
The Control of Cyclin-CDK Localisation

There are some data on how the localisation of the mitotic cyclins is specified. A region in the amino terminus of the B-type cyclins is required to retain the protein in the cytoplasm (Pines and Hunter 1994). Without this 'cytoplasmic retention signal' (CRS) cyclin B1 is transported to the nucleus at any stage of the cell cycle. However, a nuclear localisation signal (NLS) fused to cyclin B1 will override the CRS, suggesting that nuclear cyclins, such as cyclins A and E, may not have an endogenous NLS (Pines and Hunter 1994). We know relatively little about how the nuclear cyclins are properly localised. For cyclin A to be transported to the nucleus it must first bind a CDK (Maridor et al. 1993), but neither cyclin A nor CDK1/2 possess a recognisable NLS. It is possible that these complexes may be transported to the nucleus by 'piggy backing' on another protein with an NLS. For example, the requirement for cyclin A to bind to a CDK before being transported may correlate with the ability to bind nuclear proteins such as p107, p130, or the E2F-1 transcription factor.

The relocalisation of a B-type cyclin to the nucleus at prophase is conserved from starfish to man. How this translocation is triggered is unclear, but phosphorylation of cyclin B1 in the CRS appears be important (Li et al.1995, 1997; J. Pines, unpubl. observ.). The CRS region is phosphorylated during *Xenopus* oocyte maturation, and a cyclin B mutant in which all these phosphorylation sites are mutated to alanine fails to promote maturation (Li et al. 1997). In contrast, a cyclin B mutant in which the sites are mutated to glutamic acid – to mimic phosphorylation – does cause maturation.

The mechanism by which cyclin B translocates is also unknown, but there are a number of other proteins that also exhibit regulated nuclear entry at this point in the cell cycle. One of these is the *Drosophila* protein pendulin, a member of the importin β family of proteins involved in nuclear import. Pendulin has a CDK1 phosphorylation site adjacent to an NLS and is transported into the nucleus at the G2/M transition (Küssel and Frasch 1995), making pendulin and its homologues (Gorlich et al. 1994) excellent candidates to assist in the nuclear translocation of cyclin B1 at G2/M. However, no direct link with cyclin B1 nuclear relocalisation has been shown.

10 Cks Proteins

As mentioned above, the cyclin B-CDK1 complex alone does not constitute MPF. Another component is probably a Cks protein (CDK subunit), i.e. one of the small proteins (9-13 kDa) that are among the most conserved elements of the G2-M machinery. Cks proteins bind to the cyclin B-CDK complexes and are essential in yeast, but their exact role in the cell cycle remains elusive. The *Schizosacchraomyces pombe* Cks protein, Suc1, was isolated as a suppressor of a cdc2 allele (Hayles et al. 1986). Fission yeast cells with a disrupted *suc1* gene are able to enter mitosis but arrest there with a high level of H1 kinase activity (Moreno et al. 1989). In contrast, temperature-sensitive alleles of the homologous *S. cerevisiae CKS1* gene (identified as a Cdc28 kinase subunit; Hadwiger et al. 1989) show that Cks1 performs an essential role in exiting G2 phase of the cell cycle (Tang and Reed 1993). In agreement with both these findings, interphase *Xenopus* cell-free extracts without Cks1 are unable to enter mitosis and remain in interphase with CDK1 in its Y15-phosphorylated state, whereas mitotic extracts lacking Cks1 are unable to re-enter interphase, correlating with a defect in cyclin B destruction (Patra and Dunphy 1996). This suggests that Cks1 could be involved in both CDK1 dephosphorylation and cyclin B destruction. Thus the Cks proteins may perform (at least) two different roles in the cell cycle; at the end of G2, phase dephosphorylation and activation of cyclin B-CDK1, and at the end of metaphase, inactivation of cyclin B-CDK1. Two human homologues, CksHs1 and 2, that are very similar in amino acid sequence have been cloned (Richardson et al. 1990). Both are able to complement a yeast Cks, but the functional differences between human Cks1 and Cks2 have not yet been defined.

The crystal structures of fission yeast suc1 and human Cks2 have both been solved (Parge et al. 1993; Bourne et al. 1995; Endicott et al. 1995; Khazanovich et al. 1996), and show that the Cks proteins have a putative β–hinge region that could be either extended on folded back on itself. Thus the Cks proteins could take on one of two different conformations, perhaps to perform different functions. The atomic structure of the CDK2-CksHs1 complex (Bourne et al. 1996) shows that CksHs1 interacts solely with the C-terminal lobe of CDK2 and does not overlap the cyclin binding site on the N-terminal lobe. However, this is only true when the β–hinge region of the Cks folds back on itself. When the hinge is in its extended conformation the Cks is predicted to interfere with cyclin binding, perhaps precluding it altogether (Bourne et al. 1996). Thus a model for the two different roles of the Cks proteins might involve a change in the conformation of the β–hinge region.

Apart from subtle changes in conformation at the binding interface, the structures of both the Cks1 and the CDK hardly alter upon binding (Bourne et al. 1996), suggesting that Cks proteins do not directly affect CDK activity. When the CDK binds CksHs1, this brings a highly conserved, positively charged pocket of residues on Cks1 into close proximity to the catalytic cleft of CDK2. This pocket could act as a phosphate anion binding site, meaning that Cks proteins might enhance interactions between the CDK and regulators or substrates that are phosphorylated - such as, for example, activated Cdc25.

11
Cyclins upstream of cyclin B-CDK1

11.1
Cyclin A-CDK2 and cyclin E-CDK2

Thus the essential elements for the activation of Cyclin B-CDK1 to initiate mitosis are as set out in Fig. 1. However, this is by no means a complete picture. Apart from the identity of the kinase(s) that initially activates Cdc25, there is also the continuing mystery of the requirement for cyclin A.

Cyclin A is required at two points in the cell cycle; at the beginning of S phase (see Chap. 1, 2, this vol) and to enter mitosis, but although cyclin A is required for mitosis, its role in the initiation of mitosis, and indeed in mitosis itself, is unclear. Synthetic cyclin A mRNA will induce germinal vesicle breakdown and spindle formation when injected into frog oocytes (Swenson et al. 1986), and this action is synergistic with cyclin B (Knoblich and Lehner 1993). Anti-cyclin A antibodies will prevent mitosis when injected into mammalian tissue culture cells (Pagano et al. 1992), and a *Drosophila* embryo with a mutant cyclin A arrests in G2 phase of cycle 14 (Lehner and O'Farrell 1990). Alternatively, it may be that the requirement for cyclin A is more a requirement for CDK2 kinase activity. In the *Xenopus* embryo, cyclin A is exclusively bound to CDK1, whereas CDK2 is bound to cyclin E. When CDK2 kinase activity is blocked with the $p21^{Cip1/Waf1}$ CDK inhibitor, this prevents *Xenopus* extracts from entering mi-

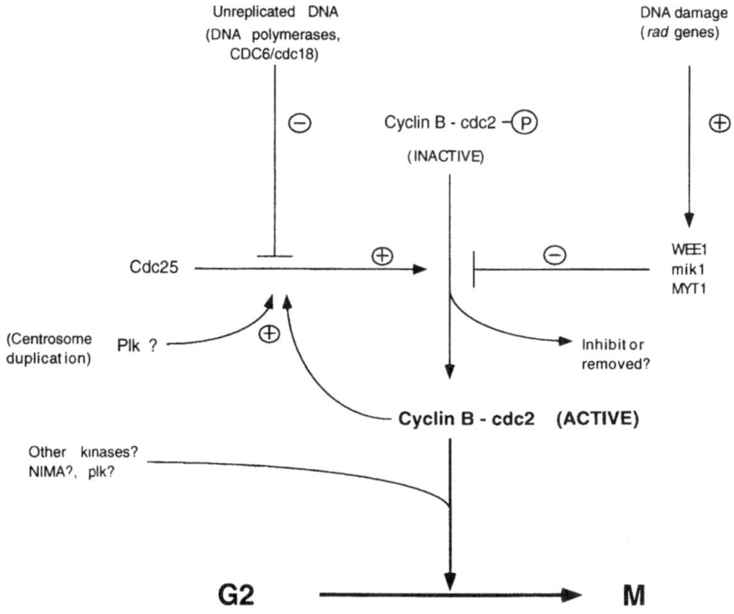

Fig 1. Essential elements for the activation of cyclin B-CDK1 to initiate mitosis

tosis (Guadagno and Newport 1996), and these extracts can be rescued by the addition of active cyclin E-CDK2.

Once cells have entered mitosis cyclin A is degraded in metaphase, before cyclin B (Pines and Hunter 1991; Hunt et al. 1992). In *Drosophila* and in cycling *Xenopus* egg extracts, a non-destructible cyclin A mutant causes cells to delay in metaphase (Luca et al. 1991; Knoblich and Lehner 1993).

Some of the normal cellular proteins that interact with cyclin A have been identified. Cyclin A is able to bind and activate two members of the CDK family; CDK2 (Pines and Hunter 1990; Elledge and Spottswood 1991; Tsai et al. 1991), and in some cell types, CDK1 (Pagano et al. 1992). Active cyclin A-CDK complexes in mitosis are also associated with the Cks proteins and with the $p21^{Cip1/Waf1}$ CDK inhibitor that is upregulated in response to activated p53 (Xiong et al. 1993). Recently, two other cyclin A associated proteins have been identified, Skp1 and Skp2, and the proportion of cyclin A associated with these proteins appears to be altered in transformed cells (Zhang et al. 1995). Skp1 and Skp2 have subsequently been implicated in targeting proteins to the ubiquitin-proteasome pathway (Bai et al. 1996). Skp1 binds a variety of proteins, including Skp2, via a common motif, the F box, and this interaction is necessary for their degradation (Bai et al. 1996). Skp2 in turn binds to cyclin A. Moreover, Skp1 has also been identified as a component of the yeast kinetochore (Connelly and Hieter 1996; Stemmann and Lechner 1996). This may have implications for the potential role of cyclin A in mitosis because cyclin A appears to be associated with condensing chromatin, and then to be degraded first on condensed chromosomes during metaphase (Pines and Hunter 1991).

11.2
Cyclin F?

The association with Skp1 links cyclin A to cyclin F, which may also have a role in G2 or M phase (Bai et al. 1994). In a similar manner to cyclin A, cyclin F levels peak in G2 phase and it is degraded around the time of mitosis. However, unlike cyclin A and B, cyclin F does not have the destruction box motif required for specific destruction by the ubiquitin degradation pathway via the 'cyclosome/APC' (see Chap. 6, this vol. for details). Rather, cyclin F has PEST sequences that are correlated with short lived proteins, and a F-box which means that it can interact directly with Skp1 (Bai et al. 1996). When cyclin F is overexpressed in cells this causes an apparent increase in the population in G2 phase, indicating that it's degradation might be linked to the start of mitosis. However, given that the CDK partner for cyclin F has not yet been isolated, the role, if any, for cyclin F in mitosis remains unclear.

12
Downstream of cyclin B-CDK1

Once cyclin B-CDK1 is activated, the cell enters the mitotic state. Cyclin B-CDK1 is the major protein kinase activity in mitosis and has a role as a 'master regulator' of downstream kinases, as well as directly phosphorylating a number of proteins responsible for altering the subcellular architecture.

12.1
Downstream kinases

12.1.1
NIMA

Although histone H1 is routinely used to assay cyclin-CDK activity, it is by no means clear whether cyclin B-CDK1 is responsible for chromosome condensation. A role for CDK1 in chromosome condensation is suggested by the observation that histones H1, H2A and H3, and the high mobility group (HMG) proteins, I, Y, and P1 are phosphorylated by cyclin/CDK1 in vitro on sites that are phosphorylated during mitosis (Meijer et al. 1991; Nissen et al. 1991). However, chromosome condensation can occur after CDK1 is inactivated in a mouse cell line with a temperature sensitive mutation in CDK1 (Hamaguchi et al. 1992), indicating that an alternative kinase can induce chromosome condensation.

One candidate is the NIMA (Never in mitosis gene A) kinase, a cell-cycle regulated Ser/Thr kinase identified in the filamentous fungus *Aspergillus nidulans* (Osmani et al. 1991; Lu et al. 1993). NIMA is required for *Aspergillus* cells to initiate mitosis and when a a stable form of NIMA kinase is expressed in fungi or human cells, it induces chromatin condensation without active CDK1 (O'Connell et al. 1994). The NIMA in cells arrested in G2 is only partially active, but is activated a further two fold after phosphorylation by CDK1. Phosphorylation by CDK1 also stabilises NIMA (Ye et al. 1995). This puts NIMA downstream of CDK1.

NIMA may also be regulated by the peptidyl-prolyl isomerase Pin1, because the lethal effects of overexpressing NIMA in yeast are suppressed by Pin1, and the two proteins colocalise when NIMA is transfected into human cells (Lu et al. 1996). As yet no NIMA homologues have been found in organisms other than filamentous fungi. The Nek2 kinase is a vertebrate Ser/Thr kinase related to NIMA in the kinase domain, whose protein levels and kinase activity peak during G2, but the physiological function of Nek2 is not yet known (Schultz et al. 1994; Fry et al. 1995).

12.1.2
MAP Kinase

The importance of MAP kinase activity in maintaing metaphase arrest in oocytes is well established. In *Xenopus* oocytes MAP kinase is activated by newly synthesised c-*mos*, and requires cyclin B-CDK1 activity (Nebreda and Hunt 1993; Posada et al. 1993; Nebreda et al. 1995). Recently a role for MAP kinase in the somatic cell cycle has been revealed. MAP kinase activity is not required for somatic tissue culture cells to enter mitosis, but it is required to maintain cells in metaphase when the spindle apparatus is perturbed (ie for a functional spindle assembly checkpoint in mitosis; Minshull et al. 1994; Takenaka et al. 1997).

12.2
Substrates

The re-organisation of the cell that occurs so rapidly and in such a coordinated fashion at mitosis is largely carried out by phosphorylation of key components of the subcellular architecture. These include; the nuclear lamins that disassemble at mitosis, the microtubules that reorganise to form the spindle apparatus, motor proteins that both organise the spindle and regulate chromosome movement, the actin microfilaments that cause the cell to round up and form the contractile ring that divides the cell into two, and components of the Golgi apparatus and the endoplasmic reticulum which cause the membrane compartment to disassemble into vesicles. A complete description of these changes is beyond the scope of this review, but it is worth mentioning that a number of the proteins involved in the reorganisation of the subcellular architecture are direct substrates for the cyclin B-CDK kinases (see Nigg 1991, 1993; Jackman and Pines 1997 for reviews). Among these substrates are as yet unidentified components of the cyclosome/APC-dependent destruction pathway, because once cyclin B-CDK1 kinase activity appears, it sets in train the process by which the mitotic cyclin destruction machinery is activated and the consequent exit from mitosis.

Note Added in Proof:
Since this review was written the Chk1 has been shown to phosphorylate Cdc25 on a conserved serine residue, causing Cdc25 to be bound by a 14-3-3 protein and functionally inactivated (Furnari B, Rhind N, Russell P (1997) Cdc25 mitotic inducer targeted by chk1 DNA damage checkpoint kinase. Science 277: 1495–1497; Peng CY, Graves PR, Thoma RS et al. (1997) Mitotic and G2 checkpoint control: regulation of 14-3-3 protein binding by phosphorylation of Cdc25C on serine-216. Science 277: 1501–1505); Sanchez Y, Wong C, Thoma RS et al. (1997) Conservation of the Chk1 checkpoint pathway in mammals: linkage of DNA damage to Cdk regulation through Cdc25. Science 277: 1497–1501).

References

Alfa CE, Ducommun B, Beach D et al. (1990) Distinct nuclear and spindle pole body populations of cyclin-cdc2 in fission yeast. Nature 347: 680–682

Amon A, Surana U, Muroff I et al. (1992) Regulation of p34CDC28 tyrosine phosphorylation is not required for entry into mitosis in *S. cerevisiae.* Nature 355: 368–371

Amon A, Irniger S, Nasmyth K (1994) Closing the cell cycle circle in yeast: G2 cyclin proteolysis initiated at mitosis persists until the activation of G1 cyclins in the next cycle. Cell 77: 1037–1050

Arion D, Meijer L, Brizuela L et al. (1988) cdc2 is a component of the M phase-specific histone H1 kinase: evidence for identity with MPF. Cell 55: 371–378

Atherton Fessler S, Parker LL, Geahlen RL et al. (1993) Mechanisms of p34cdc2 regulation. Mol Cell Biol 13: 1675–1685

Bai C, Richman R, Elledge SJ (1994) Human cyclin F. EMBO J 13: 6087–6098

Bai C, Sen P, Hofmann K et al. (1996) SKP1 connects cell cycle regulators to the ubiquitination proteolysis machinery through a novel motif, the F-box. Cell 86: 263–274

Baldin V, Ducommun B (1995) Subcellular localisation of human WEE1 kinase is regulated during the cell cycle. J Cell Sci 108: 2425–2432

Booher RN, Alfa CE, Hyams JS et al. (1989) The fission yeast cdc2/cdc13/suc1 protein kinase: regulation of catalytic activity and nuclear localization. Cell 58: 485–497

Booher RN, Deshaies RJ, Kirschner MW (1993) Properties of Saccharomyces cerevisiae wee1 and its differential regulation of $p34^{cdc28}$ in response to G_1 and G_2 cyclins. EMBO J 12: 3417–3426

Bourne Y, Arvai AS, Bernstein SL et al. (1995) Crystal structure of the cell cycle-regulatory protein suc1 reveals a beta-hinge conformational switch. Proc Natl Acad Sci USA 92: 10232–10236

Bourne Y, Watson MH, Hickey MJ et al. (1996) Crystal structure and mutational analysis of the human CDK2 kinase complex with cell cycle-regulatory protein CksHs1. Cell 84: 863–874

Brandeis M, Hunt T (1996) The proteolysis of mitotic cyclins in mammalian cells persists from the end of mitosis until the onset of S phase. EMBO J. 15: 5280–5289

Bueno A, Russell P (1992) Dual functions of CDC6: a yeast protein required for DNA replication also inhibits nuclear division. EMBO J 11: 2167–2176

Carr AM, Moudjou M, Bentley NJ et al. (1995) The chk1 pathway is required to prevent mitosis following cell-cycle arrest at 'start'. Curr Biol 5: 1179–1190

Clarke PR, Hoffmann I, Draetta G et al. (1993) Dephosphorylation of cdc25C by a type-2A protein phosphatase: specific regulation during the cell cycle in *Xenopus* egg extracts. Mol Biol Cell 4: 397–411

Coleman TR, Tang Z, Dunphy WG (1993) Negative regulation of the Wee1 protein kinase by direct action of the nim1/cdr1 mitotic inducer. Cell 73: 919–929

Conklin DS, Galaktionov K, Beach D (1995) 14-3-3 proteins associate with cdc25 phosphatases. Proc Natl Acad Sci USA 92: 7892–7896

Connelly C, Hieter P (1996) Budding yeast SKP1 encodes an evolutionarily conserved kinetochore protein required for cell cycle progression. Cell 86: 275–285

Correa Bordes J, Nurse P (1995) p25rum1 orders S phase and mitosis by acting as an inhibitor of the p34cdc2 mitotic kinase. Cell 83: 1001–1009

Dasso M, Newport JW (1990) Completion of DNA replication is monitored by a feedback system that controls the initiation of mitosis in vitro: studies in *Xenopus*. Cell 61: 811–823

Donovan JD, Toyn JH, Johnson AL et al. (1994) $P40^{SDB25}$, a putative CDK inhibitor, has a role in the M/G_1 transition in *Saccharomyces cerevisiae*. Genes Dev 8: 1640–1653

Dunphy WG, Brizuela L, Beach D et al. (1988) The *Xenopus* cdc2 protein is a component of MPF, a cytoplasmic regulator of mitosis. Cell 54: 423–431

Elledge SJ, Spottswood MR (1991) A new human p34 protein kinase, CDK2, identified by complementation of a cdc28 mutation in *Saccharomyces cerevisiae*, is a homolog of *Drosophila* Egl. EMBO J 10: 2653–2659

Endicott JA, Noble ME, Garman EF et al. (1995) The crystal structure of p13suc1, a p34cdc2-interacting cell cycle control protein. EMBO-J 14: 1004–1014

Enoch T, Nurse P (1990) Mutation of fission yeast cell cycle control genes abolishes dependence of mitosis on DNA replication. Cell 60: 665–673

Enoch T, Nurse P (1991) Coupling M phase and S phase: controls maintaining the dependence of mitosis on chromosome replication. Cell 65: 921–923

Enoch T, Carr AM, Nurse P (1992) Fission yeast genes involved in coupling mitosis to completion of DNA replication. Genes Dev 6: 2035–2046

Featherstone C, Russell P (1991) Fission yeast p107wee1 mitotic inhibitor is a tyrosine/serine kinase. Nature 349: 808–811

Fogarty P, Kalpin RF, Sullivan W (1994) The *Drosophila* maternal-effect mutation grapes causes a metaphase arrest at nuclear cycle 13. Development 120: 2131–2142 issn: 0950–1991

Fry AM, Schultz SJ, Bartek J et al. (1995) Substrate specificity and cell cycle regulation of the Nek2 protein kinase, a potential human homolog of the mitotic regulator NIMA of *Aspergillus nidulans*. J Biol Chem 270: 12899–12905

Gabrielli BG, De Souza CP, Tonks ID et al. (1996) Cytoplasmic accumulation of cdc25B phosphatase in mitosis triggers centrosomal microtubule nucleation in HeLa cells. J Cell Sci 109: 1081–1093

Galaktionov K, Beach D (1991) Specific activation of cdc25 tyrosine phosphatases by B-type cyclins: evidence for mutiple roles of mitotic cyclins. Cell 67: 1181–1194

Galaktionov K, Chen X, Beach D (1996) Cdc25 cell-cycle phosphatase as a target of c-myc. Nature 382: 511–517

Gallant P, Nigg EA (1992) Cyclin B2 undergoes cell cycle-dependent nuclear translocation and, when expressed as a non-destructible mutant, causes mitotic arrest in HeLa cells. J Cell Biol 117: 213–224

Gautier J, Minshull J, Lohka M et al. (1990) Cyclin is a component of MPF from *Xenopus*. Cell 60: 487–494

Girard F, Strausfeld U, Cavadore JC et al. (1992) cdc25 is a nuclear protein expressed constitutively throughout the cell cycle in nontransformed mammalian cells. J Cell Biol 118: 785–794

Golsteyn RM, Schultz SJ, Bartek J et al. (1994) Cell cycle analysis and chromosomal localization of human Plk1, a putative homologue of the mitotic kinases *Drosophila* polo and *Saccharomyces cerevisiae* Cdc5. J Cell Sci 107: 1509–1517

Gorlich D, Prehn S, Laskey RA et al. (1994) Isolation of a protein that is essential for the first step of nuclear protein import. Cell 79: 767–778

Gould KL, Nurse P (1989) Tyrosine phosphorylation of the fission yeast cdc2+ protein kinase regulates entry into mitosis. Nature 342: 39–45
Guadagno TM, Newport JW (1996) Cdk2 kinase is required for entry into mitosis as a positive regulator of Cdc2-cyclin B kinase activity. Cell 84: 73–82
Hamaguchi JR, Tobey RA, Pines J et al. (1992) Requirement for p34cdc2 kinase is restricted to mitosis in the mammalian cdc2 mutant FT210. J Cell Biol 117: 1041–1053
Hartwell LH, Weinert TA (1989) Checkpoints: controls that ensure the order of cell cycle events. Science 246: 629–634
Hayles J, Beach D, Durkacz B et al. (1986) The fission yeast cell cycle control gene cdc2: isolation of a sequence suc1 that suppresses cdc2 mutant function. Mol Gen Genet 202: 291–293
Heald R, McLoughlin M, McKeon F (1993) Human WEE1 maintains mitotic timing by protecting the nucleus from cytoplasmically activated Cdc2 kinase. Cell 74: 463–474
Hoffmann I, Clarke PR, Marcote MJ et al. (1993) Phosphorylation and activation of human cdc25-C by cdc2–cyclin B and its involvement in the self-amplification of MPF at mitosis. EMBO J 12: 53–63
Hoffmann I, Draetta G, Karsenti E (1994) Activation of the phosphatase activity of human cdc25A by a cdk2-cyclin E dependent phosphorylation at the G_1/S transition. EMBO J 13: 4302–4310
Honda R, Ohba Y, Nagata A et al. (1993) Dephosphorylation of human p34cdc2 kinase on both Thr-14 and Tyr-15 by human cdc25B phosphatase. FEBS Lett 318: 331–334
Hunt T, Luca FC, Ruderman JV (1992) The requirements for protein synthesis and degradation, and the control of destruction of cyclins A and B in the meiotic and mitotic cell cycles of the clam embryo. J Cell Biol 116: 707–724
Izumi T, Maller JL (1995) Phosphorylation and activation of the *Xenopus* Cdc25 phosphatase in the absence of Cdc2 and Cdk2 kinase activity. Mol Biol Cell 6: 215–226
Jackman MR, Pines JN (1997) Cyclins and the G2/M transition. In: Kastan MB (eds) Checkpoint controls and cancer. Cancer Surveys 29 Cold Spring Harbor Laboratory Press, New York, pp 47–73
Jackman MR, Firth M, Pines J (1995) Human cyclins B1 and B2 are localised to strikingly different structures: B1 to microtubules, B2 primarily to the Golgi apparatus. EMBO J 14: 1646–1654
Jallepalli PV, Kelly TJ (1996) Rum1 and Cdc18 link inhibition of cyclin dependent kinase to the initiation of DNA replication in *Schizosaccharomyces pombe*. Genes Dev 10: 541–552
Jin P, Gu Y, Morgan DO (1996) Role of inhibitory CDC2 phosphorylation in radiation-induced G2 arrest in human cells. J Cell Biol 134: 963–970
Jinno S, Suto K, Nagata A et al. (1994) Cdc25A is a novel phosphatase functioning early in the cell cycle. EMBO J. 13: 1549–1556
Johnson RT, Rao PN (1970) Mammalian cell fusion: induction of premature chromosome condensation in interphase nuclei. Nature 226: 717–722
Kakizuka A, Sebastian B, Borgmeyer U et al. (1992) A mouse cdc25 homolog is differentially and developmentally expressed. Genes Dev 6: 578–590
Kelly TJ, Martin GS, Forsburg SL et al. (1993) The fission yeast cdc18+ gene product couples S phase to START and Mitosis. Cell 74: 371–382
Khazanovich N, Bateman KS, Chernaia M et al. (1996) Crystal structure of the yeast cell cycle control Protein, p13(Suc1), in a strand exchanged dimer. Structure 4: 299–309
Knoblich JA, Lehner CF (1993) Synergistic action of *Drosophila* cyclins A and B during the G2-M transition. EMBO J 12: 65–74
Kornbluth S, Sebastian B, Hunter T et al. (1994) Membrane localization of the kinase which phosphorylates p34cdc2 on threonine 14. Mol Biol Cell 5: 273–282
Krek W, Nigg EA (1991) Differential phosphorylation of vertebrate p34cdc2 kinase at the G1/S and G2/M transitions of the cell cycle: identification of major phosphorylation sites. EMBO J 10: 305–316
Kuang J, Ashorn CL (1993) At least two kinases phosphorylate the MPM-2 epitope during *Xenopus* oocyte maturation. J Cell Biol 123: 859–868
Kumagai A, Dunphy WG (1991) The cdc25 protein controls tyrosine dephosphorylation of the cdc2 protein in a cell-free system. Cell 64: 903–914

Kumagai A, Dunphy WG (1992) Regulation of the cdc25 protein during the cell cycle in *Xenopus* extracts. Cell 70: 139–151
Kumagai A, Dunphy WG (1995) Control of the cdc2/cyclin B complex in *Xenopus* egg extracts arrested at the G2/M checkpoint with DNA synthesis inhibitors. Mol Biol Cell 6: 199–213
Kumagai A, Dunphy WG (1996) Purification and molecular cloning of Plx1, a Cdc25-regulatory kinase from *Xenopus* egg extracts. Science 273: 1377–1380
Küssel P, Frasch M (1995) Pendulin, a *Drosophila* protein with cell cycle-dependent nuclear localization, is required for normal cell proliferation. J Cell Biol 129: 1491–1507
Labbé J-C, Capony J-P, Caput D et al. (1989) MPF from starfish oocytes at first meiotic metaphase is a heterodimer containing one molecule of cdc2 and one molecule of cyclin B. EMBO J 8: 3053–3058
Lamb NJC, Fernandez A, Watrin A et al. (1990) Microinjection of $p34^{cdc2}$ kinase induces marked changes in cell shape, cytoskeletal organization and chromatin structure in quiescent and G1-phase mammalian fibroblasts. Cell 60: 151–165
Lane HA, Nigg EA (1996) Antibody microinjection reveals an essential role for human polo-like kinase 1 (Plk1) in the functional maturation of mitotic centrosomes. J Cell Biol 135: 1701–1713
Lauze E, Stoelcker B, Luca FC et al. (1995) Yeast spindle pole body duplication gene MPS1 encodes an essential dual specificity protein kinase. EMBO J 14: 1655–1663
Lee TH, Kirschner MW (1996) An inhibitor of $p34^{cdc2}$ cyclin B that regulates the G2/M transition in *Xenopus* extracts. Proc Natl Acad Sci USA 93: 352–356
Lehner CF, O'Farrell PH (1990) The roles of *Drosophila* cyclins A and B in mitotic control. Cell 61: 535–547
Lew DJ, Reed SI (1993) Morphogenesis in the yeast cell cycle: regulation by Cdc28 and cyclins. J Cell Biol 120: 1305–1320
Lew DJ, Reed SI (1995) A cell cycle checkpoint monitors cell morphogenesis in budding yeast. J Cell Biol 129: 739–749
Li J, Meyer AN, Donoghue DJ (1995) Requirement for phosphorylation of cyclin B1 for *Xenopus* oocyte maturation. Mol Biol Cell 6: 1111–1124
Li J, Meyer AN, Donoghue DJ (1997) Nuclear localization of cyclin B1 mediates its biological activity and is regulated by phosphorylation. Proc Natl Acad Sci USA 94: 502–507
Liu F, Stanton JJ, Wu Z et al. (1997) The human Myt1 kinase preferentially phosphorylates Cdc2 on threonine 14 and localizes to the endoplasmic reticulum and Golgi complex. Mol Cell Biol 17: 571–583
Llamazares S, Moreira A, Tavares A et al. (1991) polo encodes a protein kinase homolog required for mitosis in *Drosophila*. Genes Dev 5: 2153–2165
Lu KP, Osmani SA, Means AR (1993) Properties and regulation of the cell cycle-specific NIMA protein kinase of *Aspergillus nidulans*. J Biol Chem 268: 8769–8776
Lu KP, Hanes SD, Hunter T (1996) A human peptidyl-prolyl isomerase essential for regulation of mitosis. Nature 380: 544–547
Luca FC, Shibuya EK, Dohrmann CE et al. (1991) Both cyclin A delta 60 and B delta 97 are stable and arrest cells in M-phase, but only cyclin B delta 97 turns on cyclin destruction. EMBO J 10: 4311–4320
Lundgren K, Walworth N, Booher R et al. (1991) mik1 and wee1 cooperate in the inhibitory tyrosine phosphorylation of cdc2. Cell 64: 1111–1122
Maity A, McKenna WG, Muschel RJ (1995) Evidence for post-transcriptional regulation of cyclin B1 mRNA in the cell cycle and following irradiation of HeLa cells. EMBO J. 14: 603–609
Maridor G, Gallant P, Golsteyn R et al. (1993) Nuclear localisation of vertebrate cyclin A correlates with its ability to form complexes with CDK catalytic subunits. J Cell Sci. 106: 535–544
Martin Castellanos C, Labib K, Moreno S (1996) B-type cyclins regulate G1 progression in fission yeast in opposition to the p25rum1 cdk inhibitor. EMBO J 15: 839–849
Masui Y, Markert CL (1971) Cytoplasmic control of nuclear behavior during meiotic maturation of frog oocytes. J Exp Zool 177: 129–145
McGowan CH, Russell P (1993) Human Wee1 kinase inhibits cell division by phosphorylating p34cdc2 exclusively on Tyr15. EMBO J 12: 75–85
McGowan CH, Russell P (1995) Cell cycle regulation of human WEE1. EMBO J 14: 2166–2175

McGowan CH, Russell P, Reed SI (1990) Periodic biosynthesis of the human M-phase promoting factor catalytic subunit p34 during the cell cycle. Mol Cell Biol 10: 3847–3851

Meijer L, Ostvold A-C, Walaas SI et al. (1991) High mobility group (HMG) proteins I, Y and P1 as substrates of the M-phase-specific $p34^{cdc2}/cyclin^{cdc13}$ kinase. Eur J Biochem 196: 557–567

Millar JB, Blevitt J, Gerace L et al. (1991) p55CDC25 is a nuclear protein required for the initiation of mitosis in human cells. Proc Natl Acad Sci USA 88: 10500–10504

Millar JB, Lenaers G, Russell P (1992) Pyp3 PTPase acts as a mitotic inducer in fission yeast. EMBO J 11: 4933–4941

Minshull J, Sun H, Tonks NK et al. (1994) A MAP kinase-dependent spindle assembly checkpoint in *Xenopus* egg extracts. Cell 79: 475–486

Moreno S, Nurse P (1994) Regulation of progression through the G1 phase of the cell cycle by the rum1+ gene. Nature 367: 236–242

Moreno S, Hayles J, Nurse P (1989) Regulation of $p34^{cdc2}$ protein kinase during mitosis. Cell 58: 361–372

Morgan DO (1995) Principles of CDK regulation. Nature 374: 131–134

Mueller PR, Coleman TR, Dunphy WG (1995a) Cell cycle regulation of a *Xenopus* Wee1-like Kinase. Mol Biol Cell 6: 119–134

Mueller PR, Coleman TR, Kumagai A et al. (1995b) Myt1: a membrane-associated inhibitory kinase that phosphorylates Cdc2 on both threonine-14 and tyrosine-15. Science 270: 86–90

Murray A, Hunt T (1993) The cell cycle: an introduction. Freeman, New York

Murray AW, Kirschner MW (1989) Dominoes and clocks: the union of two views of the cell cycle. Science 246: 614–621

Navas TA, Zheng Z, Elledge SJ (1995) DNA polymerase ε links the DNA replication machinery to the S phase checkpoint. Cell 74: 29–39

Nebreda AR, Gannon JV, Hunt T (1995) Newly synthesized protein(s) must associate with p34cdc2 to activate MAP kinase and MPF during progesterone-induced maturation of *Xenopus* oocytes. EMBO J 14: 5597–5607

Nebreda AR, Hunt T (1993) The c-*mos* proto-oncogene protein kinase turns on and maintains the activity of MAP kinase, but not MPF, in cell-free extracts of *Xenopus* oocytes and eggs. EMBO J 12: 1979–1986

Nigg EA (1991) The substrates of the cdc2 kinase. Semin Cell Biol 2: 261–270

Nigg EA (1993) Cellular substrates of $p34^{cdc2}$ and its companion cyclin-dependent kinases. Trends Cell Biol. 3: 296–301

Nissen MS, Langan TA, Reeves R (1991) Phosphorylation by cdc2 kinase modulates DNA binding activity of high mobility group I nonhistone chromatin protein. J Biol Chem 266: 19945–19952

Norbury C, Blow J, Nurse P (1991) Regulatory phosphorylation of the $p34^{cdc2}$ protein kinase in vertebrates. EMBO J. 10: 3321–3329

O'Connell MJ, Norbury C, Nurse P (1994) Premature chromatin condensation upon accumulation of NIMA. EMBO J. 13: 4926–4937

O'Connell MJ, Raleigh JM, Verkade HM et al. (1997) Chk1 is a wee1 kinase in the G2 DNA damage checkpoint inhibiting cdc2 by Y15 phosphorylation. EMBO-J 16: 545–554

Ookata K, Hisanaga S, Okano T et al. (1992) Relocation and distinct subcellular localization of p34cdc2-cyclin B complex at meiosis reinitiation in starfish oocytes. EMBO J 11: 1763–1772

Osmani AH, McGuire SL, Osmani SA (1991) Parallel activation of the NIMA and p34cdc2 cell cycle-regulated protein kinases is required to initiate mitosis in *A nidulans*. Cell 67: 283–291

Pagano M, Pepperkok R, Verde F et al. (1992) Cyclin A is required at two points in the human cell cycle. EMBO J 11: 961–971

Parge HE, Arvai AS, Murtari DJ et al. (1993) Human CksHs2 atomic structure: a role for its hexameric assembly in cell cycle control. Science 262: 387–395

Parker LL, Piwnica-Worms H (1992) Inactivation of the p34cdc2-cyclin B complex by the human WEE1 tyrosine kinase. Science 257: 1955–1957

Parker LL, Atherton-Fessler S, Lee MS et al. (1991) Cyclin promotes the tyrosine phosphorylation of p34cdc2 in a wee1+ dependent manner. EMBO J 10: 1255–1263

Parker LL, Walter SA, Young PG et al. (1993) Phosphorylation and inactivation of the mitotic inhibitor Wee1 by the nim1/cdr1 kinase. Nature 363: 736–738

Patra D, Dunphy WG (1996) Xe-p9, a *Xenopus* sucl/CKS homolog, has multiple essential roles in cell cycle control. Genes Dev. 10: 1503–1515

Piaggio G, Farina A, Perrotti D et al. (1995) Structure and growth-dependent regulation of the human cyclin B1 promoter. Exp Cell Res 216: 396–402

Pines J, Hunter T (1989) Isolation of a human cyclin cDNA: evidence for cyclin mRNA and protein regulation in the cell cycle and for interaction with $p34^{cdc2}$. Cell 58: 833–846

Pines J, Hunter T (1990) Human cyclin A is adenovirus E1A-associated protein p60, and behaves differently from cyclin B. Nature 346: 760–763

Pines J, Hunter T (1991) Human cyclins A and B are differentially located in the cell and undergo cell cycle dependent nuclear transport. J. Cell Biol. 115: 1–17

Pines J, Hunter T (1994) The differential localization of human cyclins A and B is due to a cytoplasmic retention signal in cyclin B. EMBO J. 13: 3772–3781

Posada J, Yew N, Ahn NG et al. (1993) Mos stimulates MAP kinase in *Xenopus* oocytes and activates a MAP kinase kinase in vitro. Mol Cell Biol 13: 2546–2553

Ranganathan R, Lu KP, Hunter T et al. (1997) Structural and functional analysis of the mitotic rotamase Pinl suggests substrate recognition is phosphorylation dependent. Cell 89: 875–886

Rao PN, Johnson RT (1970) Mammalian cell fusion studies on the regulation of DNA synthesis and mitosis. Nature 225: 159–164

Rhind N, Furnari B, Russell P (1997) Cdc2 tyrosine phosphorylation is required for the DNA damage checkpoint in fission yeast. Genes-Dev 11: 504-511 issn: 0890–9369

Richardson HE, Stueland CS, Thomas J et al. (1990) Human cDNAs encoding homologs of the small p34Cdc28/Cdc2-associated protein of *Saccharomyces cerevisiae* and *Schizosaccharomyces pombe*. Genes Dev 4: 1332–1344

Rowley R, Hudson J, Young PG (1992) The weel protein kinase is required for radiation-induced mitotic delay. Nature 356: 353–355

Russell P, Nurse P (1987a) The mitotic inducer nim1+ functions in a regulatory network of protein kinase homologs controlling the initiation of mitosis. Cell 49: 569–576

Russell P, Nurse P (1987b) Negative regulation of mitosis by wee1+, a gene encoding a protein kinase homolog. Cell 49: 559–567

Schultz SJ, Fry AM, Sutterlin C et al. (1994) Cell cycle-dependent expression of Nek2, a novel human protein kinase related to the NIMA mitotic regulator of *Aspergillus nidulans*. Cell Growth Differ 5: 625–635

Schwob E, Böhm T, Mendenhall MD et al. (1994) The B-type cyclin kinase inhibitor $p40^{SIC1}$ controls the G1/S transition in *Saccharomyces cerevisiae*. Cell 79: 233–244

Sebastian B, Kakizuka A, Hunter T (1993) Cdc25M2 activation of cyclin-dependent kinases by dephosphorylation of threonine-14 and tyrosine-15. Proc Natl Acad Sci USA 90: 3521–3524

Seki T, Yamashita K, Nishitani H et al. (1992) Chromosome condensation caused by loss of RCC1 function requires the cdc25C protein that is located in the cytoplasm. Mol Biol Cell 3: 1373–1388

Sherwood SW, Rush DF, Kung AL et al. (1994) Cyclin B1 expression in HeLa S3 cells studied by flow cytometry. Exp Cell Res 211: 275–281

Sibon OCM, Stevenson VA, Theurkauf WE (1997) DNA-replication checkpoint control at the *Drosophila* midblastula transition. Nature 388: 93–97

Sorger PK, Murray AW (1992) S-phase feedback control in budding yeast independent of tyrosine phosphorylation of p34cdc28. Nature 355: 365–368

Stemmann O, Lechner J (1996) The saccharomyces cerevisiae kinetochore contains a cyclin CDK complexing homolog, as identified by *in vitro* reconstitution. EMBO J 15: 3611–3620

Swenson K, Farrell KM, Ruderman JV (1986) The clam embryo protein cyclin A induces entry into M-phase and the resumption of meiosis in *Xenopus* oocytes. Cell 47: 861–870

Takenaka K, Gotoh Y, Nishida E (1997) MAP kinase is required for the spindle assembly checkpoint but is dispensable for the normal M phase entry and exit in *Xenopus* egg cell cycle extracts. J Cell Biol 136: 1091–1097

Tang Y, Reed SI (1993) The Cdk-associated protein Cks1 functions both the G_1 and G_2 in Saccharomyces cerevisiae. Genes Dev 7: 822–832

Tang Z, Coleman TR, Dunphy WG (1993) Two distinct mechanisms for negative regulation of the Weel protein kinase. EMBO J 12: 3427–3436

Tsai L-H, Harlow E, Meyerson M (1991) Isolation of the human cdk2 gene that encodes the cyclin A and adenovirus E1A-associated p33 kinase. Nature 353: 174–177

Tyers M (1996) The cyclin dependent kinase inhibitor p40(Sic1) imposes the requirement for Cln G1 cyclin function at Start. Proc Natl Acad Sci USA 93: 7772–7776

Walworth NC, Bernards R (1996) *rad*-dependent response of the chk1-encoded protein kinase at the DNA damage checkpoint . Science 271: 353–356 issn: 0036–8075

Walworth NC, Davey S, Beach D (1993) Fission yeast chk1 protein kinase links the *rad* checkpoint pathway to cdc2. Nature 363: 368–371

Watanabe N, Broome M, Hunter T (1995) Regulation of the human WeelHu CDK tyrosine 15-kinase during the cell cycle. EMBO J. 14: 1878–1891

Weiss E, Winey M (1996) The *Saccharomyces cerevisiae* spindle pole body duplication gene *MPS1* is part of a mitotic checkpoint. J-Cell-Biol 132: 111–123

Welch PJ, Wang JY (1992) Coordinated synthesis and degradation of cdc2 in the mammalian cell cycle. Proc Natl Acad Sci USA 89: 3093–3097

Wu L, Russell P (1993) Nim1 kinase promotes mitosis by inactivating wee1 tyrosine kinase. Nature 363: 738–741

Wu L, Shiozaki K, Aligue R et al. (1996) Spatial organization of the Nim1-Wee1-Cdc2 mitotic control network in *Schizosaccharomyces pombe*. Mol-Biol-Cell 7: 1749–1758

Xiong Y, Zhang H, Beach D (1993) Subunit rearrangement of the cyclin-dependent kinases is associated with cellular transformation. Genes Dev 7: 1572–1583

Yamashita K, Yasuda H, Pines J et al. (1991) Okadaic acid, a potent inhibitor of type 1 and type 2A protein phosphatases, activates cdc2/H1 kinase and transiently induces a premature mitosis-like state in BHK21 cells. EMBO J 9: 4331–4338

Ye XS, Xu G, Pu RT et al. (1995) The NIMA protein kinase is hyperphosphorylated and activated downstream of p34cdc2-cyclin B: co-ordination of two mitosis promoting kinases. EMBO J 14: 986–994

Zhang H, Kobayashi R, Galaktionov K et al. (1995) p19Skp1 and p45Skp2 are essential elements of the cyclin A-CDK2 S phase kinase. Cell 82: 915–925

Regulation of CDKs by phosphorylation

M. J. Solomon[1] and P. Kaldis

1 Introduction

The key transitions of the cell cycle are controlled via the sequential activation and inactivation of members of the cyclin-dependent kinase (CDK) subfamily of protein kinases. The activities of these enzymes are regulated by multiple mechanisms including both activating and inactivating phosphorylations, binding to regulatory subunits termed cyclins, subcellular localization, and association with inhibitory proteins (CKIs). There are two broad classes of regulatory inputs. Regulation in response to intrinsic signals ensures the proper timing of the basic cell cycle and coordinates various cell cycle events via checkpoints that monitor the completion of each step. This regulation is responsible for the abrupt transitions between cell cycle phases as well as for the fidelity of the cell division process. Phosphorylation of CDKs is very important for such intrinsic regulation, although other mechanisms participate as well. In contrast, extrinsic signals impinge on the cell, resulting in stimulation or inhibition of the cell cycle. These signals typically affect the levels of CDK inhibitors.

In this review, we will consider the regulation of CDKs by phosphorylation. We will focus on p34^{cdc2}, the CDK responsible for inducing entry into mitosis, since the most is known about its regulation. There are three major sites of phosphorylation of p34^{cdc2}. In human p34^{cdc2} these are Thr-14 and Tyr-15, which are each inhibitory, and Thr-161, which is required for kinase activity. We will discuss other CDKs that are known to be regulated by similar mechanisms, including p33^{cdk2}, cdk4 and cdk6, which are involved in promoting cell cycle transitions, and some that regulate processes other than the cell cycle (see Table 1 for examples). Although the enzymes that phosphorylate and dephosphorylate CDKs are themselves regulated by diverse mechanisms in different organisms, we will pay particular attention to how they, in turn, are regulated by phosphorylation.

[1] Yale University School of Medicine, Department of Molecular Biophysics and Biochemistry, 333 Cedar Street, New Haven, Connecticut 06520-8024, USA, E-mail: mark.solomon@yale.edu

2 Phosphorylation During the Cyclin Activation of p34^{cdc2}

A detailed description of the activation of p34^{cdc2} during the cell cycle came from studies in *Xenopus* egg extracts (Solomon et al. 1990, 1992), although the basic process is conserved (see Fig. 1). Monomeric p34^{cdc2} is unphosphorylated and inactive. Binding to a mitotic cyclin as it accumulates during interphase induces a conformational change in p34^{cdc2} that allows it to be recognized by multiple protein kinases, resulting in its phosphorylation on Thr-14, Tyr-15, and Thr-161. The inhibitory phosphorylations (Thr-14 and Tyr-15) are dominant over the activating phosphorylation (Thr-161) and the triply phosphorylated form of p34^{cdc2} is inactive. Subsequent dephosphorylation of Thr-14 and Tyr-15 yields the active protein kinase complex that induces entry into mitosis. Dephosphorylation of p34^{cdc2} on Thr-161 occurs following the degradation of cyclin at anaphase (Lorca et al. 1992; Poon and Hunter 1995), thereby resetting the system for the next cell cycle. Two feedback loops operate during the activation of p34^{cdc2} to generate the abrupt, "autocatalytic" activation of this enzyme and entry into mitosis: active p34^{cdc2} stimulates the production of additional active complexes by both up-regulating the enzyme that dephosphorylates the inhibitory sites and down regulating the enzymes that rephosphorylate them (see Fig. 1).

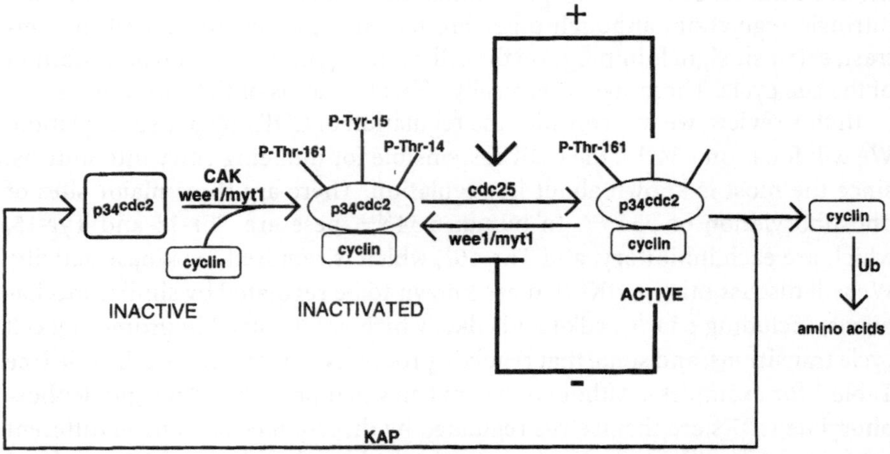

Fig. 1. Cyclin activation of p34^{cdc2}. Cyclin binding to p34^{cdc2} induces its phosphorylation on three sites. (Cyclin B is the main mitotic cyclin binding to p34^{cdc2}.) Two of these phosphorylations, Thr-14 and Tyr-15, are inhibitory whereas the third, on Thr-161, is required for kinase activity. Phosphorylation of the inhibitory sites is carried out by the Weel (Tyr-15) and Mytl (Thr-14 and Tyr-15) protein kinases, and phosphorylation of Thr-161 is carried out by the CDK-activating kinase (CAK). Dephosphorylation of the inhibitory sites is performed mainly by Cdc25, a dual-specificity phosphatase. Feedback from the active form of the cyclin-p34^{cdc2} complex stimulates Cdc25 activity and inhibits Weel/Mytl activity. This activation loop leads to a rapid, "autocatalytic" activation of p34^{cdc2} and transition into mitosis. Degradation of cyclin via the ubiquitin system (UB) inactivates p34^{cdc2} and leads to its dephosphorylation, probably by the KAP phosphatase

The first site of inhibitory phosphorylation to be identified was Tyr-15, in *Schizosaccharomyces pombe* $p34^{cdc2}$ (Gould and Nurse 1989). Tyr-15 and the adjacent Thr-14 were subsequently identified in $p34^{cdc2}$ proteins from chickens, humans and frogs (Krek and Nigg 1991a; Norbury et al. 1991; Solomon et al. 1992); phosphorylation of Thr-14 occurs to a lesser extent in *S. pombe* (Den Haese et al. 1995). Mutation of Tyr-15 to Phe-15 in *S. pombe* $p34^{cdc2}$ leads to a precocious entry into mitosis, indicating that phosphorylation of this site normally restrains cell cycle progression (Gould and Nurse 1989). Similarly a mutant form of vertebrate $p34^{cdc2}$ containing replacements of Thr-14 and Tyr-15 with Ala-14 and Phe-15 (an "AF" mutant) was activated early in a *Xenopus* egg extract (Norbury et al. 1991) and, when overexpressed in HeLa cells, led to a premature induction of mitotic events (Krek and Nigg 1991b). Inactivation of cyclin-$p34^{cdc2}$ complexes in vitro via phosphorylation on either Thr-14 or Tyr-15 (Parker and Piwnica-Worms 1992; Liu et al. 1997) firmly established the role of these phosphorylations. Interestingly, careful analyses showed that the single phosphorylations of either Thr-14 or Tyr-15 only inhibit $p34^{cdc2}$ activity by about 90 % whereas the double phosphorylation reduces the activity by about 99 % (Borgne and Meijer 1996; Liu et al. 1997), suggesting that a singly-phosphorylated species may serve as the trigger for the "autocatalytic" activation of $p34^{cdc2}$ via inhibition of Thr-14 and Tyr-15 phosphorylation and stimulation of their dephosphorylation. Genetic studies in the fission yeast *Schizosaccharomyces pombe* led the way to the identification of the first kinase acting on Tyr-15, Weel (see Sect. 6), and to the main opposing phosphatase, Cdc25 (see Sect. 7; see Forsberg and Nurse 1991 for a review of cell cycle genetics in *S. pombe*). A second Tyr-15 kinase was subsequently isolated in *S. pombe* (Lundgren et al. 1991) and a Thr-14/Tyr-15 kinase, Mytl, was identified in vertebrates (Mueller et al. 1995b; Liu et al. 1997).

In contrast, activating phosphorylation (of Thr-161 in human $p34^{cdc2}$) is required for kinase activity. This phosphorylation was first identified in *S. pombe* $p34^{cdc2}$, where the equivalent site is Thr-167 (Gould et al. 1991), and was subsequently found in other species (Krek and Nigg 1992; Solomon et al. 1992). Mutation of Thr-161 to alanine leads to the production of inactivatable and nonfunctional forms of $p34^{cdc2}$ (Booher and Beach 1986; Gould et al. 1991; Lee et al. 1991; Krek and Nigg 1992; Solomon et al. 1992; Cismowski et al. 1995). Conversely, phosphorylation of this site in vitro leads to CDK activation (Solomon et al. 1992; Fesquet et al. 1993; Poon et al. 1993; Solomon et al. 1993). The kinase acting on Thr-161 was originally named the "$p34^{cdc2}$-activating kinase" (CAK) -- though it is now known as the CDK-activating kinase to reflect its broader substrate range – and was identified biochemically (Solomon et al. 1992; Fesquet et al. 1993; Poon et al. 1993; Solomon et al. 1993). A phosphatase acting on the equivalent site in $p33^{cdk2}$ most likely also acts on $p34^{cdc2}$ and has been termed KAP (Poon and Hunter 1995).

Two minor sites of phosphorylation of p34^cdc2^ have been found, although their functions, if any, and conservation are unclear: phosphorylation of Ser-277 in chicken p34^cdc2^ peaks during G1 (Krek and Nigg 1991a), and human p34^cdc2^ is phosphorylated on Ser-39 during G1 in vivo and on this site by casein kinase II in vitro (Russo et al. 1992).

The three major sites of phosphorylation are conserved in the sequences of most cyclin-dependent kinases (see Table 1), although few of these phosphorylations have actually been demonstrated (see Sect. 5). These CDKs include enzymes involved directly in cell cycle progression such as p34^cdc2^ (G2-M transition), p33^cdk2^ (G1-S transition and progression through S phase), cdk3 (G1-S transition), cdk4 (G1 progression) and cdk6 (G1 progression). Among the others, cdk5 is involved in neuronal differentiation, cdk7/p40^MO15^, cdk8, Kin28p and Srb10p in transcription (and Cdk7/p40^MO15^ also in activation of other CDKs; see Sect. 8); Pho85p in phosphate metabolism and in G1 progression, and Cdc7p in the initiation of DNA synthesis.

Table 1. Potential phosphorylation sites in cyclin-dependent protein kinases

CDK	Major binding partner	Potential inhibitory sites	Sequence	Potential activating site	Sequence
Cdc2	Cyclin B	Thr-14, Tyr-15	G g G **T Y** G v v	Thr-161	v y **T** h e v v t l w
Cdk2	Cyclin A, E	Thr-14, Tyr-15	G e G **T Y** G v v	Thr-160	t y **T** h e v v t l w
Cdk3	?	Thr-42, Tyr-43	G e G **T Y** G v v	Thr-188	t y **T** h e v v t l w
Cdk4	Cyclin D	Tyr-17	G v G a **Y** G t v	Thr-172	a l **T** p e v v t l w
Cdk5	p35	Thr-14, Tyr-15	G e G **T Y** G t v	Ser-158	c y **S** a e v v t l w
Cdk6	Cyclin D	Tyr-24	G e G a **Y** G k v	Thr-177	a l **T** s v v v t l w
Cdk7/MO15	Cyclin H	-	G e G q f a t v	Thr-170	a y **T** h q v v t r w
Cdk8	Cyclin C	Thr-28, Tyr-29	G r G **T Y** G h v	-	d l d p v v v t f w
Cdc7p	Dbf4p	Thr-43	G e G **T** f s s v	-	k r a n r a g t r g
Kin28p	Ccl1p	Thr-17, Tyr-18	G e G **T Y** a v v	Thr-162	i l **T** s n v v t r w
Pho85p	Pho80p, Pcls	Thr-14, Tyr-15	G n G **T Y** a t v	Ser-163	t f **S** s e v v t l w
Srb10p	Srb11p	Thr-71, Tyr-72	a a G **T Y** G k v	-	t g d k v v v t i w

Selected cyclin-dependent kinases are shown. Cdc2 and cdk2 through cdk8 represent the human sequences, whereas Cdc7p, Kin28p, Pho85p, and Srb10p are from *Saccharomyces cerevisiae*. The glycines within the GxGxxG nucleotide-binding motif are shown in capitals and the potential sites of inhibitory and activating phosphorylation are shown in bold capital letters. Activating phosphorylations of cdk8 and Cdc7p seem unlikely. For cdk8, the nearest serine or threonine is at position 182, nine amino acids N-terminal to the expected position, whereas for Cdc7p the nearest sites are at positions 281 and 270, five amino acids C-terminal and six amino acids N-terminal to the expected site. Srb10p contains a potential threonine two amino acids N-terminal to the expected site, within a three amino acid insertion that might maintain the appropriate positioning of the T-loop threonine. References and accession numbers for the various proteins are as follows: Cdc2: Lee and Nurse (1987), Y00272; cdk2: Elledge et al. (1992), X61622; cdk3: Meyerson et al. (1992), X66357; cdk4: Khatib et al. (1993), S67448; cdk5 ("PSSALRE"): Meyerson et al. (1992), X66364; cdk6("PLSTIRE"): Meyerson et al. (1992), X66365; cdk7/MO15: Darbon et al. (1994), X77743 and Tassan et al. (1994), X79193; cdk8: Tassan et al. (1995b), X85753; Cdc7p: Ham et al. (1989), X15362; Kin28p: Simon et al. (1986), X04423; Pho85p: Toh-e et al. (1988), X13515; Srb10p: Liao et al. (1995), U20222.

3 Biochemical Effects of CDK Phosphorylation

Some insights into the biochemical effects of these phosphorylations come from structural studies of $p33^{cdk2}$ (reviewed in Morgan 1996), the CDK responsible for promoting entry into and through S phase. The so-called T-loop (Johnson et al. 1996) contains Thr-160, the site of activating phosphorylation in $p33^{cdk2}$ (see Table 1). In monomeric $p33^{cdk2}$, the T-loop physically blocks access of substrates to the active site of the kinase (De Bondt et al. 1993). In addition, residues in $p33^{cdk2}$ that help position the β and γ phosphates for catalysis are out of position (De Bondt et al. 1993). Binding to cyclin A leads to a repositioning of the ATP phosphates into their correct positions and to a movement of the T-loop away from the active site (Jeffrey et al. 1995). Subsequent phosphorylation of Thr-160 exerts a more subtle effect on the cyclin interface and the substrate-binding region and leads to further movement of T-loop residues by up to 7 Å (Russo et al. 1996). The structure of the active cyclin A-$p33^{cdk2}$ complex suggests that Thr-160 phosphorylation might play a greater role in substrate recognition than in catalysis (Russo et al. 1996). Biochemically, this phosphorylation is important for recognition of a basic residue three amino acids C-terminal to the phosphorylation site. In the absence of Thr-160 phosphorylation, cyclin A-$p33^{cdk2}$ complexes have about 1 % the activity of the Thr-160 phosphorylated form (Connell-Crowley et al. 1993). Although phosphorylation of a mutant substrate containing Ser-Pro-Arg-Ala by the phosphorylated enzyme is greatly reduced relative to the phosphorylation of an optimal substrate (Ser-Pro-Arg-Lys), the unphosphorylated enzyme phosphorylates these two substrates with nearly equal efficiency (J. Holmes and M.J. Solomon, unpubl. data), indicating that Thr-160 phosphorylation is only essential when there is a potential ionic interaction with the +3 position of the substrate. Other unphosphorylated cyclin-CDK complexes appear to be completely inactive, suggesting that their activating phosphorylations may have multiple effects, stimulating substrate binding as well as catalysis. This phosphorylation also strengthens the interaction between cyclin and the CDK (Ducommun et al. 1991; Desai et al. 1992).

Inhibition via phosphorylation of Thr-14 and Tyr-15 is not as well understood. These residues lie at positions 4 and 5 in the conserved GxGxxG motif involved in nucleotide binding. Surprisingly, their phosphorylation does not inhibit ATP binding to p34^{cdc2} (Atherton-Fessler et al. 1994), suggesting a more subtle effect on the positioning of ATP for catalysis. Thr-14 and Tyr-15 are buried behind the T-loop in monomeric p33^{cdk2} (De Bondt et al. 1993), providing a structural explanation for the observed cyclin-dependent phosphorylation of these residues (Solomon et al. 1990; Parker et al. 1991), which would displace the T-loop from in front of the active site (De Bondt et al. 1993; Jeffrey et al. 1995).

4
Inhibitory Phosphorylation of p34^{cdc2} During DNA Checkpoints

Checkpoints are the cellular mechanisms for ensuring that certain cell cycle events do not begin until other events have been completed (for a review see Elledge 1996). For example, entry into mitosis does not normally occur until the DNA has been fully replicated and any lesions have been repaired. It can be important to distinguish between the DNA replication checkpoint, which monitors the completion of S phase, and the DNA damage checkpoint, which can be activated even if S phase is complete. The replication checkpoint blocks cell cycle progression in response to DNA synthesis inhibitors such as hydroxyurea whereas the DNA damage checkpoint responds to UV- and X-ray-induced lesions, among others.

Although there remains some controversy surrounding this issue, it is becoming increasingly clear that inhibitory phosphorylation plays a role in both the DNA replication and the DNA damage checkpoints in most organisms. The initial work in this area was conducted in *S. pombe* by Enoch and Nurse (1990) who observed that reduction in the Tyr-15 phosphorylation of p34^{cdc2} via either mutation to phenylalanine or overexpression of *cdc25* led to progression through a DNA synthesis block. The DNA damage checkpoint in *S. pombe* also appears to require Tyr-15 phosphorylation (Rhind et al. 1997), although an earlier report suggested that it might not (Barbet and Carr 1993). Furthermore, the cell cycle arrest caused by overexpression of the Chk1 protein kinase, which is involved in transducing the checkpoint signal (Walworth et al. 1993), is relieved by mutation of *wee1*, indicating that it occurs via inhibitory phosphorylation of p34^{cdc2}.

Related studies have been conducted in diverse species. The role of inhibitory phosphorylation in *Saccharomyces cerevisiae* is quite different. In budding yeast, Tyr-19 phosphorylation does not play a significant role during a normal cell cycle, and its mutation has no effect on either the DNA replication or the DNA damage checkpoints (Amon et al. 1992; Sorger and Murray 1992). Instead, phosphorylation of Tyr-19 (mediated by the *S. cerevisiae* homologue of wee1, swelp) is essential for a morphogenesis checkpoint monitoring bud formation (Lew and Reed 1995; Sia et al. 1996) and also appears to play a role in the exit from mitosis in cells that have been held for a prolonged time at a spindle assembly checkpoint (Minshull et al. 1996). In *Aspergillus nidulans*, mutation of Tyr-15 overcomes the DNA damage checkpoint (Ye et al. 1997) as well as the cell cycle delay caused by a slowing of S phase (Ye et al. 1996). However, to override an S phase arrest requires both the prevention of tyrosine phosphorylation of $p34^{cdc2}$ and the premature activation of the NIMA protein kinase (Ye et al. 1996). In a similar vein, elimination of inhibitory phosphorylations of $p34^{cdc2}$ in human HeLa cells by expression of subphysiological levels of an "AF" mutant override the normal DNA replication and DNA damage checkpoints, thereby inducing partial mitotic events (Jin et al. 1996) and hypersensitivity to DNA damage or inhibition of DNA replication (Blasina et al. 1997). Importantly, both studies suggested that an additional, unidentified control operates in parallel to tyrosine phosphorylation during these checkpoints. Finally, the DNA replication checkpoint in Xenopus egg extracts functions even in the absence of inhibitory phosphorylations of $p34^{cdc2}$ (Kumagai and Dunphy 1995). Instead, this checkpoint appears to operate via a titratable, possibly membrane-bound, inhibitor of $p34^{cdc2}$ (Kumagai and Dunphy 1995; Lee and Kirschner 1996). Although no simple model can rationalize all of these results, a recurring theme is the presence of both phosphorylation-dependent and phosphorylation-independent (CDK-inhibitor-dependent?) forms of regulation. The relative balance between these mechanisms may vary between cell types and organisms and give rise to the observed complexity.

5 Regulation of Other CDKs by Phosphorylation

Despite the apparent conservation of sites for both activating and inhibitory phosphorylations of cyclin-dependent kinases, our knowledge of their use in the regulation of other CDKs is incomplete. Potential sites of activating phosphorylation are present in most of these enzymes. Like $p34^{cdc2}$, $p33^{cdk2}$, cdk4 and cdk6 are activated by a Thr-161-like phosphorylation event (Gu et al. 1992; Fesquet et al. 1993; Poon et al. 1993; Solomon et al. 1993; Matsuoka et al. 1994; Iavarone and Massagué 1997). Although cdk3 has not been studied extensively, it is likely, given its overall sequence similarity to $p33^{cdk2}$ and involvement in G1-S progression (van den Heuvel and Harlow 1993), that it will also be regulated by both activating and inhibitory phosphorylations. Activating phosphorylation does not seem to be necessary for cdk5, which is involved in neuronal differentiation rather than in cell cycle progression (Nikolic et al. 1996; Chae et al. 1997; Philpott et al. 1997). In fact, bacterially expressed cdk5 appears to be fully activated by binding to its regulatory partner, p35 (Qi et al. 1995; Poon et al. 1997) and is not further activated by treatment with CAK or rendered less active by mutation of the putative activating phosphorylation site Ser-259 (Poon et al. 1997). Activating phosphorylation appears to be important for the activity of yeast Pho85p since an S163A mutant is non-functional (Santos et al. 1995). This CDK is involved in both phosphate metabolism and progression through G1 (O'Neill et al. 1996; Measday et al. 1997). Activating phosphorylation also provides one route to the activation of cdk7/p40^{MO15} as a CDK-activating kinase (see Sect. 8). Yeast Kin28p, which, like cdk7/p40^{MO15}, serves as a subunit of transcription factor IIH (TFIIH; Feaver et al. 1994), undergoes activating phosphorylation on Thr-162 (J. Kimmelman, G. Laff, and M.J. Solomon, unpubl. data). Cells containing an Ala-162 substitution are viable, but the resulting protein has reduced protein kinase activity (J. Kimmelman, G. Laff, and M.J. Solomon, unpubl. data).

Inhibitory phosphorylations play important roles in the regulation of some CDKs. p33^{cdk2} is phosphorylated on both Thr-14 and Tyr-15 (Gabrielli et al. 1992; Gu et al. 1992) whereas cdk4 and cdk6 can be phosphorylated on inhibitory tyrosines (Terada et al. 1995; Iavarone and Massagué 1997). No inhibitory phosphorylations have been reported for the other CDKs listed in Table 1, though it seems likely that some of these enzymes will be so regulated, at least under the appropriate conditions. For example, although the Kin28p subunit of yeast TFIIH is normally not tyrosine-phosphorylated (J. Kimmelman, G. Laff, and M.J. Solomon, unpubl. data), inhibition of Kin28p via such phosphorylation would be an attractive way to regulate transcription at a specific stage of the transcription cycle, in response to particular cellular stresses, or following local damage to the DNA template. p33^{cdk2} is extensively phosphorylated on Thr-14 and Tyr-15 in vivo (Gu et al. 1992), yet mutation of these sites does not have a significant effect on progression through S phase (Jin et al. 1996; Poon et al. 1996). Interestingly, UV irradiation of mammalian cells leads to the inactivation of cdk4 and increases its phosphorylation on Tyr-17 (Terada et al. 1995; Poon et al. 1996). When expressed in these cells, a mutant form of cdk4 containing Phe-17 was no longer inactivated and the cells failed to arrest in G1 (Terada et al. 1995). Similarly, TGF-β treatment can cause cell cycle arrest and inhibitory tyrosine phosphorylation of both cdk4 and cdk6; mutation of the phosphorylated tyrosine prevents the inactivation of cdk6 (Iavarone and Massagué 1997).

No equivalent activating or inhibitory phosphorylations have been observed for yeast Srb10p, a CDK subunit of the RNA Pol II holoenzyme (Liao et al. 1995); for cdk8, a likely higher eukaryotic Srb10p homologue (Tassan et al. 1995b; Leclerc et al. 1996; Rickert et al. 1996), or for yeast Cdc7p, an S-phase inducer (Sclafani and Jackson 1994).

6 Inhibitory Kinases – Wee1, Mik1, Myt1

Early studies of the *wee1* gene indicated that it acted formally as a dosage-sensitive inhibitor of $p34^{cdc2}$ in *Schizosaccharomyces pombe* (Russell and Nurse 1987a; Forsberg and Nurse 1991) and its sequence suggested that it encoded a serine/threonine protein kinase (Russell and Nurse 1987a). However, since the major site of inhibitory phosphorylation on *S. pombe* $p34^{cdc2}$ is on Tyr-15 (Gould and Nurse 1989), it was not at all clear that Wee1 could inhibit $p34^{cdc2}$ directly. It was thus surprising when Wee1 was found to phosphorylate $p34^{cdc2}$, but only on Tyr-15, not on Thr-14 (Parker et al. 1992). Even human Wee1 could only phosphorylate Tyr-15 (Parker and Piwnica-Worms 1992; McGowan and Russell 1993), despite the fact that human $p34^{cdc2}$ is phosphorylated on both sites in vivo. Interestingly, overexpression of *wee1* in *S. pombe* leads to significant phosphorylation of Thr-14; a low level of Thr-14 phosphorylation occurs normally during S phase in a *wee1*-dependent manner (Den Haese et al. 1995). This phosphorylation requires the prior phosphorylation of $p34^{cdc2}$ on Tyr-15. These observations indicate that *wee1* controls the phosphorylation of *S. pombe* $p34^{cdc2}$ on Thr-14, but they do not indicate that the effect is necessarily a direct one. Nevertheless, they raise the possibility that Wee1 proteins may be able to phosphorylate Thr-14, at least under certain circumstances.

wee1 has been conserved in diverse organisms (see Table 2). A functional fragment of a human homologue was identified by complementation in *S. pombe* (Igarashi et al. 1991) followed by the isolation of the intact gene (McGowan and Russell 1995; Parker et al. 1995; Watanabe et al. 1995) and may be somewhat more similar to *S. pombe* Mik1 (see below) than to Wee1 (Watanabe et al. 1995). This enzyme can phosphorylate $p33^{cdk2}$ as well as $p34^{cdc2}$ (Watanabe et al. 1995). The *S. cerevisiae* homologue has been termed Swe1p. It can phosphorylate and inactivate Cdc28p (the *S. cerevisiae* equivalent of $p34^{cdc2}$), and the cell cycle arrest caused by its overexpression can be suppressed by a Y19F mutation of its phosphorylation site on Cdc28p (Booher et al. 1993). A *Drosophila* homologue (Dwee1) was isolated by complementation in *S. pombe* (Campbell et al. 1995) and a *Xenopus* homologue was isolated by PCR (Mueller et al. 1995a).

The identification of Mik1, an *S. pombe* cousin of Wee1, helped to explain how the $p34^{cdc2}$ in cells lacking Wee1 function could nevertheless contain phosphorylated Tyr-15 (Lundgren et al. 1991). These related proteins (48 % identity in their catalytic domains) act redundantly, with Wee1 activity predominating. Cells mutant for both genes lack tyrosine phosphorylation of $p34^{cdc2}$ and undergo a catastrophic premature entry into mitosis (Lundgren et al. 1991). Like Wee1, Mik1 can directly phosphorylate $p34^{cdc2}$ on Tyr-15 (Lee et al. 1994).

Table 2. Wee1-related proteins

Species	Protein	Length (aa)
H. sapiens	Wee1	664
X. laevis	Wee1	555
S. pombe	Wee1	877
S. cerevisiae	Swe1p	819
D. melanogaster	Dwee1	618
S. pombe	Mik1	581
H. sapiens	Myt1	499
X. laevis	Myt1	548

The first five Wee1 proteins as well as Mik1 are known or suspected of phosphorylating CDKs solely on Tyr-15. The Myt1 proteins are transmembrane proteins that can phosphorylate both Thr-14 and Tyr-15. References and accession numbers for the various proteins are as follows: *H. sapiens* Wee1: Watanabe et al. (1995), U10564; *X. laevis* Wee1: Mueller et al. (1995a), U13962; *S. pombe* Wee1: Russell and Nurse (1987a), M16508; *S. cerevisiae* Swe1p: Booher et al. (1993), X73966; *D. melanogaster* Dwee1: Campbell et al. (1995), U17223; *S. pombe* Mik1: Lundgren et al. (1991), M60834; *H. sapiens* Myt1: Liu et al. (1997), U56816; *X. laevis* Myt1: Mueller et al. (1995b), U28931.

A second inhibitory kinase has also been found in vertebrates, but it differs from Wee1 in some quite interesting ways. Although Wee1 proteins can only phosphorylate $p34^{cdc2}$ on Tyr-15, crude *Xenopus* egg extracts can phosphorylate both Thr-14 and Tyr-15 (Solomon et al. 1990; Atherton-Fessler et al. 1994; Kornbluth et al. 1994). Simple fractionations separated a cytoplasmic activity that was specific for Tyr-15 (i.e., Wee1) and a membrane-bound activity that could phosphorylate both sites (Atherton-Fessler et al. 1994; Kornbluth et al. 1994). The novel Thr-14 and Tyr-15 kinase activities co-purified through a number of steps, indicating that they resided in a single, dual-specificity kinase (Atherton-Fessler et al. 1994). The gene for a *Xenopus* protein kinase with these properties was subsequently isolated during a PCR-based screen for *wee1*-like genes (Mueller et al. 1995b). The encoded protein was termed "Myt1", for membrane-associated, tyrosine- and threonine-specific, Cdc2 inhibitory kinase, and was predicted to contain a single transmembrane domain (Mueller et al. 1995b). Although *Xenopus* Myt1 and *Xenopus* Wee1 can both phosphorylate $p34^{cdc2}$ on Tyr-15, the kinase domains of the two proteins are only 39 % identical, whereas the Xenopus and human Wee1 proteins are 72 % identical over the same regions (Mueller et al. 1995b). Human Myt1, which is 46 % identical to Xenopus Myt1, localizes to the endoplasmic reticulum and the Golgi complex (Liu et al. 1997). Unlike Xenopus Myt1, human Myt1 preferentially phosphorylates Thr-14 (Liu et al. 1997).

6.1 Regulation of Inhibitory Kinases

The observation that the rate of inhibitory phosphorylation of $p34^{cdc2}$ declines dramatically as a *Xenopus* egg extract passes from interphase into mitosis (Solomon et al. 1990; Smythe and Newport 1992; Atherton-Fessler et al. 1994; Kornbluth et al. 1994; Kumagai and Dunphy 1995) suggested that enzymes such as Wee1 and Myt1 would be cell-cycle regulated. Such feedback regulation makes intuitive as well as theoretical (Novak and Tyson 1993; Tyson et al. 1996) sense as it can, when coupled to a parallel activation of Cdc25, lead to an abrupt, "autocatalytic" activation of $p34^{cdc2}$ and transition into mitosis. A strong inhibition of Wee1 during mitosis has been observed in many organisms including humans (McGowan and Russell 1995; Parker et al. 1995; Watanabe et al. 1995), mice (Honda et al. 1995), and frogs (Mueller et al. 1995a). Inhibition of Wee1 correlates with its phosphorylation and can be reversed by treatment with phosphatases (Tang et al. 1993; Mueller et al. 1995a; Watanabe et al. 1995). The activity of the dual-specificity Myt1 protein kinase also declines during mitosis in *Xenopus* egg extracts (Atherton-Fessler et al. 1994; Kornbluth et al. 1994; Mueller et al. 1995b).

Because the DNA replication checkpoint acts in large part through tyrosine-phosphorylation of $p34^{cdc2}$ in most organisms, it was logical to consider whether the checkpoint had an effect on Wee1 activity. Indeed, an early report suggested that Wee1 activity increased five- to ten-fold when *Xenopus* egg extracts were blocked in S phase with a DNA synthesis inhibitor (Smythe and Newport 1992). However, subsequent studies in both this system and in mammalian tissue culture cells indicated that the overall rate of Tyr-15 phosphorylation as well as the activities of Wee1 and Myt1 are unaltered by the presence of incompletely replicated DNA (Atherton-Fessler et al. 1994; Kumagai and Dunphy 1995; McGowan and Russell 1995; Mueller et al. 1995a, b). The DNA replication checkpoint may instead act by delaying the inactivation of these enzymes as cells approach mitosis, effectively increasing the threshold concentration of cyclin necessary for induction of mitosis (Novak and Tyson 1993; Tyson et al. 1996).

A number of enzymes can phosphorylate Wee1 proteins. *S. pombe* Wee1 contains a C-terminal catalytic domain and an N-terminal regulatory domain with clusters of potential sites of phosphorylation by $p34^{cdc2}$ (Aligue et al. 1997). Phosphorylation of the regulatory domain in a *Xenopus* egg extract inactivates *S. pombe* Wee1 (Tang et al. 1993). At least two kinases can phosphorylate *Xenopus* Wee1 in this region, $p34^{cdc2}$ and "kinase X" (Mueller et al. 1995a). $p34^{cdc2}$ can directly phosphorylate and inhibit Wee1, providing an important feedback loop during the entry into mitosis. It seems unlikely that $p34^{cdc2}$ would be the trigger that initiates Wee1 inhibition; an unidentified "kinase X" may be a better candidate.

Genetic studies identified *nim1* as a negative regulator of *wee1* and its sequence suggested that it encoded a protein kinase (Russell and Nurse 1987b; Forsberg and Nurse 1991). Subsequent studies using purified proteins showed that Nim1 directly phosphorylates and inhibits Wee1 (Coleman et al. 1993; Parker et al. 1993; Wu and Russell 1993). Unlike the phosphorylation of Wee1 in a *Xenopus* egg extract, Nim1 phosphorylates Wee1 within its C-terminal catalytic domain. *nim1* is nonessential in *S. pombe*, indicating that it may be only one component in the cell cycle regulation of Wee1. Nim1 functions to relay nutritional signals to the mitotic machinery (Feilotter et al. 1991); other "Wee1 kinases" may transduce different inputs or participate in the initial inactivation of Wee1 at the G2-M transition. An apparent *S. cerevisiae* homologue of *nim1* regulates *SWE1* and is called *HSL1* (Ma et al. 1996).

Wee1 in *S. pombe* is also phosphorylated by the Chk1 DNA checkpoint sensor protein kinase in vitro (O'Connell et al. 1997). Chk1 lies on a signal transduction pathway leading from damaged DNA to cell cycle arrest (Walworth et al. 1993). Its own phosphorylation is induced by DNA damage (Walworth and Bernards 1996) and it may also respond to unreplicated DNA (Francesconi et al. 1997). UV irradiation or overexpression of *chk1* arrests cells in G2 with hyperphosphorylated Wee1 and Tyr-15-phosphorylated $p34^{cdc2}$ (O'Connell et al. 1997). Cell cycle arrest is abrogated in the absence of *wee1* (O'Connell et al. 1997). Phosphorylation by Chk1 has no obvious effect on Wee1 activity in vitro (O'Connell et al. 1997), which is consistent with the unaltered Tyr-15 kinase activity during DNA checkpoints in other systems (Atherton-Fessler et al. 1994; Kumagai and Dunphy 1995; McGowan and Russell 1995; Mueller et al. 1995a, b). This phosphorylation may instead prevent the inhibition of Wee1 that normally occurs at the G2-M border, regulate the subcellular localization of Wee1, affect the interaction of Wee1 with $p34^{cdc2}$ or some other protein, or affect an unappreciated aspect of Wee1 function. Chk1 also appears to regulate Cdc25 function (see below).

7 Activating Phosphatases – Cdc25

Although genetic studies indicated that *cdc25* functioned as an activator of *cdc2* and in opposition to *wee1*, its sequence gave no hint that it might be a dual-specificity phosphatase acting on Thr-14 and Tyr-15 (Russell and Nurse 1986; Forsberg and Nurse 1991). In addition, initial experiments found that recombinant Cdc25 had no phosphatase activity toward model substrates and later that it was insensitive to N-ethylmaleimide, which inactivates the catalytic cysteine found in all protein tyrosine phosphatases. It was only after some unusual protein tyrosine phosphatases had been identified and the consensus motifs for this class of enzymes had shrunk to just a few well-placed amino acids – including the critical HCxxxxxR sequence surrounding the catalytic cysteine residue (Fauman and Saper 1996; see Table 3) – that sequence analysis suggested that Cdc25 might actually be this essential activating phosphatase (Gautier et al. 1991; Moreno and Nurse 1991). The key biochemical observations included the detection of phosphatase activity toward synthetic model compounds (Gautier et al. 1991; Dunphy and Kumagai 1991; Lee et al. 1992) and the use of pure proteins from frogs, humans and fission yeast to demonstrate direct dephosphorylation of $p34^{cdc2}$ that was dependent on the presence of the putative catalytic cysteine (Dunphy and Kumagai 1991; Gautier et al. 1991; Millar et al. 1991; Strausfeld et al. 1991; Lee et al. 1992).

cdc25 homologues are ubiquitous in eukaryotes (see Table 3). Proteins with similar biological and/or biochemical properties have been identified in budding yeast (*MIH1*; Russell et al. 1989), *Drosophila* (*string* and *twine*; Edgar and O'Farrell 1989; Alphey et al. 1992), frogs (Kumagai and Dunphy 1992), and humans (Cdc25A, Cdc25B, Cdc25C; Sadhu et al. 1990; Galaktionov and Beach 1991). Though the N-terminal regulatory domains of these proteins have diverged, their C-terminal catalytic domains are clearly related: This region of *S. pombe* Cdc25 is 40 % identical to *S. cerevisiae* Mih1p and 37 % identical to human Cdc25C (Russell et al. 1989; Sadhu et al. 1990). Although each of the three vertebrate Cdc25 proteins can dephosphorylate a variety of CDKs in vitro, they are expressed and activated at different times during the cell cycle and have

Table 3. Cdc25 proteins

Species	Protein	Length (aa)	Catalytic motif	Pos. of Arg
H. sapiens	Cdc25A	523	**H C** e f s s e **R**	-436
H. sapiens	Cdc25B	565	**H C** e f s s e **R**	-479
H. sapiens	Cdc25C	473	**H C** e f s s e **R**	-383
X. laevis	Cdc25C	572	**H C** e f s s h **R**	-482
S. pombe	Cdc25	580	**H C** e h s a h **R**	-486
S. cerevisiae	Mih 1p	474	**H C** e f s s h **R**	-326
D. melanogaster	string	479	**H C** e f s s e **R**	-385
D. melanogaster	twine	426	**H C** e f s s e **R**	-324

Shown in bold capital letters are the conserved residues surrounding the catalytic cysteine in protein tyrosine phosphatases. The amino acid position of the terminal Arg is indicated. References and accession numbers for the various proteins are as follows: *H. sapiens* Cdc25A: Galaktionov and Beach (1991), M81933; *H. sapiens* Cdc25B: Galaktionov and Beach (1991), M81934; *H. sapiens* Cdc25C: Sadhu et al. (1990), M34065; *X. laevis* Cdc25C: Kumagai and Dunphy (1992), M94264; *S. pombe* Cdc25: Russell and Nurse (1986), M13158; *S. cerevisiae* Mih1p: Russell et al. (1989), J04846; *D. melanogaster* string: Jimenez et al. (1990), X57495; *D. melanogaster* twine: Alphey et al. (1992), M94064.

been proposed to act predominantly on different cyclin-CDK complexes, with Cdc25C functioning at the G2-M transition, and Cdc25A and Cdc25B functioning to promote S phase (Galaktionov and Beach 1991; Gabrielli et al. 1992; Sebastian et al. 1993; Hoffmann et al. 1994). In addition, inhibition of Cdc25A in vivo prevents cells from entering S phase (Hoffmann et al. 1994). These enzymes are related to each other, but clearly form subfamilies. For example, human Cdc25A and Cdc25C are 48 % identical and Cdc25B and Cdc25C are 43 % identical over their catalytic regions, which is not much greater than the identity between any of these proteins and *Drosophila* homologues (Galaktionov and Beach 1991). The *S. pombe* Pyp3 protein tyrosine phosphatase cooperates with Cdc25 in vivo and can directly dephosphorylate $p34^{cdc2}$ in vitro, though its physiological effect is modest (Millar et al. 1992).

7.1 Regulation of Activating Phosphatases

Stimulation of Cdc25C activity at the G2-M transition, like the parallel inhibition of Wee1, is essential for the concerted activation of $p34^{cdc2}$ and abrupt transition into mitosis (Solomon et al. 1990; Izumi et al. 1992; Kumagai and Dunphy 1992; Hoffmann et al. 1993). This five- to seven fold activation is accompanied by a dramatic electrophoretic retardation of the protein caused by its phosphorylation; dephosphorylation in vitro restores Cdc25 activity to its low interphase level (Izumi et al. 1992; Kumagai and Dunphy 1992). Cdc25C activity remains low during a DNA replication checkpoint (Kumagai and Dunphy 1995) although this checkpoint could, in principle, act partly by postponing the normal cell cycle activation of Cdc25.

The direct phosphorylation of Cdc25C by $p34^{cdc2}$ in vitro can stimulate Cdc25C activity to near mitotic levels (Hoffmann et al. 1993; Izumi and Maller 1993; Strausfeld et al. 1994). This finding provides a very attractive explanation for the "autocatalytic" amplification of $p34^{cdc2}$ activity whereby a low level of active $p34^{cdc2}$ can stimulate Cdc25C (and presumably inhibit Wee1), leading to the further activation of $p34^{cdc2}$ complexes etc. (see Fig. 1). Part of this amplification loop could be provided by the observed sequential dephosphorylation of $p34^{cdc2}$ first on Thr-14 and then on Tyr-15, both in vivo (in starfish) and in vitro by Cdc25 (Borgne and Meijer 1996), and by the low but detectable activity of $p34^{cdc2}$ singly phosphorylated on Tyr-15 (Borgne and Meijer 1996; Liu et al. 1997). $p33^{cdk2}$ complexes can also phosphorylate and activate Cdc25C (Izumi and Maller 1995), which is especially interesting in light of the requirement of $p33^{cdk2}$ activity for $p34^{cdc2}$ activation in *Xenopus* egg extracts (Guadagno and Newport 1996). Mutation of five potential $p34^{cdc2}$ (or $p33^{cdk2}$) phosphorylation sites that are conserved in the N-terminal halves of *Xenopus* and human Cdc25 proteins reduced the ability of $p34^{cdc2}$ to stimulate Cdc25C, although the basal activity of the mutated Cdc25C was low (Izumi and Maller 1993). An additional

Cdc25C activating kinase was detected in *Xenopus* egg extracts depleted of both $p34^{cdc2}$ and $p33^{cdk2}$ to which phosphatase inhibitors had been added to stabilize phosphoproteins (Izumi and Maller 1995). The identity of this kinase has not been determined, but it may correspond to Plx1 (see below). Phosphorylation and activation of Cdc25A at the G1-S transition is controlled by cyclin E-$p33^{cdk2}$ (Hoffmann et al. 1994), suggesting that amplification loops may be general phenomena.

Another *Xenopus* kinase that can activate Cdc25 is Plx1 (Kumagai and Dunphy 1996). This enzyme binds to and phosphorylates the N-terminal regulatory domain of Cdc25 and was purified from *Xenopus* egg extracts. Plx1 activity is higher in M phase than in interphase, consistent with a role in the activation loop. Plx1 is so named because of its similarity to the *polo* gene product of *Drosophila* (Fenton and Glover 1993). *polo* homologues have been found in numerous other species including humans (*Plk1*), *S. pombe* (*plo1*), and *S. cerevisiae* (*CDC5*) (Pringle and Hartwell 1981; Golsteyn et al. 1995; Ohkura et al. 1995). Both Plx1 and $p34^{cdc2}$ phosphorylate Cdc25 on multiple sites in vitro, all of which appear to be phosphorylated in vivo as well; some sites are phosphorylated by both Plx1 and $p34^{cdc2}$ (Kumagai and Dunphy 1996). These observations raise a number of important questions, including which sites are actually responsible for the stimulation of Cdc25 activity and whether Plx1, $p34^{cdc2}$, and any yet-to-be-identified Cdc25 kinases act through overlapping or distinct phosphorylation sites.

One perplexing aspect of this finding is that the activities of polo proteins peak during mitosis, rather than at the G2-M transition (Fenton and Glover 1993; Golsteyn et al. 1995). Moreover, genetic analyses indicate that the yeast homologues *plo1* in *S. pombe* and *CDC5* in *S. cerevisiae* function late in mitosis. Thus neither Plx1 nor $p34^{cdc2}$ seems well positioned to be the trigger for the Cdc25-$p34^{cdc2}$ activation loop since each appears to become active after the function of Cdc25 is complete. However, Plx1 may have multiple roles, with a unique (i.e. non-redundant) function late in mitosis. Alternatively, a low level of $p34^{cdc2}$ activity may be the trigger. In either case, full activation of both Cdc25 and $p34^{cdc2}$ would ensure that the process is unidirectional.

Three recent reports indicate that Cdc25 is a direct target of the Chk1 protein kinase that is induced by DNA damage in *S. pombe* and in human cells (Furnari et al. 1997; Peng et al. 1997; Sanchez et al. 1997). Chk1 is identical to a kinase activity that had been partially purified and that phosphorylates the major phosphorylation site in Cdc25 during asynchronous growth of HeLa cells, serine-216 (Ogg et al. 1994). This phosphorylation creates a binding site for 14-3-3 proteins (Peng et al. 1997). Mutation of serine-216 prevents 14-3-3 binding and expression of this mutant protein abrogates the DNA-replication and DNA-damage checkpoints, indicating that this phosphorylation serves to down-regulate Cdc25 function. Although Cdc25 and Wee1 have been proposed as targets of Chk1 phosphorylation, the case for Cdc25 is somewhat stronger

(Furnari et al. 1997), though Chk1 could act through both enzymes. The observation that *wee1* mutants eliminated the cell cycle arrest caused by overexpression of *chk1* (O'Connell et al. 1997) was interpreted as indicating that *wee1* lies downstream of *chk1*, though it is equally consistent with Cdc25 being the target of Chk1 since elimination of Wee1 could compensate for the down-regulation of Cdc25 by Chk1.

8 The CDK-Activating Kinase (CAK)

Since phosphorylation on the equivalent of Thr-161 in human $p34^{cdc2}$ is essential for CDK activity, it was important to identify the responsible kinase. Although it had been suggested that $p34^{cdc2}$ might autophosphorylate (Nigg et al. 1991), subsequent experiments demonstrated the existence of a distinct CDK-activating kinase, or CAK (Desai et al. 1992; Solomon et al. 1992). There were no genetic candidates for this activity and CAK was identified biochemically. An important first step was the development of a sensitive assay in which CAK could be detected by its ability to activate a CDK as a histone H1 kinase (Solomon et al. 1992). Purification of CAKs from *Xenopus*, starfish, and later humans identified its catalytic subunit as $p40^{MO15}$ (also called cdk7; Fesquet et al. 1993; Poon et al. 1993; Solomon et al. 1993; Tassan et al. 1994). *MO15* was originally identified in a PCR-based screen for $p34^{cdc2}$-related genes in *Xenopus* and was suggested to be a negative regulator of meiotic maturation in oocytes (Shuttleworth et al. 1990). Homologues have been detected in diverse organisms (see Table 4 for a partial listing).

In addition to its catalytic subunit, CAK also contains regulatory subunits. An essential regulatory subunit is cyclin H, a member of the cyclin C family (Fisher and Morgan 1994; Mäkelä et al. 1994). A cyclin H-$p40^{MO15}$ dimer is the minimal CAK unit and is active so long as $p40^{MO15}$ is phosphorylated on Thr-170 (in human $p40^{MO15}$), a site equivalent to activating phosphorylation sites in other cyclin-dependent kinases (Table 1; Fisher et al. 1995). The three-dimensional structure of cyclin H is similar to that of cyclin A with additional required helices at the N- and C-termini (Kim et al. 1996; Andersen et al. 1997). An additional regulatory subunit termed MAT1 (Devault et al. 1995; Fisher et al. 1995; Tassan et al. 1995a) was originally identified by its co-immunoprecipitation with $p40^{MO15}$ and cyclin H from human cell extracts (Tassan et al. 1994). MAT1 lacks extensive similarity to known proteins except for a C3HC4 putative zinc-binding domain or RING-finger (Devault et al. 1995; Fisher et al. 1995; Tassan et al. 1995). MAT1 promotes the assembly of $p40^{MO15}$ and cyclin H, producing an active trimeric complex even in the absence of Thr-170 phosphorylation (Devault et al. 1995; Fisher et al. 1995; Tassan et al. 1995a). The function of the RING-finger domain remains unclear as it is dispensable in vitro and does not seem to bind DNA (Tassan et al. 1995a).

Table 4. CAK-like proteins

Organism	Protein Name	Identity to human MO15 (%)	GxGxxG region
H. sapiens	MO15/Cdk7	100	**G** e **G** q f **A**
X. laevis	MO15	86	**G** e **G** q f **A**
S. pombe	Crk1/Mop1	50	**G** e **G** t y **A**
S. cerevisiae	Kin28p	47	**G** e **G** t y **A**
D. melanogaster	MO15	67	**G** e **G** q f **A**
S. pombe	Csk1	28	**D** g **T** i s **E**
S. cerevisiae	Cak1p	20	**D** i **T** h c **Q**

The three MO15 proteins as well as Crk1/Mop1 have CAK activity in vitro and are known or suspected of being subunits of TFIIH. Kin28p, though not a CAK, is also a subunit of TFIIH. Instead, Cak1p is the physiological CAK in budding yeast, and is not part of TFIIH. Csk1 is a CAK in *S. pombe* acting on Crk1/Mop1. Residues within the GxGxxG nucleotide-binding motif are shown in bold capital letters. References and accession numbers for the various proteins are as follows: *H. sapiens* MO15/cdk7: Darbon et al. (1994), X77743 and Tassan et al. (1994), X79193; *X. laevis* MO15: Shuttleworth et al. (1990), X53962; *S. pombe* Crk1/Mop1: Buck et al. (1995), X91239 and Damagnez et al. (1995), L47353; *S. cerevisiae* Kin28p: Simon et al. (1986), X04423; *D. melanogaster* MO15: unpublished, U56661; *S. pombe* Csk1: Molz and Beach (1993), S59896; *S. cerevisiae* Cak1p: Kaldis et al. (1996), U60192.

8.1 CAK and Transcription

A major surprise came with the realization that all three CAK subunits are also subunits of basal transcription factor IIH (TFIIH; Feaver et al. 1994; Roy et al. 1994; Serizawa et al. 1995; Shiekhattar et al. 1995; Adamczewski et al. 1996) where they phosphorylate the C-terminal domain (CTD) of the large subunit of RNA polymerase II. In addition to their role in transcriptional initiation, some subunits of TFIIH also function in the nucleotide-excision repair form of DNA repair (Orphanides et al. 1996). CAK exists in at least three distinct complexes in mammalian cells: (1) Free CAK (trimeric form), (2) CAK-ERCC2/XPD (Drapkin et al. 1996; Reardon et al. 1996), and (3) TFIIH. All three complexes contain $p40^{MO15}$, cyclin H, and MAT1. The linkage between CAK and the remainder of TFIIH appears to be mediated via XPB/ERCC3 and XPD/ERCC2, and trimeric CAK can be dissociated from TFIIH by low salt concentrations (Marinoni et al. 1997).

Despite its presence in TFIIH, there is some controversy over whether CAK activity is required for transcription. The first question is whether CTD phosphorylation is necessary for transcription. Although the CTD is clearly essential, its phosphorylation in vitro is either essential or dispensable depending on the purity of the transcription system and the promoter used (for review see Orphanides et al. 1996). Furthermore, some experiments indicate that the kinase activity of $p40^{MO15}$ is dispensable for transcription (Mäkelä et al. 1995) whereas others indicate that it is essential (Akoulitchev et al. 1995). The situation may be clearer in yeast, where Kin28p (Simon et al. 1986) is the equivalent protein kinase subunit of TFIIH (Feaver et al. 1994). Transcriptional initiation ceases rapidly (Cismowski et al. 1995) and CTD phosphorylation declines (Valay et al. 1995) in *kin28-ts* mutants at the non-permissive temperature. Moreover, the catalytic activity of Kin28p is necessary for cell growth (J. Kimmelman, G. Laff, M.J. Solomon, unpubl. data).

8.2 Substrate Specificity of CAK

CAK was purified based on its ability to phosphorylate and activate $p34^{cdc2}$ and $p33^{cdk2}$, but it can also activate cdk4 (Matsuoka et al. 1994) and cdk6 (Iavarone and Massagué 1997). Despite the similarity between the activating phosphorylation site in $p40^{MO15}$ and those in its substrates, it does not appear that $p40^{MO15}$ can autophosphorylate (but see Fisher and Morgan 1994). Phosphorylation by CAK is stimulated by cyclin bound to the CDK and at least $p34^{cdc2}$ is not phosphorylated at all as a monomer (Fisher and Morgan 1994). It is unclear if CAK will be responsible for activating phosphorylations of all CDKs or if other CAKs remain to be identified. Besides CDKs, CAK can phosphorylate sites within the coiled coil CTD of RNA polymerase II (see above). The structure of these sites and the functions of this phosphorylation differ fundamentally from the activation of CDKs [for a review see Dahmus 1996].

As a component of TFIIH, CAK can also phosphorylate a growing number of transcription components. These include the basal transcription factors TFIIE and TFIIF (Rossignol et al. 1997; Yankulov and Bentley 1997). Phosphorylation by CAK of the retinoic acid receptor α on Ser-77, results in its activation (Rochette-Egly et al. 1997). This transcriptional transactivator binds to free CAK as well as to TFIIH. The region surrounding Ser-77 (VPSPP<u>S</u>PPPLPR) bears no obvious similarity to the activating phosphorylation site in $p34^{cdc2}$ (Y<u>T</u>HEVV) or to the repeated heptad motif (YSPTSPS) of the CTD. CAK can also phosphorylate one or more sites in p53, resulting in its enhanced binding to DNA (Lu et al. 1997). Future CAK substrates may include additional CDKs (in particular, cdk3) and transcription factors. Also worth considering are MAP kinases (for example, see Wagner et al. 1997) since they also require activating phosphorylation of T-loop residues (Hanks and Hunter 1995), though in most cases the responsible MAP kinase kinase (MEK) is already known.

8.3 Regulation of CAK

Although CAK activity, protein levels, and subunit composition are constant throughout the cell cycle (Solomon et al. 1992; Brown et al. 1994; Poon et al. 1994; Tassan et al. 1994; Bartkova et al. 1996), CAK activity and function can be affected by phosphorylation and by the assembly factor, MAT1. Human $p40^{MO15}$ is phosphorylated on two sites, Ser-164 and Thr-170. Phosphorylation of Thr-170, which is equivalent to Thr-161 in $p34^{cdc2}$, promotes cyclin H binding and is necessary for the activity of the dimeric cyclin H-$p40^{MO15}$ form of CAK (Labbé et al. 1994; Poon et al. 1994; Fisher et al. 1995; Martinez et al. 1997). Mutations of Ser-164 do not have any obvious effect on CAK activity (Labbé et al. 1994; Poon et al. 1994). Although the physiological CAK-activating kinase (CAKAK) is unknown, $p33^{cdk2}$ and $p34^{cdc2}$ can phosphorylate Thr-170 in vitro (Fisher et al. 1995; Martinez et al. 1997). In the presence of MAT1 cyclin H can bind to and activate $p40^{MO15}$ even in the absence of Thr-170 phosphorylation (Fisher et al. 1995). The presence of MAT1 affects the substrate specificity of CAK by favoring CTD kinase activity over CDK activation. Assembly into TFIIH also allows CAK to phosphorylate TFIIE and TFIIF, proteins that are not substrates for free CAK (Rossignol et al. 1997; Yankulov and Bentley 1997).

$p40^{MO15}$ is also regulated by nuclear localization which is important for its activation and is dependent on an identified nuclear localization sequence (Darbon et al. 1994; Labbé et al. 1994; Tassan et al. 1994; Bartkova et al. 1996). Transport into the nucleus makes sense for the transcription (TFIIH) functions of CAK, but it raises unanswered questions about how and where various cytoplasmic CDKs are phosphorylated by CAK.

8.4 The Physiological CAK in Budding Yeast

The dual functions of vertebrate CAK in transcription and in cell cycle control are intriguing, but are not conserved in budding yeast. The budding yeast protein that is most closely related to $p40^{MO15}$ is Kin28p (Simon et al. 1986), which, like $p40^{MO15}$, functions as the CTD kinase subunit of TFIIH (Feaver et al. 1994). Kin28p phosphorylates RNA polymerase II (Valay et al. 1995) and is required for mRNA transcription in vivo (Cismowski et al. 1995; Valay et al. 1995), but is devoid of CAK activity (Cismowski et al. 1995). Thus *S. cerevisiae* has distinct enzymes for the transcriptional (Kin28p) and cell cycle (Cak1p, see below) functions apparently both carried out by $p40^{MO15}$ in vertebrates.

The CDK-activating kinase from budding yeast has been purified and named Cak1p (or Civ1p; Espinoza et al. 1996; Kaldis et al. 1996; Thuret et al. 1996). Cak1p is an unusual protein kinase that lacks all of the glycines in the GxGxxG nucleotide-binding motif (see Table 4) and contains non-conservative replacements at four of 15 nearly invariant amino acid positions in protein kinases. Its sequence is only 20-25 % identical to $p40^{MO15}$ proteins and CDKs, which are its closest relatives. In contrast, yeast Kin28p, which is not a CAK, is about 45 % identical to p40^{MO15} proteins, reflecting their common function as components of TFIIH. Unlike p40^{MO15}, Cak1p functions as a monomer, is active when expressed in *E. coli*, and lacks CTD kinase activity (Kaldis et al. 1996). Expression of p40^{MO15} fails to rescue a temperature-sensitive *CAK1* mutant (Kaldis et al. 1996). Cak1p is responsible for essentially all of the CAK activity in yeast lysates (Espinoza et al. 1996; Kaldis et al. 1996) and its inactivation in vivo leads to the inactivation of Cdc28p and a block to its phosphorylation on Thr-169 (Kaldis et al. 1996; Thuret et al. 1996). Genetic interactions with *CLB2*, which encodes the major mitotic cyclin in budding yeast, firmly established the function of Cak1p as the physiological yeast CAK (Kaldis et al. 1996). Although the amount of CAK activity, *CAK1* mRNA, and Cak1p are constant throughout the mitotic cell cycle (Espinoza et al. 1996; Sutton and Freiman 1997), *CAK1* mRNA levels oscillate during meiosis (Wagner et al. 1997). *CAK1* is required for spore wall morphogenesis during meiosis and its overexpression can suppress a conditional mutant of the Smk1p MAP kinase that is defective for spore wall morphogenesis, suggesting that Cak1p may have interesting non-CDK substrates.

8.5
Cak1p-like CAKs in Other Species

Since the cell cycle is generally well conserved from yeast to man, the identification of Cak1p as the physiological *S. cerevisiae* CAK raises some questions about whether p40^{MO15} actually functions as a CAK. Although p40^{MO15} was identified as a CAK because it is responsible for the bulk of the detectable CAK activity in cell extracts, there is no evidence that it has this activity in vivo. On the other hand, there are no reports of even a minor non-p40^{MO15} CAK activity from vertebrate cells. If a Cak1p-like CAK exists that functions in addition to, or instead of, p40^{MO15} as the physiological human CAK, it should not be long before it is identified either biochemically or by random cDNA sequencing efforts.

The situation in *S. pombe* appears to be intermediate between those in budding yeast and humans. A protein usually referred to as either Crk1 or Mop1 has CAK activity (Buck et al. 1995; Damagnez et al. 1995). Its sequence, binding to a cyclin-like regulatory protein (Mcs2; Molz and Beach 1993), and associated CTD kinase activity predict that it will, like p40^{MO15}, also function as a subunit of TFIIH. *S. pombe* also contains a second CAK, Csk1 (Hermand et al. 1997). Csk1 exhibits limited sequence similarity to Cak1p (for example, see Table 4) and, like Cak1p, is active as a monomer. Although it is non-essential it can directly phosphorylate and activate the Mop1 CDK. (The equivalent relationship does not appear to exist in *S. cerevisiae*, that is Cak1p is probably not the activating kinase for Kin28p; J Kimmelman and MJ Solomon, unpublished data.) These results indicate that there can be multiple CAKs in one cell and suggest that a search for Csk1/Cak1p-related proteins in higher eukaryotes could be particularly interesting.

9
Removal of the Activating Phosphorylation

Despite the rapidly growing literature on CAKs, much less is known about the protein phosphatase that counters this activating phosphorylation. A *Xenopus* type 2A phosphatase can dephosphorylate Thr-161 in a cyclin-$p34^{cdc2}$ complex, but very high concentrations of this enzyme are required and its physiological significance is unclear (Lee et al. 1991). Inhibition of a type 1 phosphatase can prevent the dephosphorylation of $p34^{cdc2}$ in *Xenopus* egg extracts (Lorca et al. 1992), although indirect vs. direct effects could not be distinguished. More recently, a phosphatase termed KAP, Cdi1, or Cip2 that was identified in two-hybrid screens for $p33^{cdk2}$-binding proteins (Gyuris et al. 1993; Harper et al. 1993; Hannon et al. 1994) was found to specifically dephosphorylate Thr-160 of human $p33^{cdk2}$ and to be responsible for ~45 % of such phosphatase activity in HeLa cell extracts (Poon and Hunter 1995). KAP is a dual-specificity phosphatase and has motifs characteristic of protein tyrosine phosphatases. These enzymes typically dephosphorylate T-loop residues in MAP kinases (Fauman and Saper 1996), a reaction that is structurally similar to the dephosphorylation of the T-loop activating site in CDKs. KAP dephosphorylates $p33^{cdk2}$ only in the absence of cyclin, which agrees with experimental observations (Lorca et al. 1992) and makes biological sense. An indication of its specificity is that it can dephosphorylate native, but not denatured $p33^{cdk2}$, in contrast to protein phosphatase 2A which can dephosphorylate both forms of monomeric $p33^{cdk2}$.

An intriguing candidate for a KAP homologue in *S. cerevisiae* is Cdc14p. Little is known about *CDC14* other than that it encodes a dual-specificity phosphatase (Wan et al. 1992; Fauman and Saper 1996) mutations in which give a late-anaphase arrest (Culotti and Hartwell 1971). This phenotype is plausible for a lack of Thr-169 dephosphorylation in Cdc28p and is similar to that produced following expression of a non-degradable cyclin (Surana et al. 1993). Cdc14p is the yeast phosphatase that is most similar to KAP, although the similarity is limited (25 % identity and ~50 % similarity over a stretch of 96 amino acids).

10
Concluding Remarks

The cell cycle field has come a long way in the last decade toward constructing a complete parts list for the fundamental cell cycle machinery and understanding how these pieces fit together and adjust to each other. Yet even at this most basic level much will be learned in the coming years about how cyclin-dependent protein kinases are regulated by phosphorylation. Will the structural model of $p33^{cdk2}$ regulation apply directly to other CDKs? Some differences seem likely given that the cyclin A-$p33^{cdk2}$ complex is a somewhat special case, with detectable activity even in the absence of activating phosphorylation. A large structural unknown concerns the precise biochemical roles of Thr-14 and Tyr-15 phosphorylation.

Despite recent advances, additional kinases and phosphatases acting on CDKs, including $p34^{cdc2}$, will probably be identified. It was only 2 years ago that Myt1, the membrane-bound Thr-14/Tyr-15 kinase, was discovered. Other surprises are likely. One area of particular ignorance concerns the phosphatase acting on Thr-161. KAP can perform this reaction on $p33^{cdk2}$ in vitro, and may do so for $p34^{cdc2}$ as well. The identification of the equivalent phosphatase in one of the yeasts would help to establish its physiological role. The greatest uncertainty right now is the identity of the vertebrate CAK. The discovery that budding yeast has two enzymes, Kin28p and Cak1p, for the transcriptional and CAK functions reported to be performed by $p40^{MO15}$ in vertebrates has raised serious concerns that $p40^{MO15}$ may not be a physiological CAK.

It is generally assumed that most CDKs will be regulated according to the $p34^{cdc2}$ paradigm, though this has only been established for some aspects of the regulation of some CDKs. Are all potential sites of inhibitory or stimulatory phosphorylation in other CDKs used for those purposes? Are these sites phosphorylated by the known machinery including Wee1, Cdc25 and CAK, or are there other kinases and phosphatases, perhaps related to those acting on $p34^{cdc2}$, that act preferentially on other CDKs? The issue of specificity will prove difficult to address given the apparent promiscuity of the known CDK modifying enzymes.

Much remains to be learned about the regulation of Wee1, Myt1, Cdc25, and possibly CAK. Current efforts are only beginning to define their specific sites of phosphorylation and the roles of these phosphorylations in regulating the activities of these enzymes during a normal cell cycle or in response to checkpoints. The high degree of phosphorylation of Wee1 and Cdc25 in vivo will prolong this analysis. No doubt more kinases acting on these enzymes will be identified. Some of them will serve apparently redundant functions but others may transduce specific signals much as the *S. pombe* Chk1 protein relays checkpoint signals to Wee1 and/or Cdc25. Closing the activation loop will require a more complete understanding of what brings about the initial changes in Wee1 and Cdc25 activities that lead to the rapid activation of $p34^{cdc2}$ via reinforcing feedback mechanisms. The extent to which such circuitry also applies to the regulation of other CDKs remains open.

Finally, hints of a reductionist biochemist's nightmare are emerging concerning enzymatic reactions localized to intracellular membranes. The presence of Myt1 in the endoplasmic reticulum and Golgi membranes is intriguing, if for no other reason than that it came as a complete surprise. Is there a profound reason that these reactions could not simply occur "free in solution"? More aspects of cell cycle control may well be found occurring in two dimensions or at discrete sites within the cell. Identifying them will be one thing, understanding them biochemically may be quite another.

Acknowledgments. The authors are grateful to members of the lab for discussion. Related work in the authors' lab is supported by a fellowship from Donaghue Medical Research Foundation (to P.K.) and by grant GM47830 from the National Institutes of Health (to M.J.S.). M.J.S. is a scholar of the Leukemia Society of America.

References

Adamczewski JP, Rossignol M, Tassan J-P, Nigg EA, Moncollin V, Egly J-M (1996) MAT1, cdk7 and cyclin H form a kinase complex which is UV light-sensitive upon association with TFIIH. EMBO J 15:1877–1884

Akoulitchev S, Mäkelä TP, Weinberg RA, Reinberg D (1995) Requirement for TFIIH kinase activity in transcription by RNA polymerase II. Nature 377:557–560

Aligue R, Wu L, Russell P (1997) Regulation of *Schizosaccharomyces pombe* Weel tyrosine kinase. J Biol Chem 272:13320–13325

Alphey L, Jimenez J, White-Cooper H, Dawson I, Nurse P, Glover DM (1992) twine, a cdc25 homolog that functions in the male and female germline of Drosophila. Cell 69:977–988

Amon A, Surana U, Muroff I, Nasmyth K (1992) Regulation of $p34^{CDC28}$ tyrosine phosphorylation is not required for entry into mitosis in *S. cerevisiae*. Nature 355:368–371

Andersen G, Busso D, Poterszman A, Hwang JR, Wurtz J-M, Ripp R, Thierry J-C, Egly J-M, Moras D (1997) The structure of cyclin H: common mode of kinase activation and specific features. EMBO J 16:958–967

Atherton-Fessler S, Liu F, Gabrielli B, Lee MS, Peng CY, Piwnica-Worms H (1994) Cell cycle regulation of the $p34^{cdc2}$ inhibitory kinases. Mol Biol Cell 5:989–1001

Barbet NC, Carr AM (1993) Fission yeast weel protein kinase is not required for DNA damage-dependent mitotic arrest. Nature 364:824–827

Bartkova J, Zemanova M, Bartek J (1996) Expression of cdk7/CAK in normal and tumor cells of diverse histogenesis, cell-cycle position and differentiation. Int J Cancer 66:732–737

Blasina A, Paegle ES, McGowan CH (1997) The role of inhibitory phosphorylation of Cdc2 following DNA replication block and radiation-induced damage in human cells. Mol Biol Cell 8:1013–1023

Booher R, Beach D (1986) Site-specific mutagenesis of *cdc2*$^{+}$, a cell cycle control gene of the fission yeast *Schizosaccharomyces pombe*. Mol Cell Biol 6:3523–3530

Booher RN, Deshaies RJ, Kirschner MW (1993) Properties of *Saccharomyces cerevisiae* weel and its differential regulation of $p34^{CDC28}$ in response to G1 and G2 cyclins. EMBO J 12:3417–3426

Borgne A, Meijer L (1996) Sequential dephosphorylation of $p34^{cdc2}$ on Thr-14 and Tyr-15 at the prophase/metaphase transition. J Biol Chem 271:27847–27854

Brown AJ, Jones T, Shuttleworth J (1994) Expression and activity of $p40^{MO15}$, the catalytic subunit of cdk-activating kinase, during Xenopus oogenesis and embryogenesis. Mol Biol Cell 1994 5:921–932

Buck V, Russell P, Millar JBA (1995) Identification of a cdk-activating kinase in fission yeast. EMBO J 14:6173–6183

Campbell SD, Sprenger F, Edgar BA, O'Farrell PH (1995) Drosophila Weel kinase rescues fission yeast from mitotic catastrophe and phosphorylates Drosophila Cdc2 in vitro. Mol Biol Cell 6:1333–1347

Chae T, Kwon YT, Bronson R, Dikkes P, Li E, Tsai LH (1997) Mice lacking p35, a neuronal specific activator of cdk5, display cortical lamination defects, seizures, and adult lethality. Neuron 18:29–42

Cismowski MJ, Laff GM, Solomon MJ, Reed SI (1995) *KIN28* encodes a C-terminal domain kinase that controls mRNA transcription in *Saccharomyces cerevisiae* but lacks cyclin-dependent kinase-activating kinase (CAK) activity. Mol Cell Biol 15:2983–2992

Coleman TR, Dunphy WG (1994) Cdc2 regulatory factors. Curr Opin Cell Biol 6:877–882

Coleman TR, Tang Z, Dunphy WG (1993) Negative regulation of the weel protein kinase by direct action of the niml/cdrl mitotic inducer. Cell 72:919–929

Connell-Crowley L, Solomon MJ, Wei N, Harper JW (1993) Phosphorylation-independent activation of human cyclin-dependent kinase 2 by cyclin A in vitro. Mol Biol Cell 4:79–92

Culotti J, Hartwell LH (1971) Genetic control of the cell division cycle in yeast III. Seven genes controlling nuclear division. Exp Cell Res 67:389–401

Dahmus ME (1996) Reversible phosphorylation of the C-terminal domain of RNA polymerase II. J Biol Chem 271:19009–19012

Damagnez V, Mäkelä TP, Cottarel G (1995) *Schizosaccharomyces pombe* Mop1-Mcs2 is related to mammalian CAK. EMBO J 14:6164–6172

Darbon J-M, Devault A, Taviaux S, Fesquet D, Martinez A-M, Galas S, Cavadore J-C, Dorée M, Blanchard J-M (1994) Cloning, expression and subcellular localization of the human homolog of $p40^{MO15}$ catalytic subunit of cdk-activating kinase. Oncogene 9:3127–3138

De Bondt HL, Rosenblatt J, Jancarik J, Jones HD, Morgan DO, Kim S-H (1993) Crystal structure of cyclin-dependent kinase 2. Nature 363:595–602

Den Haese GJ, Walworth N, Carr AM, Gould KL (1995) The Weel protein kinase regulates T14 phosphorylation of fission yeast Cdc2. Mol Biol Cell 6:371–385

Desai D, Gu Y, Morgan DO (1992) Activation of human cyclin-dependent kinases in vitro. Mol Biol Cell 3:571–582

Devault A, Martinez A-M, Fesquet D, Labbé J-C, Morin N, Tassan J-P, Nigg EA, Cavadore J-C, Dorée M (1995) MAT1 ('menage à trois') a new RING finger protein subunit stabilizing cyclin H-cdk7 complexes in starfish and *Xenopus* CAK. EMBO J 14:5027–5036

Drapkin R, Le Roy G, Cho H, Akoulitchev S, Reinberg D (1996) Human cyclin-dependent kinase-activating kinase exists in three distinct complexes. Proc Natl Acad Sci USA 93:6488–6493

Ducommun B, Brambilla P, Félix MA, Franza BR Jr, Karsenti E, Draetta G (1991) cdc2 phosphorylation is required for its interaction with cyclin. EMBO J 10:3311–3319

Dunphy WG, Kumagai A (1991) The cdc25 protein contains an intrinsic phosphatase activity. Cell 67:189–196

Edgar BA, O'Farrell PH (1989) Genetic control of cell division patterns in the *Drosophila* embryo. Cell 57:177–187

Elledge SJ (1996) Cell cycle checkpoints: preventing an identity crisis. Science 724:1664–1672

Elledge SJ, Richman R, Hall FL, Williams RT, Lodgson N, Harper JW (1992) CDK2 encodes a 33-kDa cyclin A-associated protein kinase and is expressed before CDC2 in the cell cycle. Proc Natl Acad Sci USA 89:2907–2911

Enoch T, Nurse P (1990) Mutation of fission yeast cell cycle control genes abolishes dependence of mitosis on DNA replication. Cell 60:665–673

Espinoza FH, Farrell A, Erdjument-Bromage H, Tempst P, Morgan DO (1996) A cyclin-dependent kinase-activating kinase (CAK) in budding yeast unrelated to vertebrate CAK. Science 273:1714–1717

Fauman EB, Saper MA (1996) Structure and function of the protein tyrosine phosphatases. Trends Biochem Sci 21:413–417

Feaver WJ, Svejstrup JQ, Henry NL, Kornberg RD (1994) Relationship of CDK-activating kinase and RNA polymerase II CTD kinase TFIIH/TFIIK. Cell 79:1103–1109

Feilotter H, Nurse P, Young PG (1991) Genetic and molecular analysis of cdrl/niml in *Schizosaccharomyces pombe*. Genetics 127:309–318

Fenton B, Glover DM (1993) A conserved mitotic kinase active at late anaphase-telophase in syncytial *Drosophila* embryos. Nature 363:637–640

Fesquet D, Labbé J-C, Derancourt J, Capony J-P, Galas S, Girard F, Lorca T, Shuttleworth J, Dorée M, Cavadore J-C (1993) The *MO15* gene encodes the catalytic subunit of a protein kinase that activates cdc2 and other cyclin-dependent kinases (CDKs) through phosphorylation of Thr161 and its homologues. EMBO J 12:3111–3121

Fisher RP, Morgan DO (1994) A novel cyclin associates with MO15/CDK7 to form the CDK-activating kinase. Cell 78:713–724

Fisher RP, Jin P, Chamberlin HM, Morgan DO (1995) Alternative mechanisms of CAK assembly require an assembly factor or an activating kinase. Cell 83:47–57

Forsberg SL, Nurse P (1991) Cell cycle regulation in the yeast *Saccharomyces cerevisiae* and *Schizosaccharomyces pombe*. Annu Rev Cell Biol 7:227–256

Francesconi S, Grenon M, Bouvier D, Baldacci G (1997) $p56^{chk1}$ protein kinase is required for the DNA replication checkpoint at 37 °C in fission yeast. EMBO J 16:1332–1341

Furnari B, Rhind N, Russell P (1997) Cdc25 mitotic inducer targeted by Chk1 DNA damage checkpoint kinase. Science 277:1495–1497

Gabrielli BG, Lee MS, Walker DH, Piwnica-Worms H, Maller JL (1992) Cdc25 regulates the phosphorylation and activity of the *Xenopus* cdk2 protein kinase complex. J Biol Chem 267:18040–18046

Gabrielli BG, De Souza CP, Tonks ID, Clark JM, Hayward NK, Ellem KA (1996) Cytoplasmic accumulation of cdc25B phosphatase in mitosis triggers centrosomal microtubule nucleation in HeLa cells. J Cell Sci 109:1081–1093

Galaktionov K, Beach D (1991) Specific activation of cdc25 tyrosine phosphatases by B-type cyclins: evidence for multiple roles of mitotic cyclins. Cell 67:1181–1194

Gautier J, Solomon MJ, Booher RN, Bazan JF, Kirschner MW (1991) Cdc25 is a specific tyrosine phosphatase that directly activates $p34^{cdc2}$. Cell 67:197–211

Golsteyn RM, Mundt KE, Fry AM, Nigg EA (1995) Cell cycle regulation of the activity and subcellular localization of Plk1, a human protein kinase implicated in mitotic spindle function. J Cell Biol 129:1617–1628

Gould KL, Nurse P (1989) Tyrosine phosphorylation of the fission yeast *cdc2*$^{+}$ protein kinase regulates entry into mitosis. Nature 342:39–45

Gould KL, Moreno S, Owen SJ, Sazer S, Nurse P (1991) Phosphorylation at Thr167 is required for *Schizosaccharomyces pombe* $p34^{cdc2}$ function. EMBO J 10:3297–3309

Gu Y, Rosenblatt J, Morgan DO (1992) Cell cycle regulation of CDK2 activity by phosphorylation of Thr160 and Tyr15. EMBO J 11:3995–4005

Guadagno TM, Newport JW (1996) Cdk2 kinase is required for entry into mitosis as a positive regulator of Cdc2-cyclin B kinase activity. Cell 84:73–82

Gyuris J, Golemis E, Chertkov, H, Brent R (1993) Cdi1, a human G1 and S phase protein phosphatase that associates with Cdk2. Cell 75:791–803

Ham J, Moore D, Rosamond J, Johnston IR (1989) Transcriptional analysis of the *CDC7* protein kinase gene of *Saccharomyces cerevisiae*. Nucleic Acids Res 17:5781–5792

Hanks SK, Hunter T (1995) The eukaryotic protein kinase superfamily: kinase (catalytic) domain structure and classification. FASEB J 9:576–596

Hannon GJ, Casso D, Beach D (1994) KAP: a dual specificity phosphatase that interacts with cyclin-dependent kinases. Proc Natl Acad Sci USA 91:1731–1735

Harper JW, Adami GR, Wei N, Keyomarsi K, Elledge SJ (1993) The p21 Cdk-interacting protein Cip1 is a potent inhibitor of G1 cyclin-dependent kinases. Cell 75:805–816

Hermand D, Damagnez V, Cottarel G, Vandenhaute J, Mäkelä TP (1997) *Schizosaccharomyces pombe* csk1 is an in vivo Cdk-activating kinase (CAK). Yeast 13:S39

Hoffmann I, Clarke PR, Marcote MJ, Karsenti E, Draetta G (1993) Phosphorylation and activation of human cdc25-C by cdc2-cyclin B and its involvement in the self-amplification of MPF at mitosis. EMBO J 12:53–63

Hoffmann I, Draetta G, Karsenti E (1994) Activation of the phosphatase activity of human cdc25A by a cdk2-cyclin E dependent phosphorylation at the G1/S transition. EMBO J 13:4302–4310

Honda R, Tanaka H, Ohba Y, Yasuda H (1995) Mouse $p87^{wee1}$ kinase is regulated by M-phase specific phosphorylation. Chromosome Res 3:300–308

Iavarone A, Massagué J (1997) Repression of the CDK activator Cdc25A and cell-cycle arrest by cytokine TGF-beta in cells lacking the CDK inhibitor p15. Nature 387:417–422

Igarashi M, Nagata A, Jinno S, Suto K, Okayama H (1991) Wee1^{+}-like gene in human cells. Nature 353:80–83

Izumi T, Maller JL (1993) Elimination of cdc2 phosphorylation sites in the cdc25 phosphatase blocks initiation of M-phase. Mol Biol Cell 4:1337–1350
Izumi T, Maller JL (1995) Phosphorylation and activation of the *Xenopus* Cdc25 phosphatase in the absence of Cdc2 and Cdk2 kinase activity. Mol Biol Cell 6:215–226
Izumi T, Walker DH, Maller JL (1992) Periodic changes in phosphorylation of the *Xenopus* cdc25 phosphatase regulate its activity. Mol Biol Cell 3:927–939
Jeffrey PD, Russo AA, Polyak K, Gibbs E, Hurwitz J, Massagué J, Pavletich NP (1995) Mechanisms of CDK activation revealed by the structure of a cyclin A-CDK2 complex. Nature 376:313–320
Jimenez J, Alphey L, Nurse P, Glover DM (1990) Complementation of fission yeast *cdc2ts* and *cdc25ts* mutants identifies two cell cycle genes from *Drosophila*: a cdc2 homologue and *string*. EMBO J 9:3565–3571
Jin P, Gu Y, Morgan DO (1996) Role of inhibitory CDC2 phosphorylation in radiation-induced G2 arrest in human cells. J Cell Biol 134:963–970
Johnson LN, Noble ME, Owen DJ (1996) Active and inactive protein kinases: structural basis for regulation. Cell 85:149–158
Kaldis P, Sutton A, Solomon MJ (1996) The cdk-activating kinase (CAK) from budding yeast. Cell 86:553–564
Khatib ZA, Matsushime H, Valentine M, Shapiro DN, Sherr CJ, Look AT (1993) Coamplification of the CDK4 gene with MDM2 and GLI in human sarcomas. Cancer Res 53:5535–5541
Kim KK, Chamberlin HM, Morgan DO, Kim S-H (1996) Three-dimensional structure of human cyclin H, a positive regulator of the CDK-activating kinase. Nature [Struct Biol] 3:849–855
Kornbluth S, Sebastian B, Hunter T, Newport J (1994) Membrane localization of the kinase which phosphorylates $p34^{cdc2}$ on threonine 14. Mol Biol Cell 5:273–282
Krek W, Nigg EA (1991a) Differential phosphorylation of vertebrate $p34^{cdc2}$ kinase at the G1/S and G2/M transitions of the cell cycle: identification of major phosphorylation sites. EMBO J 10:305–316
Krek W, Nigg EA (1991b) Mutations of $p34^{cdc2}$ phosphorylation sites induce premature mitotic events in HeLa cells: evidence for a double block to $p34^{cdc2}$ kinase activation in vertebrates. EMBO J 10:3331–3341
Krek W, Nigg EA (1992) Cell cycle regulation of vertebrate $p34^{cdc2}$ activity: identification of Thr-161 as an essential in vivo phosphorylation site. New Biol 4:323–329
Kumagai A, Dunphy WG (1992) Regulation of the cdc25 protein during the cell cycle in *Xenopus* extracts. Cell 70:139–151
Kumagai A, Dunphy WG (1995) Control of the Cdc2/cyclin B complex in *Xenopus* egg extracts arrested at a G2/M checkpoint with DNA synthesis inhibitors. Mol Biol Cell 6:199–213
Kumagai A, Dunphy WG (1996) Purification and molecular cloning of Plx1, a Cdc25-regulatory kinase from *Xenopus* egg extracts. Science 273:1377–1380
Labbé J-C, Martinez A-M, Fesquet D, Capony J-P, Darbon J-M, Derancourt J, Devault A, Morin N, Cavadore J-C, Dorée M (1994) $p40^{MO15}$ associates with a p36 subunit and requires both nuclear translation and Thr176 phosphorylation to generate cdk-activating kinase activity in *Xenopus* oocytes. EMBO J 13:5155–5164
Leclerc V, Tassan JP, O'Farrell PH, Nigg EA, Leopold P (1996) *Drosophila* Cdk8, a kinase partner of cyclin C that interacts with the large subunit of RNA polymerase II. Mol Biol Cell 7:505-513
Lee MG, Nurse P (1987) Complementation used to clone a human homologue of the fission yeast cell cycle control gene cdc2. Nature 327:31–35
Lee MS, Ogg S, Xu M, Parker LL, Donoghue DJ, Maller JL, Piwnica-Worms H (1992) *cdc25⁺* encodes a protein phosphatase that dephosphorylates $p34^{cdc2}$. Mol Biol Cell 3:73–84
Lee MS, Enoch T, Piwnica-Worms H (1994) *mik1⁺* encodes a tyrosine kinase that phosphorylates $p34^{cdc2}$ on tyrosine 15. J Biol Chem 269:30530–30537
Lee TH, Kirschner MW (1996) An inhibitor of $p34^{cdc2}$/cyclin B that regulates the G2/M transition in *Xenopus* extracts. Proc Natl Acad Sci USA 93:352–356
Lee TH, Solomon MJ, Mumby MC, Kirschner MW (1991) INH, a negative regulator of MPF, is a form of protein phosphatase 2A. Cell 64:415–423

Levedakou EN, He M, Baptist EW, Craven RJ, Cance WG, Welcsh, PL, Simmons A, Naylor SL, Leach RJ, Lewis TB, Liu ET (1994) Two novel human serine/threonine kinases with homologies to the cell cycle regulating *Xenopus* MO15, and NIMA kinases: cloning and characterization of their expression pattern. Oncogene 9:1977–1988

Lew DJ, Reed SI (1995) A cell cycle checkpoint monitors cell morphogenesis in budding yeast. J Cell Biol 129:739–749

Liao SM, Zhang J, Jeffery DA, Koleske AJ, Thompson CM, Chao DM, Viljoen M, van Vuuren HJ, Young RA (1995) A kinase-cyclin pair in the RNA polymerase II holoenzyme. Nature 374:193–196

Liu F, Stanton JJ, Wu Z, Piwnica-Worms H (1997) The human Myt1 kinase preferentially phosphorylates Cdc2 on threonine 14 and localizes to the endoplasmic reticulum and Golgi complex. Mol Cell Biol 17:571–583

Lorca T, Labbé JC, Devault A, Fesquet D, Capony JP, Cavadore JC, Le Bouffant F, Dorée M (1992) Dephosphorylation of cdc2 on threonine 161 is required for cdc2 kinase inactivation and normal anaphase. EMBO J 11:2381–2390

Lu H, Fisher RP, Bailey P, Levine AJ (1997) The CDK7-cycH-p36 complex of transcription factor IIH phosphorylates p53, enhancing its sequence-specific DNA binding activity in vitro. Mol Cell Biol 17:5923–5934

Lundgren K, Walworth N, Booher R, Dembski M, Kirschner M, Beach D (1991) mik1 and wee1 cooperate in the inhibitory tyrosine phosphorylation of cdc2. Cell 64:1111–1122

Ma XJ, Lu Q, Grunstein M (1996) A search for proteins that interact genetically with histone H3 and H4 amino termini uncovers novel regulators of the Swe1 kinase in *Saccharomyces cerevisiae*. Genes Dev 10:1327–1340

Mäkelä TP, Tassan J-P, Nigg EA, Frutiger S, Hughes GJ, Weinberg RA (1994) A cyclin associated with the CDK-activating kinase MO15. Nature 371:254–257

Mäkelä TP, Parvin JD, Kim J, Huber LJ, Sharp PA, Weinberg RA (1995) A kinase-deficient transcription factor TFIIH is functional in basal and activated transcription. Proc Natl Acad Sci USA 92:5174–5178

Marinoni J-C, Roy R, Vermeulen W, Miniou P, Lutz Y, Weeda G, Seroz T, Molina Gomez D, Hoeijmakers JHJ, Egly J-M (1997) Cloning and characterization of p52, the fifth subunit of the core of the transcription/DNA repair factor TFIIH. EMBO J 16:1093–1102

Martinez A-M, Afshar M, Martin F, Cavadore J-C, Labbé J-C, Dorée M (1997) Dual phosphorylation of the T-loop in cdk7: its role in controlling cyclin H binding and CAK activity. EMBO J 16:343–354

Matsuoka M, Kato J-Y, Fisher RP, Morgan DO, Sherr CJ (1994) Activation of cyclin-dependent kinase 4 (cdk4) by mouse MO15-associated kinase. Mol Cell Biol 14:7265–7275

McGowan CH, Russell P (1993) Human Wee1 kinase inhibits cell division by phosphorylating $p34^{cdc2}$ exclusively on Tyr15. EMBO J 12:75–85

McGowan CH, Russell P (1995) Cell cycle regulation of human WEE1. EMBO J 14:2166–2175

Measday V, Moore L, Retnakaran R, Lee J, Donoviel M, Neiman AM, Andrews B (1997) A family of cyclin-like proteins that interact with the Pho85 cyclin-dependent kinase. Mol Cell Biol 17:1212–1223

Meyerson M, Enders GH, Wu CL, Su LK, Gorka C, Nelson C, Harlow E, Tsai L-H (1992) A family of human cdc2-related protein kinases. EMBO J 11:2909–2917

Millar JBA, McGowan CH, Lenaers G, Jones R, Russell P (1991) $p80^{cdc25}$ mitotic inducer is the tyrosine phosphatase that activates $p34^{cdc2}$ kinase in fission yeast. EMBO J 10:4301–4309

Millar JBA, Lenaers G, Russell P (1992) Pyp3 PTPase acts as a mitotic inducer in fission yeast. EMBO J 11:4933–4941

Minshull J, Straight A, Rudner AD, Dernburg AF, Belmont A, Murray AW (1996) Protein phosphatase 2A regulates MPF activity and sister chromatid cohesion in budding yeast. Curr Biol 6:1609–1620

Molz L, Beach D (1993) Characterization of the fission yeast mcs2 cyclin and its associated protein kinase activity. EMBO J 12:1723–1732

Moreno S, Nurse P (1991) Clues to action of cdc25 protein. Nature 351:194

Morgan DO (1996) The dynamics of cyclin dependent kinase structure. Curr Opin Cell Biol 8:767–772

Mueller PR, Coleman TR, Dunphy WG (1995a) Cell cycle regulation of a *Xenopus* Weel-like kinase. Mol Biol Cell 6:119–134

Mueller PR, Coleman TR, Kumagai A, Dunphy WG (1995b) Mytl: a membrane-associated inhibitory kinase that phosphorylates Cdc2 on both threonine-14 and tyrosine-15. Science 270:86–90

Nigg EA, Krek W, Peter M (1991) Vertebrate cdc2 kinase: its regulation by phosphorylation and its mitotic targets. Cold Spring Harb Symp Quant Biol 56:539–547

Nikolic M, Dudek H, Kwon YT, Ramos YF, Tsai LH (1996) The cdk5/p35 kinase is essential for neurite outgrowth during neuronal differentiation. Genes Dev 10:816–825

Norbury C, Blow J, Nurse P (1991) Regulatory phosphorylation of the $p34^{cdc2}$ protein kinase in vertebrates. EMBO J 10:3321–3329

Novak B, Tyson JJ (1993) Numerical analysis of a comprehensive model of M-phase control in *Xenopus* oocyte extracts and intact embryos. J Cell Sci 106:1153–1168

O'Connell MJ, Raleigh JM, Verkade HM, Nurse P (1997) Chkl is a weel kinase in the G2 DNA damage checkpoint inhibiting cdc2 by Y15 phosphorylation. EMBO J 16:545–554

O'Neill EM, Kaffman A, Jolly ER, O'Shea EK (1996) Regulation of Pho4 nuclear localization by the Pho80-Pho85 cyclin-cdk complex. Science 271:209–212

Ogg S, Gabrielli B, Piwnica-Worms H (1994) Purification of a serine kinase that associates with and phosphorylates human Cdc25C on serine 216. J Biol Chem 269:30461–30469

Ohkura H, Hagan IM, Glover DM (1995) The conserved *Schizosaccharomyces pombe* kinase plol, required to form a bipolar spindle, the actin ring, and septum, can drive septum formation in Gl and G2 cells. Genes Dev 9:1059–1073

Orphanides G, Lagrange T, Reinberg D (1996) The general transcription factors of RNA polymerase II. Genes Dev 10:2657–2683

Parker LL, Piwnica-Worms H (1992) Inactivation of the $p34^{cdc2}$-cyclin B complex by the human WEE1 tyrosine kinase. Science 257:1955–1957

Parker LL, Atherton-Fessler S, Lee MS, Ogg S, Falk JL, Swenson KI, Piwnica-Worms H (1991) Cyclin promotes the tyrosine phosphorylation of $p34^{cdc2}$ in a $weel^{+}$ dependent manner. EMBO J 10: 1255–1263

Parker LL, Atherton-Fessler S, Piwnica-Worms H (1992) $p107^{weel}$ is a dual-specificity kinase that phosphorylates $p34^{cdc2}$ on tyrosine 15. Proc Natl Acad Sci USA 89:2917–2921

Parker LL, Walter SA, Young PG, Piwnica-Worms H (1993) Phosphorylation and inactivation of the mitotic inhibitor Weel by the niml/cdrl kinase. Nature 363:736–738

Parker LL, Sylvestre PJ, Byrnes MJ III, Liu F, Piwnica-Worms H (1995) Identification of a 95-kDa WEE1-like tyrosine kinase in HeLa cells. Proc Natl Acad Sci USA 92:9638–9642

Peng C-Y, Graves PR, Thoma RS, Wu A, Shaw AS, Piwnica-Worms H (1997) Mitotic and G2 checkpoint control: regulation of 14-3-3 protein binding by phosphorylation of Cdc25C on serine-216. Science 277:1501–1505

Philpott A, Porro EB, Kirschner MW, Tsai LH (1997) The role of cyclin-dependent kinase 5 and a novel regulatory subunit in regulating muscle differentiation and patterning. Genes Dev 11:1409–1421

Poon RYC, Hunter T (1995) Dephosphorylation of Cdk2 Thr160 by the cyclin-dependent kinase-interacting phosphatase KAP in the absence of cyclin. Science 270:90–93

Poon RYC, Yamashita K, Adamczewski JP, Hunt T, Shuttleworth J (1993) The cdc2-related protein $p40^{MO15}$ is the catalytic subunit of a protein kinase that can activate $p33^{cdk2}$ and $p34^{cdc2}$. EMBO J 12:3123–3132

Poon RYC, Yamashita K, Howell M, Ershler MA, Belyavsky A, Hunt T (1994) Cell cycle regulation of the $p34^{cdc2}/p33^{cdk2}$-activating kinase $p40^{MO15}$. J Cell Sci 107:2789–2799

Poon RYC, Jiang W, Toyoshima H, Hunter T (1996) Cyclin-dependent kinases are inactivated by a combination of p21 and Thr-14/Tyr-15 phosphorylation after UV-induced DNA damage. J Biol Chem 271:13283–13291

Poon RYC, Lew, J, Hunter T (1997) Identification of functional domains in the neuronal Cdk5 activator protein. J Biol Chem 272:5703–5708

Pringle JR, Hartwell LH (1981) The *Saccharomyces cerevisiae* cell cycle. In: Strathern J, Jones E, Broach J (eds) The molecular biology of the yeast Saccharomyces. Cold Spring Harbor Laboratory Press, Cold Spring Harbor, pp 97–142

Qi Z, Huang Q-Q, Lee K-Y, Lew J, Wang JH (1995) Reconstitution of neuronal Cdc2-like kinase from bacteria-expressed Cdk5 and an active fragment of the brain-specific activator: kinase activation in the absence of Cdk5 phosphorylation. J Biol Chem 270:10847–10854

Reardon JT, Ge H, Gibbs E, Sancar A, Hurwitz J, Pan Z-Q (1996) Isolation and characterization of two human transcription factor IIH (TFIIH)-related complexes: ERCC2/CAK and TFIIH. Proc Natl Acad Sci USA 93:6482–6487

Rhind N, Furnari B, Russell P (1997) Cdc2 tyrosine phosphorylation is required for the DNA damage checkpoint in fission yeast. Genes Dev 11:504–511

Rickert P, Seghezzi W, Shanahan F, Cho H, Lees E (1996) Cyclin C/Cdk8 is a novel CTD kinase associated with RNA polymerase II. Oncogene 12:2631–2640

Rochette-Egly C, Adam S, Rossignol M, Egly J-M, Chambon P (1997) Stimulation of RARα activation function AF-1 through binding to the general transcription factor TFIIH and phosphorylation by CDK7. Cell 90:97–107

Rossignol M, Kolb-Cheynel I, Egly J-M (1997) Substrate specificity of the cdk-activating kinase (CAK) is altered upon association with TFIIH. EMBO J 16:1628–1637

Roy R, Adamczewski JP, Seroz T, Vermeulen W, Tassan J-P, Schaeffer L, Nigg EA, Hoeijmakers JHJ, Egly J-M (1994) The MO15 cell cycle kinase is associated with the TFIIH transcription-DNA repair factor. Cell 79:1093–1101

Russell P, Nurse P (1986) *cdc25*$^{+}$ functions as an inducer in the mitotic control of fission yeast. Cell 45:145–153

Russell P, Nurse P (1987a) Negative regulation of mitosis by *wee1*$^{+}$, a gene encoding a protein kinase homolog. Cell 49:559–567

Russell P, Nurse P (1987b) The mitotic inducer *nim1*$^{+}$ functions in a regulatory network of protein kinase homologs controlling the initiation of mitosis. Cell 49:569–576

Russell P, Moreno S, Reed SI (1989) Conservation of mitotic controls in fission and budding yeasts. Cell 57:295–303

Russo AA, Jeffrey PD, Pavletich NP (1996) Structural basis of cyclin-dependent kinase activation by phosphorylation. Nature [Struct Biol] 3:696–700

Russo GL, Vandenberk MT, Yu IJ, Bae Y-S, Franza BR Jr, Marshak DR (1992) Casein kinase II phosphorylates p34^{cdc2} kinase in G1 phase of the HeLa cell division cycle. J Biol Chem 267:20317–20325

Sadhu K, Reed SI, Richardson H, Russell P (1990) Human homolog of fission yeast cdc25 mitotic inducer is predominantly expressed in G2. Proc Natl Acad Sci USA 87:5139–5143

Sanchez Y, Wong C, Thoma RS, Richman R, Wu Z, Piwnica-Worms H, Elledge SJ (1997) Conservation of the Chk1 checkpoint pathway in mammals: linkage of DNA damage to CDK regulation through Cdc25. Science 277:1497–1501

Santos RC, Waters NC, Creasy CL, Bergman LW (1995) Structure-function relationships of the yeast cyclin-dependent kinase Pho85. Mol Cell Biol 15:5482–5491

Sclafani RA, Jackson AL (1994) Cdc7 protein kinase for DNA metabolism comes of age. Mol Microbiol 11:805–810

Sebastian B, Kakizuka A, Hunter T (1993) Cdc25M2 activation of cyclin-dependent kinases by dephosphorylation of threonine-14 and tyrosine-15. Proc Natl Acad Sci USA 90:3521–3524

Serizawa H, Mäkelä TP, Conaway JW, Conaway RC, Weinberg RA, Young RA (1995) Association of Cdk-activating kinase subunits with transcription factor TFIIH. Nature 374:280–282

Shiekhattar R, Mermelstein F, Fisher RP, Drapkin R, Dynlacht B, Wessling HC, Morgan DO, Reinberg D (1995) Cdk-activating kinase complex is a component of human transcription factor TFIIH. Nature 374:283–287

Shuttleworth J, Godfrey R, Colman A (1990) p40^{MO15}, a *cdc2*-related protein kinase involved in negative regulation of meiotic maturation of *Xenopus* oocytes. EMBO J 9:3233–3240

Sia RA, Herald HA, Lew DJ (1996) Cdc28 tyrosine phosphorylation and the morphogenesis checkpoint in budding yeast. Mol Biol Cell 7:1657–1666

Simon M, Seraphin B, Faye G (1986) Kin28, a yeast split gene coding for a putative protein kinase homologous to CDC28. EMBO J 5:2697–2701

Smythe C, Newport JW (1992) Coupling of mitosis to the completion of S phase in *Xenopus* occurs via modulation of the tyrosine kinase that phosphorylates $p34^{cdc2}$. Cell 68:787–797

Solomon MJ, Glotzer M, Lee TH, Philippe M, Kirschner MW (1990) Cyclin activation of $p34^{cdc2}$. Cell 63:1013–1024

Solomon MJ, Lee T, Kirschner MW (1992) Role of phosphorylation in $p34^{cdc2}$ activation: identification of an activating kinase. Mol Biol Cell 3:13–27

Solomon MJ, Harper JW, Shuttleworth J (1993) CAK, the $p34^{cdc2}$ activating kinase, contains a protein identical or closely related to $p40^{MO15}$. EMBO J 12:3133–3142

Sorger PK, Murray AW (1992) S-phase feedback control in budding yeast independent of tyrosine phosphorylation of $p34^{cdc2}$. Nature 355:365–368

Strausfeld U, Labbé JC, Fesquet D, Cavadore JC, Picard A, Sadhu K, Russell P, Dorée M (1991) Dephosphorylation and activation of a $p34^{cdc2}$/cyclin B complex in vitro by human CDC25 protein. Nature 351:242–245

Strausfeld U, Fernandez A, Capony JP, Girard F, Lautredou N, Derancourt J, Labbé JC, Lamb NJ (1994) Activation of $p34^{cdc2}$ protein kinase by microinjection of human cdc25C into mammalian cells. Requirement for prior phosphorylation of cdc25C by $p34^{cdc2}$ on sites phosphorylated at mitosis. J Biol Chem 269:5989–6000

Surana U, Amon A, Dowzer C, McGrew J, Byers B, Nasmyth K (1993) Destruction of the CDC28/CLB mitotic kinase is not required for the metaphase to anaphase transition in budding yeast. EMBO J 12:1969–1978

Sutton A, Freiman R (1997) The Cak1p protein kinase is required at G1/S and G2/M in the budding yeast cell cycle. Genetics 147:57–71

Tang Z, Coleman, TR, Dunphy WG (1993) Two distinct mechanisms for negative regulation of the Wee1 protein kinase. EMBO J 12:3427–3436

Tassan J-P, Schultz SJ, Bartek J, Nigg EA (1994) Cell cycle analysis of the activity, subcellular localization, and subunit composition of human CAK (CDK-activating kinase). J Cell Biol 127:467–478

Tassan J-P, Jaquenoud M, Fry AM, Frutiger S, Hughes GJ, Nigg EA (1995a) In vitro assembly of a functional human CDK7-cyclin H complex requires MAT1, a novel 36 kDa RING finger protein. EMBO J 14:5608–5617

Tassan JP, Jaquenoud M, Leopold P, Schultz SJ, Nigg EA (1995b) Identification of human cyclin-dependent kinase 8, a putative protein kinase partner for cyclin C. Proc Natl Acad Sci USA 92:8871–8875

Terada Y, Tatsuka M, Jinno S, Okayama H (1995) Requirement for tyrosine phosphorylation of Cdk4 in G1 arrest induced by ultraviolet irradiation. Nature 376:358–362

Thuret J-Y, Valay J-G, Faye G, Mann C (1996) Civ1 (CAK in vivo), a novel Cdk-activating kinase. Cell 86:565–576

Toh-e A, Tanaka K, Uesono Y, Wickner RB (1988) *PHO85*, a negative regulator of the PHO system, is a homolog of the protein kinase gene, *CDC28*, of *Saccharomyces cerevisiae*. Mol Gen Genet 214:162–164

Tyson JJ, Novak B, Odell GM, Chen K, Thron CD (1996). Chemical kinetic theory: understanding cell-cycle regulation. Trends Biochem Sci 21:89–96

Valay J-G, Simon M, Dubois M-F, Bensaude O, Facca C, Faye G (1995) The *KIN28* gene is required both for RNA polymerase II mediated transcription and phosphorylation of the Rpb1p CTD. J Mol Biol 249:535–544

van den Heuvel S, Harlow E (1993) Distinct roles for cyclin-dependent kinases in cell cycle control. Science 262:2050–2054

Wagner M, Pierce M, Winter E (1997) The CDK-activating kinase *CAK1* can dosage suppress sporulation defects of *smk1* MAP kinase mutants and is required for spore wall morphogenesis in *Saccharomyces cerevisiae*. EMBO J 16:1305–1317

Walworth NC, Bernards R (1996) rad-dependent response of the chk1-encoded protein kinase at the DNA damage checkpoint. Science 271:353–356

Walworth N, Davey S, Beach D (1993) Fission yeast chk1 protein kinase links the rad checkpoint pathway to cdc2. Nature 363:368–371
Wan J, Xu H, Grunstein M (1992) *CDC14* of *Saccharomyces cerevisiae*. Cloning, sequence analysis, and transcription during the cell cycle. J Biol Chem 267:11274–11280
Watanabe N, Broome M, Hunter T (1995) Regulation of the human WEE1Hu CDK tyrosine 15-kinase during the cell cycle. EMBO J 14:1878–1891
Wu L, Russell P (1993) Nim1 kinase promotes mitosis by inactivating Wee1 tyrosine kinase. Nature 363:738–741
Yankulov KY, Bentley DL (1997) Regulation of CDK7 substrate specificity by MAT1 and TFIIH. EMBO J 16:1638–1646
Ye XS, Fincher RR, Tang A, O'Donnell K, Osmani SA (1996) Two S-phase checkpoint systems, one involving the function of both BIME and Tyr15 phosphorylation of $p34^{cdc2}$, inhibit NIMA and prevent premature mitosis. EMBO J 15:3599–3610
Ye XS, Fincher RR, Tang A, Osmani SA (1997) The G2/M DNA damage checkpoint inhibits mitosis through Tyr15 phosphorylation of $p34^{cdc2}$ in *Aspergillus nidulans*. EMBO J 16:182–192

Regulation of the cell cycle by CDK inhibitors

T. J. Soos, M. Park, H. Kiyokawa, and A. Koff[1]

1
Introduction

There are at least four possible ways in which inhibitors of cyclin-dependent kinases (CDKs) might regulate cell cycle progression (Fig. 1). First, they might determine the duration of G1 phase by setting a threshold above which G1 CDK activity must accumulate before initiation of S-phase. In this threshold role, steady-state levels of CDK inhibitors (CKIs) counteract the temporal accumulation of G1 cyclin/CDK complexes. Second, they might act as checkpoints inducing cell cycle arrest transiently when genomic fidelity is threatened. Chromosome damage or inappropriate segregation of chromosomes may induce cell cycle arrest and initiate DNA repair mechanisms or apoptotic programs. Third, CKIs might allow expedient withdrawal of cells from the cell cycle and a concomitant change in cell fate. In some cases, CKIs might not only regulate the withdrawal from the cell cycle but also affect lineage-specific functions. Fourth, they might function as a gatekeeper of quiescence, keeping the G0 cell

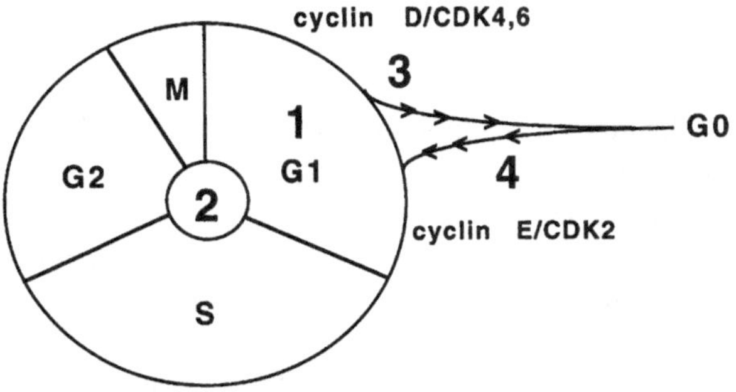

Fig. 1. Four transitions in the cell cycle that can be controlled, in part, by the CKI. The phases of the cell cycle are indicated (*G1*, *S*, *G2*, *M*, and *G0*) and the G1 cyclin/CDK complexes are placed according to the time of their first activities during the mitotic cell cycle. The four transitions are the passage of a cell to DNA replication directly from mitosis *(1)*, any time in the cell cycle that has the ability to respond to threats to genomic fidelity (*2*), during withdrawal from the cell cycle *(3)*, and during re-entry into the cell cycle *(4)*

[1] Program in Molecular Biology, Memorial Sloan-Kettering Cancer Center, RRL 917D, Box 207, 1275 York Avenue, New York, New York 10021, E-mail: a-koff@ski.mskcc.org

from inappropriately resuming proliferation. In some cases, tumors arise from well differentiated and non-mitotic cells that inexplicably return to the cell cycle. Over the last 5 years, biochemical, cellular, and molecular analyses of CKIs have combined with genetic analysis of mutant mice deficient in each CKI to address these possible functions. Our current knowledge of the role of CKI in these decisions and speculations about their roles in developmentally regulated cell cycles are discussed below.

2 Identification of CKI

Two classes of CKI were identified nearly simultaneously (Fig. 2) (Sherr and Roberts 1995). Cyclin D/CDK4,6 specific inhibitors exemplified by p16MTS1/Ink4a, p15Ink4B, p18, and p19 comprise the Ink class of inhibitors. Promiscuous CKI, i.e., those that target both CDK4,6 and CDK2, exemplified by p21Waf1/Cip1/Sdi1 and followed rapidly by p27Kip1 and p57Kip2 are members of the Cip/Kip class of inhibitors. Within these two classes of inhibitors, so far only p16 appears to be a bona fide tumor suppressor (Serrano et al. 1996); however, this might be due to the presence of an overlapping reading frame encoding another protein, p19ARF1, that when overexpressed can itself inhibit cell proliferation through an undefined mechanism (Quelle et al. 1995a). Significantly more information has accumulated on the role of Cip/Kip class inhibitors in cell cycle control than on the Ink class. This may be due in part to the implicit role that a cyclin D-specific inhibitor has affecting the expression of the D-type cyclin and its functions in cell cycle, i.e., as a rate-limiting component of S-phase entry. Consequently, we will focus mostly on the Cip/Kip-class of inhibitors and will only cursorily point out the Inks and their possible roles when deemed appropriate.

The identification of the inhibitors of the CDK4 (Ink) family occurred simultaneously with a hunt for a melanoma tumor suppressor, MTS1 (Kamb et al. 1994), and identification of proteins that associated with CDK4 (Xiong et al. 1993b). These studies indicated that a 16 kDa protein, p16INK4A, was involved in both. p16 competed with cyclin D for the CDK4 subunit (Xiong et al. 1993b), raising the possibility that the tumor suppressive function of this protein was due to inhibition of cyclin D/CDK4 activity, which is required for inactivation of the tumor suppressor Rb (Kato et al. 1993; Medema et al. 1995). Two-hybrid screens and sequence homology cloning led to the identification of p15, p18, and p19 (Serrano et al. 1993; Hannon and Beach 1994; Guan et al. 1994, 1996; Hirai et al 1995). Interestingly, p19 interacts with an orphan steroid receptor, Nur77 (Chan et al. 1995). Recently, a cyclin D1/estrogen receptor complex has also been identified, raising the interesting, albeit speculative notion that Inks and D-type cyclins may compete for steroid receptors, as well as CDKs, with unknown consequences.

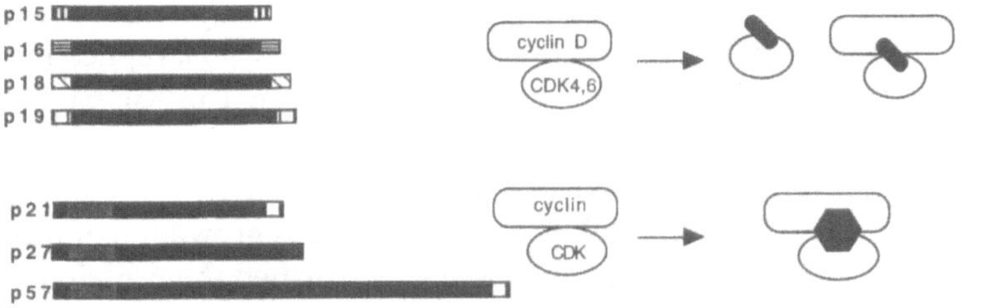

Kip subclass of Cip/Kip inhibitors (with FY domain)

mouse p27	MDARQADHPKPSACR	NLFGPVNHEE---LT	RDLEKHCRDMEEASQ	RKWNFDFQNHKPLEG	---RYEWQEVERGSL	PEFYYRPPRPP
rat p27	MDARQTEHPKPSACR	NLFGPVNHEE---LT	RDLEKHCRDMEEASQ	RKWNFDFQNHKPLEG	---RYEWQEVERGSL	PEFYYRPPRPP
human p27	MDARQAEHPKPSACR	NLFGPVDHEE---LT	RDLEKHCRDMEEASQ	RKWNFDFQNHKPLEG	---KYEWQEVEKGSL	PEFYYRPPRPP
hamster p27	MDARQAEHPKPSACR	NLFGPVNHEE---LT	RDLEKHCRDMEEASQ	RKWNFDFQNHNPLEG	---RYQWQEVDKGSL	PEFYYRPPRPP
human p57	ARGTFPVLVRTSACR	SLFGPVDHEE---LS	RELQARLAELNAEDQ	NRWDYDFQQDMPLRG	P-GRLQWTEVDSDSV	PAFYRETVQ--
mouse p57	SSDTFPVIARSSACR	SLFGPVDHEE---LG	RELRMRLAELNAEDQ	NRWDFNFQQDVPLRG	P-GRLQWMEVDSESV	PAFYRETVQ--
Xenopus p28	VLPRLSAGTGRGACR	NLFGPIDHDE---MR	SELKRQLKEIQASDC	QRWNFDFETGTPLKG	---IFCWEPVESKDM	PSFYSQNRSIA
Xenopus XICL	ALPRLSLGTGRGACR	NLFGPIDHDE---LR	SELKRQLKEIQASDC	QRWNFDFESGTPLKG	---TFCWEPVETKDV	PSFYSPSRSLA
C. elegans T05A6.2	MAATTAGDGKRKAAR	CLFGKPDPEE---VS	RQLNSSLEEMYKKDS	RKFNFDFSGGVPIVG	SRGDYEFESISASEV	PSFYREKI---
C. elegans T05A6.1	---------MSSARR	CLFGRPTPEQRSRTR	IWLEDAVKRMRQEES	QKWGFDFELETPLPS	S-AGFVYEVIPENCV	PEFYRT-----

Cip subclass of Cip/Kip inhibitors (without FY domain)

mouse p21	PQDVRPVPHRSKVCR	CLFGPVDSEQ---LR	RDCDALMAGCLQEAR	ERWNFDFVTETPLEG	---NFVWERVRSLGL	PKVYLSPGSRS
rat p21	PGDVRPVPHRSKVCR	RLFGPVDSEQ---LS	RDCDALMASCLQEAR	ERWNFDFATETPLEG	---NYVWERVRSPGL	PKIYLSPGSRR
human p21	AGDVRQNPCGSKACR	RLFGPVDSEQ---LS	RDCDALMAGCIQEAR	ERWNFDFVTETPLEG	---DFAWERVRGLGL	PKLYLPTGPRR
D. melanogaster dacapo	SRNLACGRQLNRIKR	DLFGSSKSAEGTANK	TPFNSELERHQELAT	QKWGFDFRAGCPLAE	K-SPYTWERVS---F	QESTFAPEM--

CONSENSUS

$$X_{1\ 3}\overline{CRXLFG}PX_2\overline{H}X^{E}_{Q}X_7LX_{1\ 4}\overline{WXFDFX_4PLXGX_6}WX_2VX_5PX\overline{FY}$$

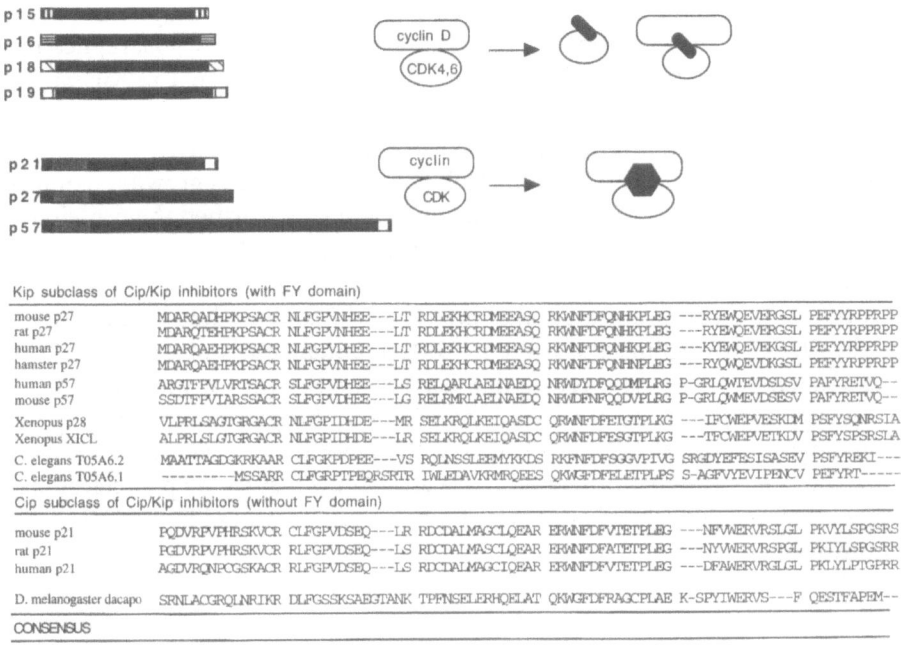

Fig. 2 A, B. Sequence homologies of the CKI. **A** Mechanism of inhibition. The four Ink-class inhibitors, *p15, p16, p18* and *p19*, have a common ankyrin repeat motif flanked by unique sequences. These inhibitors are specific for the D-type cyclin-associated kinase CDK4 and CDK6. Mechanistically they interfere with CDK activity either by competing with cyclin D for the CDK subunit, or by binding to the complex. The three mammalian Cip/Kip class inhibitors, *p21, p27* and *p57*, share a conserved 60 amino acid cyclin/CDK binding domain (*gray*), and *p21* and *p57* have a PCNA binding domain (*white*). These inhibitors are promiscuous and inhibit the activity of both CDK4/6 and CDK2 kinases, at least in vitro. Mechanistically they interfere by interacting with the complex. **B** The cyclin/CDK binding domain of the Cip/Kip-class of inhibitors is conserved from *Drosophilia* and *C. elegans* to mammals. We have aligned the known and presumptive (*C. elegans* open reading frames) cyclin/CDK binding domains of the proteins indicated on the *left*. Protein sequences have been aligned by introducing gaps marked by *dashes*. The presence of an FY motif in the carboxyl region of the motif was used to discriminate between Kip-type and Cip-type inhibitors. A consensus sequence was defined by the eight known Kip-type inhibitors (excluding the *C. elegans* presumptive inhibitors) and then compared to the Cip-type inhibitors. Sequence with identity between the classes is noted by the *bar* above the residues

p21 was identified in five laboratories where very different aspects of G1 control were studied. By examining proteins that physically interact with cyclin/CDK complexes, the groups of Beach (Xiong et al. 1993a) and Morgan (Gu et al. 1993) independently identified p21 in human and murine fibroblasts, respectively. Looking for proteins that interact with CDKs, Harper et al. (1993) identified p21. cDNA screens to identify gene products important in senescence led to the cloning of p21 (Noda et al. 1994), and screens to identify genes induced by p53 also led to p21 (El-Deiry et al. 1993). Consequently, putative roles for p21 in DNA damage-induced cell cycle arrest mediated by p53, replicative senescence, and in the cell cycle has made it the most studied of Cip/Kip family members.

Identification of p27 followed p21. We first identified an activity responsible for inhibition of cyclin E/CDK2 in mink lung epithelial cells grown to confluence or treated with TGFβ (Koff et al. 1993), and later purified the protein by its ability to bind to cyclin E/CDK2 complexes (Polyak et al. 1994a, b). This provided the first direct evidence of signal transduction pathways converging on the cell cycle through a protein that interacts with and modulates the activity of the CDKs. Toyoshima and Hunter (1994) simultaneously identified p27 using a modified two-hybrid screen that used cyclin D/CDK4 as the bait.

Finally, identification of p57 occurred in two laboratories; one looking for homologues of p21 (Lee et al. 1995), and the other using a two-hybrid screen (Matsuoka et al. 1995). Consequently, investigations focused on p53, cell aging, signal transduction, and the cell cycle found common ground in the CKI of the Cip/Kip class.

3 Kips Inhibit All G1 CDKs, Although the Mechanisms May Differ

Unlike the Ink inhibitors that compete with the D-type cyclins to preferentially interact with the CDK4,6 subunit (although ternary complexes have recently been noted in cells (Reynisdottir and Massague 1997); the Cip/Kips preferentially bind to cyclin/CDK complexes (Fig. 2). The Cip/Kips are highly homologous in the N terminus but differ widely at the C terminus. A region of approximately 60 amino acids is conserved between p21, p27, and p57 and encompasses the cyclin and CDK interaction domain (Fig. 2B; J. Chen et al. 1995, 1996; I. T. Chen et al. 1996; Russo et al. 1996). Similar domains have been identified in a number of invertebrate proteins. The solution of the cyclin A/CDK2/p27 domain structure indicates that p27 binds to the cyclin subunit and orients so that the inhibitor domain interacts with the kinase subunit (Russo et al. 1996). The conserved LFG motif in the inhibitor makes contact with residues in the peptide groove of the cyclin (Russo et al. 1996). Mutation of the leucine and phenylalanine severely reduce the ability of p27 to bind and inhibit cyclin E/CDK2 complexes (Luo et al. 1995).

Following association of p27 with the complex, inhibition of kinase activity might occur as a result of structural changes in the CDK subunit. p27 binding induces a dramatic conformational change in the N terminal lobe of CDK2 and disrupts the catalytic cleft of the kinase (Russo et al. 1996). Furthermore, the change in the structure of the cleft might destabilize bound ATP, and subsequent occupancy of the ATP binding site by the FY domain of the inhibitor prevents ATP binding (Russo et al. 1996). The FY motif may "lock" the complex together and inhibit activity by competing with ATP. However, although mutational analysis indicates the necessity of the LFG motif, at this time there is little direct evidence to support the necessity of the FY domain. First, the FY motif is less conserved than the LFG motif, and second, the interactions of p27 with CDK2 involve the region of CDK2 that forms the "roof" of the ATP binding cleft. Thus, perturbing this portion of CDK2 may be sufficient to inhibit activity. However, the reader should be cautious about generalizing this mechanism to other Cip/Kip-class inhibitors. For example, although it is tempting to speculate that these proteins share a common mechanism of inhibitory action, due in part to the amino acid homology in the cyclin/CDK binding domain, there is scarce biochemical evidence to support this notion. The crystal structure of p27/cyclin A/CDK2 suggests a stoichiometry of 1:1:1 for inhibition (Russo et al. 1996). In contrast, similar stoichiometries with p21 in place of p27 are not inhibitory, but inhibition occurs if the stoichiometry of p21 increases to 2:1:1 (Zhang et al. 1994; Harper et al. 1995). This suggests that the inhibitory ability of a CKI might be modulated by non-conserved elements.

Another possible mechanism by which CDK inhibitors might function is suggested by the observation that the cyclin domain interacting with the LFG motif is also highly conserved (Russo et al. 1996). In cyclin A this motif contains the MRAIL domain, at the amino-terminus of the cyclin box, specifically the M, I, and L, as well as amino acids W217, E220, L253, and T285. Consequently, the binding of Cip/Kips to the cyclin might limit the interactions of the cyclin/CDK complex with other proteins, including potential substrates. Likely candidates may be E2F1, E2F2, E2F3, p107 and p130, all of which possess cyclin binding motifs that share some homology with the Cip/Kips (Adams et al. 1996).

The ability of Cip/Kips to promiscuously inhibit the G1-CDKs is observed in vitro and in transient transfection experiments. In contrast, in vivo studies suggest that the ability of an inhibitor to affect a particular cyclin/CDK complex is dependent on the G1 state: Is the cell going from M to S, G0 to S, or G1 to G0? The functional consequence of p21 or p27 binding to each cyclin/CDK complex varies under different in vivo conditions (Zhang et al. 1994; Harper et al. 1995; Poon et al. 1995; Agrawal et al. 1996; Florenes et al. 1996; Soos et al. 1996; Hauser et al. 1997). In cells entering or exiting quiescence (transitions 3 and 4 in Fig. 1), the association of Kips with cyclin D and cyclin E is inhibitory (Fig. 3), consistent with the putative threshold function of the inhibitor. In addition, there is a possibility that the transient interaction of Kips with a cyclin or CDK may facilitate assembly of the CDK complex (LaBaer et al. 1997). In contrast, in mitotically growing cells (transition 1 in figure 1), p21 interacts with CDK2 in a non-inhibitory fashion. The effect on CDK4 or CDK6 complexes associated with p21 has not been determined in mitotically growing cells. In MANCA cells (Soos et al. 1996) and keratinocytes (Hauser et al. 1997) going from M to S phase (transition 1 in Fig. 1), p27 interacts with cyclin D3/CDK4 in a non-inhibitory

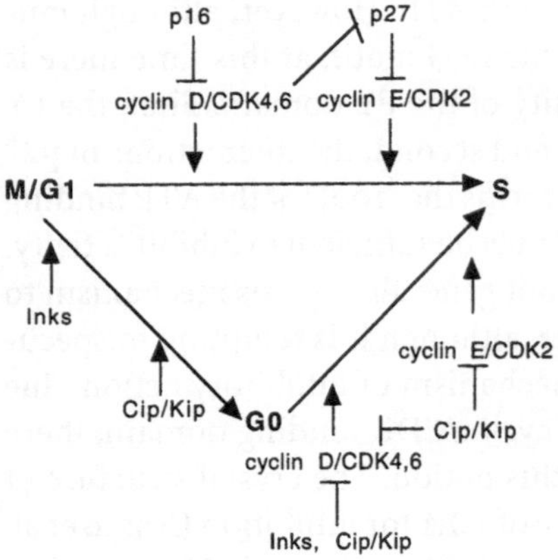

Fig. 3. The inhibitory properties of the CKIs are dependent on the *G1* transition examined. Three different *G1* transitions have been described, the transition of cells from *M*-phase into *S*-phase (*top*), the transition of cells leaving the cell cycle (*M->G0, left*), and the transition of cell re-entering the cell cycle (*G0->S, right*). The inhibitory ability of the p27 protein, and possibly others, changes depending on the conditions

fashion; however, in these cells p27 does inhibit associated CDK2 complexes (Fig. 3). Consistent with these observations, inhibition of cyclin E,A/CDK2 requires less p27 than inhibition of cyclin D2/CDK4 when normalized for activity on an Rb substrate in vitro (Polyak et al. 1994b; Harper et al. 1995). Reciprocally, p21 is a more efficient inhibitor of cyclin D/CDK4 kinase activity than cyclin A,E/CDK2 (Harper et al. 1995). It is important to insert a note of caution when interpreting these results because CKIs act stoichiometrically, and inactive cyclin/CDK complexes might affect the interpretation of these results; furthermore, the normalization of cyclin/CDK complexes by kinase activity to a specific substrate is also problematic. However, with this caveat given, the inhibitory preference of each Cip/Kip for specific G1 CDK complexes might be reconciled with distinct in vivo roles for the inhibitors regulating different transitions.

4
Expression of Inhibitors During the Cell Cycle

To determine the role of a protein, the expression pattern is often informative. Studies on tissue culture cells that retain the ability to withdraw from the cell cycle to a quiescent or differentiated state indicate that CDK inhibitors are more abundant in non-proliferating cells. Amalgamation of these studies suggests a simple interpretation. Signals from the environment establish a balance between cyclin/CDK complexes and the inhibitors by altering the steady-state levels of these classes of proteins individually or simultaneously (Fig. 4). Furthermore, the expression of these proteins may affect each other (Hatakayama and Weinberg 1995). Mitogenic signaling increases the cumulative steady state level of cyclin/CDK complexes and decreases the cumulative steady state level of CKI. Anti-mitogens have the opposite effect. Consistent with this model, quiescent T-cells return to the cell cycle following antigen and IL-2 stimulation.

Changes in both the cyclin/CDK and CKI amounts complete this two-step activation (Firpo et al. 1994). Likewise, TGFβ treatment of epithelial cells stimulates Ink expression (Reynisdottir et al. 1995) and the loss of CDK4 protein (Ewen et al. 1993).

The inhibitor expressed during growth arrest is dependent on the cell type and inductive signal. In many cases, both p21 and p27 increase, in other cases one or the other may increase. Concomitant increases in the Inks may or may not be observed. There is no evidence for induction of p57 in tissue culture cell lines. Experimental support has been gained for the hypothesis that the Inks and Cip/Kip class of inhibitors cooperate to mediate G1 arrest by TGFb; however, the choice of the Ink and Cip/Kip appear to remain cell type dependent, and in some cells p21 is used while in other cells p27 dominates. The conditions governing this choice remain to be elucidated.

In cell strains, the expression of Inks is variable. Only certain cells contain detectable levels of Inks during the cell cycle. In cycling IMR90 cells (human diploid fibroblasts), p16 levels increase during G1 phase, become maximal at S phase and remain high during G2 and M-phase (Tam et al. 1994). Similar patterns are observed for p18 and p19 in BAC1.2F5 macrophage cells (Hirai et al. 1995). These data suggest that Inks may be responsible for inactivation of cyclin D-associated kinases in S+G2+M cells. However, in normal human breast epithelium, the amounts of p15 and p16 protein are constant during the cell cycle (Gray-Bablin et al. 1997).

To find a cell type in which to identify the biological role of inhibitors, we and others examined the expression of CKI in the animal. There have been a number of histological studies on the expression of CDK inhibitors in the mouse.

Inks are expressed in a tissue-specific manner. In mice, p16 mRNA is highest in spleen and lungs, whereas p15 mRNA is highest in colon, brain, spleen, kidneys, lung and liver (Quelle et al. 1995b). p19 expression is highest in bone marrow and spleen (Hirai et al. 1995), suggesting a role in hematopoiesis; however,

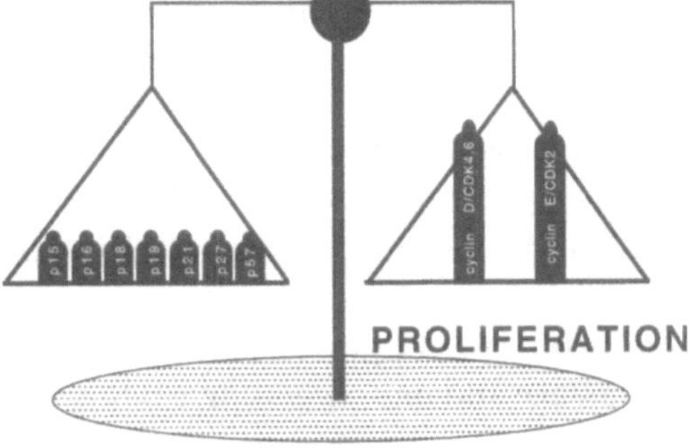

Fig. 4. The balance between CKIs and cyclin/CDK complexes determines the proliferation of cells. On the *left* we show the differnt CKIs and on the *right* the different G1 cyclin/CDK complexes. Depending on the cell type and the inductive signal, the amounts of these proteins may change; however, the stoichiometric relationship between the proteins suggests that when the amount of cyclin/CDK exceeds the amount of cyclin/CDK exceeds the amount of CKI, proliferation occurs

in humans expression is more general and the transcripts can be detected in spleen, thymus, testes, heart, brain, and skeletal muscle (Guan et al. 1996). p18 is complex and there are at least two mRNA species and possibly as many as five. In mice, highest expression is in the bone marrow, spleen, skin, and lung, and in humans in the skeletal muscle, pancreas, thymus, ovaries and kidneys (Guan et al. 1994, 1996; Hirai et al. 1995).

The expression of p21 and p57 has been studied most extensively using in situ hybridization (Lee et al. 1995; Matsuoka et al. 1995; Parker et al. 1995). These two inhibitors co-localize in skeletal muscle, cartilage, and tongue muscle, while they have distinct patterns of expression in other tissues. In general, the expression of p21 and p57 in mouse embryos is correlated with exit from the cell cycle during terminal differentiation. In adult mice, mRNA encoding these proteins were detected in postmitotic cells of the muscle, liver and kidneys, as well as in epithelial cells that are continuously turning over. These observations imply that these inhibitors might establish quiescence during differentiation programs and/or maintain the postmitotic status (transitions 3 and 4 in Fig. 1). As the regulation of p27 expression is mostly post-transcriptional (both protein translation and degradation levels), it is more informative to examine the tissue distribution of p27 protein. p27 protein was detected in extracts derived from all organs of adult mice, however, the amount of p27 varied widely. Immunohistochemistry has shown that during thymic development, p27 protein level decreases, as thymocytes resume proliferation following successful recombination of the T cell receptor β chain (Hoffman et al. 1996), and increase again as cells undergo positive and negative selection. In adult ovaries, the amount of p27 protein markedly increases as the granulosa cells in the follicle differentiate into luteal cells in correlation with cessation of proliferation (Fero et al. 1996; Kiyokawa et al. 1996). In mice lacking p27, these cells do not withdraw from the cell cycle. These observations are consistent with the hypothesis that CKIs play a role in regulating entry into and withdrawal from the cell cycle during postpartum determination of cell fate.

In addition, Cip/Kip family members may operate within the cell cycle to establish the rate of G1 passage. Alterations in G1 duration can affect the ability of cells to respond to anti-mitogenic signals that normally withdraw cells from the cell cycle. The commitment machinery is the arbiter of the choice between proliferation and withdrawal from the cell cycle. Overexpression of Cln3 in yeast (Cross et al. 1988), overexpression of cyclin E in mammalian cells (Ohtsubo and Roberts.1993), cul-1 deficiencies in *C elegans* (Kipreos et al. 1996), deficiency of p27 in mice (Fero et al. 1996; Kiyokawa et al. 1996; Nakayama et al. 1996), and deficiency of dacapo, a Kip-related CKI, in *Drosophilia* (Lane et al. 1996) diminish the response of cells to environmental signals. Consequently, it is difficult to distinguish experimentally between a CKI regulating the rate of G1 passage and one regulating withdrawal from the cell cycle.

5
p21, a Checkpoint Regulator of the Cell Cycle

In primary quiescent human T-cells there is no detectable p21 protein, although p27 is abundant (Firpo et al. 1994). The amount of p21 increases as cells return to the cell cycle following stimulation of the T-cell receptor in the presence of IL-2. This suggests that p21, at least in T-cells, may either act as an assembly factor for cyclin/CDK complexes, as a threshold regulator or titratable buffer of cyclin/CDK complexes in cycling cells, or allow rapid utilization of checkpoints.

Support for a checkpoint function is threefold. First, all of CDK2 in proliferating WI38 cells is associated with p21, but is not inhibited suggesting that the cell is "buffered" for inhibition by small changes in the steady state amount of p21 (Harper et al. 1995). Second, at high concentrations p21 associates with PCNA, a processivity factor for DNA polymerase δ enzyme (Zhang et al. 1993). Experiments have suggested that the association of p21 with PCNA decreases the processivity of the PCNA/polymerase δ complex (Waga et al. 1994; Pan et al. 1995). Furthermore, the carboxyl-terminal domain of p21 in the absence of the CDK binding portion of the inhibitor is sufficient to arrest cells in G1 (Luo et al. 1995). These data suggest that p21 has the ability to convert the replicative mode of the polymerase into a repair-mode. Third, p21-deficient mice display neither developmental phenotype nor spontaneous tumorigenesis; however, embryonic fibroblasts isolated from p21-deficient mice, like p53-deficient mice, are partially refractory to G1 arrest induced by DNA damage or nucleotide pool perturbation (Brugarolas et al. 1995; Deng et al. 1995). Together these observations suggest that a major function of p21 might be to mediate checkpoint function.

The absence of developmental phenotype can not exclude a role for p21 in withdrawal from the cell cycle and a change in cell phenotype, differentiation, or senescence. Other factors could be acting in a redundant manner, either directly replacing p21 or on other pathways obscuring the requirement for p21. In embryos the expression of p21 mRNA is well correlated with terminal differentiation of muscle (Parker et al. 1995). In addition, introduction of the muscle-specific transcription factor myoD leads to muscle differentiation and increased p21 mRNA (Halevy et al. 1995), suggesting that the lack of developmental defects, at least in muscle, might be due to overlapping function of other CKI such as p27 (Halevy et al. 1995) or p18 (Franklin and Xiong 1996).

6
p27, a Regulator of G1/G0 Transition, and Possibly G1/S.

Many of the phenotypes of p27-deficient mice suggest that p27 plays an important role in the timely withdrawal of cells from the cell cycle. p27-deficient mice exhibit enhancement of postnatal growth without increases in growth-regulatory hormones such as GH, IGF-I and IGF-II. Although the effect of p27 deficiency mimics proportional gigantism, it shows some tissue specificity. The weight of tissues that normally express the highest levels of p27, such as thymus and spleen, is more affected by the gene disruption than other organs (Fero et al. 1996; Kiyokawa et al. 1996; Nakayama et al. 1996). The disruption of p27 in mice leads to an increase in the number of thymocytes at virtually all stages of development, resulting in a markedly enlarged thymus. Hyperproliferation during thymic maturation appears to contribute to the increased production of T lymphocytes. In addition, p27 deficient mice have an increase in the number of O2A cells, precursors to the oligodendrocyte in the central nervous system (Casaccia-Bonnefil et al. 1997).

Increased numbers of cells might be interpreted to reflect a role in regulating withdrawal from the cell cycle, increasing proliferation of cells, or decreasing cell death. Apoptosis and the requirements for cytokines are not disturbed by p27 deficiency in either O2A (Casaccia-Bonnefil et al. 1997) or thymocyte (Fero et al. 1996; Kiyokawa et al. 1996; Nakayama et al. 1996) systems. In some cell lines, and importantly in primary cells, the expression pattern of Cip/Kips suggests a more direct role in the cell cycle, perhaps in regulating the rate of S phase entry. Disruption of p27 expression in 3T3 fibroblasts reduced the percentage of cells in G1 with a corresponding increase in S and G2 phases, suggesting that p27 regulated the rate of G1 progression (Coats et al. 1996). Furthermore, both O2A cells and thymocytes obtained from p27-deficient mice have an increase in the proportion of S-phase cells with a reduction in the amount of G1 cells (Fero et al. 1996; Casaccia-Bonnefil et al. 1997).

Consequently, the ability to act as a rate-limiting regulator of S phase entry and to prevent timely withdrawal suggests that p27 has a role in the critical period of time when cells decide their fate vis-a-vis the restriction point. However, neither fibroblasts (Fero et al. 1996; Nakayama et al. 1996), keratinocytes (Missero et al . 1996) nor activated T-cells (Fero et al. 1996; Kiyokawa et al. 1996; Nakayama et al. 1996) isolated from p27-deficient mice display a significant alteration in S phase entry during the mitotic cell cycle. In contrast, these cells are partially resistant to growth-inhibitory agents such as rapamycin and TGF-β (Luo et al. 1996; Missero et al. 1996). Thus, we conclude that p27 is an instrumental part of commitment, where the decision to commit to S phase or withdraw from the cell cycle is made.

7
p57, a Specific Regulator for Specific Cells

The human p57 gene maps to chromosome position 11p15.5, which is linked to Wilms' tumors and Beckwith-Wiedemann syndrome (Matsuoka et al. 1995) characterized by organ overgrowth and embryonic tumors. This locus is imprinted, and the maternal allele is expressed. Targeted disruption of p57 demonstrates that it is critical for endochondral ossification and development of the palate (Yan et al. 1997; Zhang et al. 1997). Most of p57-deficient mice die perinatally within the first day, displaying short limbs, a cleft palate, and an inflated gastrointestinal tract. Defective or delayed differentiation of chondrocytes in the epiphysis of longitudinal bones may be responsible for the short limb phenotype, presumably causing a decrease in the space in which ossification occurs. This hypothesis is supported by the observations that hypertrophic chondrocytes undergoing differentiation in wild-type bones express high levels of p57, and intramembranous ossification that does not rely on the chondrocyte function is not affected in the mutants. Consequently, p57, like p27, has a phenotype wherein the progenitor cell remains in the cycle, thus perturbing differentiation; however, in contrast to p27, this occurs in a limited number of tissues.

In the normal intestine, p57 expression is restricted to differentiated epithelial cells of the apical region of the villi as well as smooth muscle cells of the intestinal wall. Development of the palate and gastrointestinal tract may require p57-dependent control of proliferation in epithelial and mesenchymal cells, although the exact mechanism is not clear at present. In addition, increased apoptosis is observed in the palate of the mutant animals, leading to the speculation that loss of p57 function delays the decision of cells to exit from the cell cycle (possibly due to increased proliferation) with the result that they are in a state of conflicting signals and choose to die.

Finally, like p21, human p57 has a domain at the C-terminus that contains residues appropriately spaced for interaction with PCNA (Gulbis et al. 1996). The crystal structure of PCNA/p21 indicates the importance of some p21 residues and in human p57, some appropriately spaced residues can be identified. Consequently, it is tempting to speculate that human p57 may also interact with PCNA.

8
p16, Rate-Limiting for Proliferation and Required to Maintain G0

Tumors often develop from non-proliferative cells, suggesting that a key component of the tumorigenic process is the inability to maintain the non-proliferative state. The cyclin D-specific inhibitor p16 is a bona fide tumor suppressor gene. In a large body of tumor-derived cell lines and clinical tumors, inactivation of p16 is accomplished by mutations, deletions, or methylation (Hirama and Koeffler 1995). Furthermore, spontaneous tumorigenesis observed in p16 knockout mice supports the tumor-suppressive function of this molecule, with some caveats.

p16-deficient mice are viable without major defects in development, but often develop spontaneous tumors, predominantly lymphoma and fibrosarcoma (Serrano et al. 1996). Increased sensitivity to DMBA- and UV-induced carcinogenesis also suggests a major role for p16 in tumor promotion. Interestingly, the spectrum of tumors in p16-/- mice is different from that of human tumors with alterations of the p16 gene: melanoma, glioblastoma, esophageal cancer, pancreatic cancer, and acute lymphoblastic leukemia. This may be related to the observation that deletion of the p16 gene in human tumors is often, but not always, accompanied by deletion of $p15^{Ink4B}$ located only 25 kb from the p16 gene. Phenotypes of targeted disruption of p15 have not been reported yet, and it is unknown if double deficiency in p16 and p15 in mice will cause a different spectrum of tumors than observed in p16-/- mice. In addition, it is important to note that the p16 locus encodes two different transcripts from different reading frames, p16 and p19ARF (Quelle et al. 1995a). Both p16 and p19ARF are growth suppressive in vitro, and both have been disrupted in the p16-/- mice; thus, the contribution of p19ARF deficiency to tumorigenesis in the p16 knockout mice remains to be clarified.

Supporting a role for p16 in maintaining quiescence, progressive accumulation of p16 in primary cells undergoing terminal cell divisions after many passages suggested a role for p16 in senescence (Hara et al. 1996). This is supported by the results that fibroblasts obtained from p16-deficient mice fail to undergo senescence in vitro. These cells grow more rapidly in culture than wild-type cells, suggesting a role for p16 in determination of G1 duration. Moreover, fibroblasts isolated from p16-deficient mice are transformed by Ha-ras (Serrano et al. 1996), indicating that the loss of p16 is functionally equivalent to the activation of myc, an immortalizing oncogene, at least with respect to cell transformation.

Finally, p16 may also affect the entry of cells into S phase by virtue of its ability to antagonize cyclin D. Consistent with this, p16-/- mouse embryonic fibroblasts (MEFs) have a shorter doubling time.

9 Phase Models for the Relationship Between CKI and Various Cell Cycle Transitions.

No CKI discussed had a single transition function, at least as we could experimentally define, thus we suggest that the cell cycle is more similar to a phase diagram along which the cell exists in a particular state where two environmen-

Fig. 5. Phase diagram of the CKIs. We propose that there are three classes of CKIs, exemplified by p16, p21 and p27, which regulate the state of the cell. p16-type inhibitors (inks) operate both in cycling cells and in cell entering or exiting the cell cycle. p21-type inhibitors are important in cycling cells. p27-type inhibitors operate both in cycling cells and in cells withdrawing from the cycle

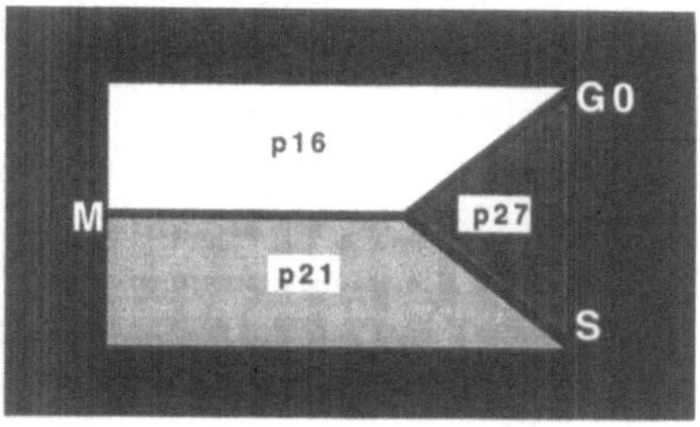

tally controlled conditions come together. Thus, we advance that the CKI proteins should be divided into three "prototypical" functions on the basis of their properties currently defined in mouse knockout studies and, among the Cip/Kip family, the conservation of the FY motif at the carboxyl end of the cyclin/CDK inhibition domain (Fig. 3).

In Fig. 5, we divided the G1 phase of the cell cycle into a phase diagram. We suggest that the cells reside in each "state," and each state is affected predominantly, although not exclusively, by a particular CKI type. Thus, Inks, exemplified by p16, play an important role in G1 to G0 and G0 to G1. Kips – more similar to p21 than p27 – play an important role in cycling cells those going from G1 to S. Finally, those Kip – more similar to p27 than p21– play a role in G1 to S as well as in withdrawing cells from the cell cycle in G1 to G0. However, as any model, this one carries the seeds of its own destruction, and from that destruction we hope to be able to rebuild to a better understanding of the mitotic cell cycle.

10
The Call of Developmental Biology – A Cell Cycle with a Twist

Cell cycle analysis during early development introduces a unique perspective to that derived from studies in cell culture (Edgar and Lehner 1996). The need to coordinate massive cell expansion with morphogenesis and differentiation in organisms raises the following question: how is proliferation halted to allow patterning to occur? In metazoans, the cell cycle begins with simple oscillations between replication and division, and as development proceeds, proliferation undergoes spatially and temporally defined restrictions in coordination with morphogenesis. Ultimately, differentiation programs ensue and cells cease proliferating. In contrast, early development in the mouse starts more slowly, perhaps to gain time for the uterine tissue to prepare for implantation. The mouse embryo divides relatively slowly without any increase in mass until E4.5. Once implantation occurs, growth and proliferation are dramatically increased as the developing fetus begins gastrulation and continues onwards to organogenesis. It is at this early time that size regulation is first observed. Thus, fused blastomeres do not develop into supersize mice because cell number can be "normalized" if necessary. However, the signal leading to size regulation is currently unknown, nor do we understand the mechanism by which cell number is restored (i.e., lack of proliferation or increase in cell death). Later, if cell number is limiting or inappropriately low, size regulation occurs again – this time increasing the number of cells. Again, we know little about the signals and mechanisms, but it is reasonable to speculate that regulation has to do with controlling proliferation or apoptosis in some manner. The embryonic program of cell proliferation presents a number of events that may rely on alternative modes of cell cycle regulation. For simplicity, we categorize these into three different types of proliferative cycles that we believe provide exciting prospects for future studies. first, there is the endoreduplicative cycle of trophectoderm which could arguably be the first differentiation event in the mouse embryo; second, during gastrulation there is a burst of rapid proliferation; and third, the transition of germ cells from mitosis to meiosis requires two subsequent divisions with no possibility of DNA replication.

11
The Importance of CKI in Drosophilia Development

Recent studies in *Drosophila* that identified dacapo (dap), a CKI of the Cip/Kip class, suggest a role for these proteins in early embryonic development (de Nooj et al. 1996; Lane et al. 1996). The dap gene encodes a novel *Drosophila* CKI bearing limited similarity to Kip family members in the N-terminal cyclin/CDK binding domain, and biochemical analyses have confirmed that dap binds and inhibits cyclin E/CDK2 complexes. Dap function is essential for normal embryonic development; transient high levels of dap expression are observed when cells exit from the mitotic cycle. In the epidermis and PNS of dap mutant embryos, cells fail to arrest in a timely fashion and progress through a complete additional cell cycle before withdrawing. Thus, dap appears to be necessary to achieve exit from the cell cycle at precisely the right stage during development. Without arrest the embryo has diminished viability. A failure to arrest cell proliferation at the correct developmental stage has also been described for the *C. elegans* mutant *cul-1* (Kipreos et al. 1996); null mutations cause hyperplasia of all larval tissues reminiscent of the hyperproliferative effect seen in the adult organs of p27-deficient mice.

G1 arrest of the epidermal and PNS cells is dependent on the timely inactivation of cyclin E/CDK2 activity (Knoblich et al. 1994). This is achieved by decreasing cyclin E transcription and increasing dap transcription. Although the downregulation of cyclin E occurs in dap mutants, low levels of cyclin E remain after the final mitotic division. Similar amounts are seen in wild-type embryos, suggesting that the activity of the residual cyclin E-associated kinase is sufficient to trigger an additional cycle in the absence of dap. Thus, redundant pathways and not simply the substitution of one inhibitor for another might explain why deficiencies of the mammalian CDK inhibitors in knockout mice have not demonstrated severe embryonic lethality singly, but may do so in combination.

The essential requirement for the negative regulators of cell proliferation during the embryonic development of the lower metazoans *Drosophila* and *C. elegans* strongly suggests that the three additional types of cell cycle we postulate in early mouse development will be affected by CKIs. A closer examination of these early proliferative cycles may provide further insight as to how CKI may play a role in the temporally controlled execution of these cell cycle transitions at the correct developmental stages.

12 Trophoblast Endoreduplication

One of the first differentiation events to occur during embryogenesis is the distinction of the trophoblast cells from the inner cell mass at the 16 cell stage. The embryo is mostly derived from the descendants of the inner cell mass, whereas the external cells, the trophoblasts or trophectoderm, produce no embryonic structures (Gardner 1983; Hogan et al. 1994). These cells form the embryonic contribution to the placenta and chorion. These cell types form during or following compaction of the embryo and are complete when the embryo forms a blastocyst or a hollow trophectoderm vessel surrounding a fluid cavity (the blastocoel) and a few cells, the inner cell mass. The trophectoderm looks like an epithelial cell layer complete with an apical and basal aspect. Furthermore, there is a distinction among the trophectoderm cells – the ones aligning the blastocoel are defined as mural trophectoderm, and the ones adjacent to the inner cell mass are the polar trophectoderm. We are concerned with the mural trophectoderm. These cells cease dividing, but continue to replicate their genomes. These endoreduplicating cells can polytene chromosomes containing 1000 times the DNA as a diploid cell. Interestingly, polar trophectoderm does not undergo endoreduplication, and in the absence of inner cell mass, trophectoderm cells become hyperploid, suggesting that being adjacent to the inner cell mass suppresses endoreduplication. It is not yet clear what signals are involved in this event, nor how the cell cycle is regulated in this manner in mammalian systems.

The involvement of CKI in endoreduplication is consistent with the observation that the product of the *Schizosaccharomyces pombe* rum1 locus is a potent B-type cyclin inhibitor (Correa-Bordes and Nurse 1995). When rum1 is overexpressed cells undergo successive S phases without intervening mitosis (Moreno and Nurse 1994). During G1 phase, the rum1 protein suppresses residual cyclin B/CDC2 activity, preventing the cell from undergoing mitosis until the DNA is replicated. Thus rum1, an inhibitor of a CDK activity, regulates the activation state of the B-type cyclins and can affect the ordering of S and M phases (Labib et al. 1995). This suggests that an analogous situation might exist in mammalian cells.

13 Rapid Proliferative Cycles During Gastrulation

Gastrulation in the mouse takes place at day 6.5 of gestation when the somatic cell and germ cell lineages are segregated from the epiblast to give rise to three germ layers and primordial germ cells. Unlike most lower vertebrates, mouse gastrulation is associated with rapid cell proliferation. Both counting of total cell number and mitotic index show that epiblast cells divide extremely rapidly between embryonic days 5.5 and 7.5 (Snow 1977). A wild-type embryo at E5.5 contains 120 epiblast cells, whereas at E6.5 and E7.5 this number increases to 660 and 14290, respectively. A decrease in cell cycle length from 11.5 h at E5.5 to 6.5 h at E6.5 accounts for this. Histological determination has demonstrated that this growth is not uniform (Snow 1977), only a small region, about 10 % of the whole epiblast, is designated as the "proliferative zone." In this region the cell generation time is 2–3 h, which is significantly shorter than the 6.5 h cell cycle for regions of the epiblast outside of the proliferative zone. Cells generated in the proliferative zone constitute the ectoderm of later stage embryos.

An intriguing characteristic of cell proliferation during gastrulation is the sudden onset of rapid cell division that persists for 24 h. We know neither the mechanism responsible for initiating the sudden proliferative burst, nor what regulatory mechanisms ensure that it does not persist longer than necessary for proper development. However, there is some hint that BRCA1 might be involved. Homozygous *BRCA1* mutant mice have been shown to die before day 7.5 of embryogenesis (Hakem et al. 1996). Reduced cell proliferation in these mutant mice and impaired cell growth of mutant blastocysts in vitro suggest a positive role in the control of cell growth in the epiblast during gastrulation. Furthermore, the amount of mdm2, a negative regulator of p53 (i.e., a positive regulator of p21), is reduced in *BRCA1* mutant mice. Consistent with this, the expression of p21 increases in E4 and E6.5 *BRCA1* mutants, and DNA replication as evidenced by BrdU incorporation is decreased. These data suggest that the imposition of a p21-mediated cell cycle block may lead to embryonic lethality during gastrulation. An analysis of p21 expression patterns, as well as of the remaining CDK inhibitors in epiblasts derived from wild-type gastrulating embryos might help define the nature of cell cycle regulation at this early stage of mouse development.

14 Primordial Germ Cells: Transition From Mitosis to Meiosis

The embryonic precursors of the ovum and sperm of the adult mouse are called primordial germ cells (PGCs) (Buehr 1997). Establishment of the germ line of the mouse occurs during embryogenesis and involves the segregation of PGCs from the somatic lineages of the epiblast. Following segregation, proliferation of PGCs is coincident with migration to the genital ridge, wherein they undergo further proliferation and eventually differentiation. The cells of the genital ridge are precursors to the adult gonad. PGCs arise from the epiblast at post-implantation day 7.5 and are first detectable as a small migratory population of approximately eight cells distinguished by their high expression levels of alkaline phosphatase (Falconer and Avery 1978; Gardner et al. 1985; Hahnel et al. 1990). Alkaline phosphatase staining is a useful marker to follow the migration pathway of the PGCs from the base of the allantois to the genital ridge (Clark and Eddy 1975; Eddy et al. 1981; Eddy and Hahnel 1983). At E8.5 the small group of PGCs begins to disperse and proliferate as it begins its migratory path through the yolk sac along the epithelium of the hindgut, traversing the dorsal mesentery to finally colonize the genital ridge. The first PGCs reach the genital ridge by E11.5 and continue to proliferate with a doubling time of approximately 16 h (Tam and Snow 1981), until E13.5 when their cell number has expanded to about 25000.

The proliferation of the primordial germ cells during migration to the genital ridge as well as their continued proliferation upon colonization of the genital ridge provides an ideal model system to study mechanisms of cell cycle control. The expansion in the size of the testes and ovaries in p27-deficient adult mice implicates a role for the CDK inhibitors in regulation of PGC number during proliferation and maturation. A striking feature of PGC development is the sex dependent fates of PGC proliferation – when does the switch to meiosis occur? Female germ cells cease proliferation and enter meiosis, becoming arrested at prophase of the first meiotic division during embryogenesis. In contrast, male germ cells entering prospective testes continue to proliferate until mitotic arrest occurs at E14.5. The differences in the signals and control mechanisms which direct PGCs in the female gonad to enter the meiotic cycle and male PGCs to simply undergo mitotic arrest are unclear. What regulatory mechanisms limit the number of mitotic divisions that take place before male PGCs arrest, and what mechanisms lead to the meiotic transition in female PGCs? Are these mechanisms intrinsic to the PGCs or mediated by the gonadal environment? Do CKI play a significant role in any of these events during embryonic germ cell development? The multitude of unanswered questions arising from cell cycle transitions limited to embryogenesis holds exciting promise for future studies.

Acknowledgement We like to thank S. Sean Millard for reading this manuscript during its evolution. Furthermore, we apologize to those whose work is not cited in the review due to space constraints. Work in this laboratory is supported by the NIH, Memorial Sloan-Kettering Cancer Center, and the Pew Foundation.

References

Adams PD, Sellers WR, Sharma SK, Wu AD, Nalin CM, Kaelin WG (1996) Identification of a cyclin-cdk2 recognition motif present in substrates and p21-like cyclin-dependent kinase inhibitors. Mol. Cell Biol. 16: 6623–6633

Agrawal D, Hauser P, McPherson F, Dong F, Garcia A, Pledger WJ (1996) Repression of p27^{Kip1} synthesis by platelet-derived growth factor in BALB/c 3T3 cells. Mol. Cell Biol. 16: 4327–4336

Brugarolas J, Chandrasekaran C, Gordon JI, Beach D, Jacks T, Hannon GJ (1995) Radiation-induced cell cycle arrest compromised by p21 deficiency. Nature 377: 552–557

Buehr M (1997) The primordial germ cells of mammals: some current perspectives. Exp Cell Res 232: 194–207

Casaccia-Bonnefil P, Tikoo R, Kiyokawa H, Friedrich V, Chao MV, Koff A (1997) Oligodendrocyte precursor differentiation is perturbed in the absence of the cyclin-dependent kinase inhibitor p27Kip1. Genes Dev 11: 2335–2346

Chan FKM, Zhang J, Cheng L, Shapiro DN, Winoto A (1995) Identification of human and mouse p19, a novel CDK4 and CDK6 inhibitor with homology to p16^{ink4}. Mol Cell Biol 15: 2682–2688

Chen IT, Akamatsu M, Smith ML, Lung FDT, Duba D, Roller PR, Fornace AJ, O'Connor PM (1996) Characterization of p21 (Cip1/Waf1) peptide domains required for cyclin E/cdk2 and PCNA interaction. Oncogene 12: 595–607

Chen J, Jackson PK, Kirschner MW, Dutta A (1995) Separate domains of p21 involved in the inhibition of Cdk kinase and PCNA. Nature 374: 386–388

Chen J, Saha P, Kornbluth S, Dynlacht BD, Dutta A (1996) Cyclin-binding motifs are essential for the function of p21^{CIP1}. Mol Cell Biol 16: 4673–4682

Clark JM, Eddy EM (1975) Fine structural observations on the origin and association of primordial germ cells in the mouse. Dev Biol 47: 136–155

Coats S, Flanagan WM, Nourse J, Roberts JM (1996) Requirement of p27Kip1 for restriction point control of the fibroblast cell cycle. Science 272: 877–880

Correa-Bordes J, Nurse P (1995) p25rum1 orders S phase and mitosis by acting as an inhibitor of the p34cdc2 mitotic kinase. Cell 83: 1001–1009

Cross FR (1988) DAF1, a mutant gene affecting size control, pheromone arrest, and cell cycle kinetics of *Saccharomyces cerevisiae*. Mol Cell Biol 8: 4675–4684.

de Nooj JC, Letendre MA, Kariharan IK (1996) A cyclin-dependent kinase inhibitor, Dacapo, is necessary for timely exit from the cell cycle during *Drosophila* embryogenesis. Cell 87: 1237–1247

Deng C, Zhang P, Harper JW, Elledge SJ, Leader P (1995) Mice lacking p21$^{CIP1/WAF1}$ undergo normal development, but are defective in G1 checkpoint control. Cell 82: 675–684

Eddy EM, Hahnel AC (1983) Establishment of the germ cell line in mammals. In: (eds) McLaren A, Wylie CC. Current problems in germ cell differentiation. Cambridge University Press, Cambridge, pp 41–70

Eddy EM, Clark JM, Gong D, Fenderson BA (1981) Origin and migration of primordial germ cells in mammals. Gamete Res 4: 333–362

Edgar B, Lehner CF (1996) Developmental control of cell cylce regulators: a fly's perspective. Science 274: 1646–1652

El-Deiry WS, Tokino T, Velculescu VE, Levy DB, Parsons R, Trent JM, Lin D, Mercer WE, Kinzler KW, Vogelstein B (1993) WAF1, a potent mediator of p53 tumor suppression. Cell 75: 817–825

Ewen ME, Sluss HK, Whitehouse LL, Livingston DM (1993) TGF beta inhibition of Cdk4 synthesis is linked to cell cycle arrest. Cell 74 : 1009–1020

Falconer DS, Avery PJ (1978) Variability of chimeras and mosaics. J Embryol Exp Morphol 43: 195–219

Fero MI, Rivkin M, Tasch M, Porter P, Carow CE, Firpo E, Polyak K, Tsai L-H, Broudy V, Perlmutter RM, Kaushansky K, Roberts JM (1996) A syndrome of multi-organ hyperplasia with features of gigantism, tumorigenesis and female sterility in p27Kip1-deficient mice. Cell 85: 733–744

Firpo EJ, Koff A, Soloman MJ, Roberts JM (1994) Inactivation of a Cdk2 inhibitor during interleukin 2-induced proliferation of human T lymphocytes. Mol Cell Biol 14: 4889–4901

Florenes VA, Bhattacharya N, Bani MR, Ben-David Y, Kerbel RS, Slingerland JM (1996) TGF-β mediated G1 arrest in a human melanoma cell line lacking p15^{INK4B}: evidence for cooperation between p21$^{Cip1/WAF1}$ and p27^{Kip1}. Oncogene 13: 2447–2457

Franklin DS, Xiong Y (1996) Induction of p18INK4c and its predominant association with CDK4 and CDK6 during myogenic differentiation. Mol Biol Cell 7: 1587–1599

Gardner RL (1983) Origin and differentiation of extraembryonic tissues in the mouse. Int Rev Exp Pathol 24: 63–133

Gardner RL, Lyon MF, Evans EP, Burtenshaw MD (1985) Conal analysis of X-chromosome inactivation and the origin of the germ line in the mouse embryo. J Embryol Exp Morphol 88: 349–363

Gray-Bablin J, Rao S, Keyomarsi K (1997) Lovastatin induction of cyclin-dependent kinase inhibitors in human breast cells occurs in a cell cycle-independent fashion. Cancer Res 57: 604–609

Gu Y, Turck CW, Morgan DO (1993) Inhibition of CDK2 activity in vivo by an associated 20K regulatory subunit. Nature 366: 707–710

Guan KL, Jenkins CW, Li Y, Nichols MA, Wu X, O'Keefe CL, Matera AG, Xiong Y (1994) Growth suppression by p18, a p16$^{INK4/MTS1}$- and p14$^{INK4B/MTS2}$-related CDK6 inhibitor, correlates with wild-type pRb function. Genes Dev 8: 2939–2952

Guan KL, Jenkins CW, Li Y, O'Keefe CL, Noh S, Wu X, Zariwala M, Matera AG, Xiong Y (1996) Isolation and characterization of p19^{INK4d}, a p16-related inhibitor specific to CDK6 and CDK4. Mol Biol Cell 7: 57–70

Gulbis JM, Kelman Z, Hurwitz J, O'Donnell M, Kuriyan J (1996) Structure of the C-terminal region of p21(WAF1/CIP1) complexed with human PCNA. Cell 87: 297–306

Hahnel AC, Rappolee DA, Millan JL, Manes T, Ziomek CA, Theodosiou NG, Werb Z, Pederson RA, Schultz GA (1990) Two alkaline phosphatase genes are expressed during early development in the mouse embryo. Development 110: 555–564

Hakem R, de la Pompa JL, Sirard C, Mo R, Woo M, Hakem A, Wakeham A, Potter J, Reitmair A, Billia F, Firpo E, Hui CC, Roberts J, Rossant J, Mak TW (1996) The tumor suppressor gene *Brca1* is required for embryonic cellular proliferation in the mouse. Cell 85: 1009–1023

Halevy O, Novitch BG, Spicer DB, Skapek SX, Rhee J, Hannon GJ, Beach D, Lassar AB (1995) Correlation of terminal cell cycle arrest of skeletal muscle with induction of p21 by MyoD. Science 267: 1018–1021

Hannon GJ, Beach D (1994) p15^{INK4B} is a potential effector of TGF-b-induced cell cycle arrest. Nature 371: 257–261

Hara E, Smith R, Parry D, Tahara H, Stone S, Peters G (1996) Regulation of p16^{CDKN2} expression and its implications for cell immortalization and senescence. Mol Cell Biol 16: 859–867.

Harper JW, Adami GR, Wei N, Keyomarsi K, Elledge SJ (1993) The p21 Cdk-interacting protein Cip1 is a potent inhibitor of G1 cyclin-dependent kinases. Cell 75: 805–816

Harper JW, Elledge SJ, Keyomarsi K, Dynlacht B, Tsai LH, Zhang P, Dobrowolski S, Bai C, Connell-Crowley L, Swindell E, Fox MP, Wei N (1995) Inhibition of cyclin-dependent kinases by p21. Mol Biol Cell 6: 387–400

Hatekayama H, Weinberg RA (1995) The role of Rb in cell cycle control. In: Meijier L, Guidet S, and Tung HYL (eds) Progress in cell cycle research, vol. 1, Plenum Press, New York, pp 9–19

Hauser PJ, Agrawal D, Flanagan M, Pledger WJ (1997) The role of p27^{Kip1} in the in vitro differentiation of murine keratinocytes. Cell Growth Diff 8: 203–211

Hirai H, Roussel MF, Kato JY, Ashmun RA, Sherr CJ (1995) Novel INK4 proteins, p19 and p18, are specific inhibitors of the cyclin D-dependent kinases CDK4 and CDK6. Mol Cell Biol 15: 2672–2681

Hirama T, Koeffler HP (1995) Roles of the cyclin-dependent kinase inhibitors in the development of cancer. Blood 86: 841–854

Hoffman ES, Passoni L, Crompton T, Leu TMJ, Schatz DG, Koff A, Owen MJ, Hayday AC (1996) Productive T-cell receptor β-chain gene rearrangement: coincident regulation of cell cycle and clonality during development in vivo. Genes Dev 10: 948–962

Hogan B, Beddington R, Constantini F, Lacy E. (1994) Manipulating the mouse embryo: a laboratory manual. Cold Spring Harbor Laboratory Press, Cold Spring Harbor, NY
Kamb A, Shattuck-Eidens D, Eeles R, Liu Q, Gruis NA, Ding W, Hussey C, Tran T, Miki Y, Weaver-Feldhaus J, et al. (1994) Analysis of the p16 gene (CDKN2) as a candidate for the chromosome 9p melanoma susceptibility locus. Nature Genet 8: 23–26
Kato J, Matsushime H, Hiebert SW, Ewen ME, Sherr CJ (1993) Direct binding of cyclin D to the retinoblastoma gene product (pRb) and pRb phosphorylation by the cyclin D-dependent kinase CDK4. Genes Dev 7: 331–342
Kipreos ET, Lander LE, Wing JP, Wu HW, Hedgecock EM (1996) *cul-1* is required for cell cycle exit in *C. elegans* and identifies a novel gene family. Cell 85: 829–839
Kiyokawa H, Kineman RD, Manova-Todorova KO, Soares VC, Hoffman E, Ono M, Khanam D, Hayday AC, Frohman LA, Koff A (1996) Enhanced growth of mice lacking the cyclin-dependent kinase inhibitor function of $p27^{Kip1}$. Cell 85: 721–732.
Knoblich JA, Sauer K, Jones L, Richardson H, Saint R, Lehner CF (1994) Cyclin E controls S phase progression and its down-regulation during *Drosophila* embryogenesis is required for the arrest of cell proliferation. Cell 77: 107–120
Koff A, Ohtsuki M, Polyak K, Roberts JM, Massaque J (1993) Negative regulation of G1 in mammalian cells: inhibition of cyclin E-dependent kinase by TGF-β. Science 260: 536–538
LaBaer J, Garrett MD, Stevenson LF, Slingerland JM, Sandhu C, Chou HS, Fattaey A, Harlow E (1997) New functional activities for the p21 family of CDK inhibitors. Genes Dev 11:847–862
Labib K, Moreno S, Nurse P (1995) Interaction of cdc2 and rum 1 regulates Start and S phase in fission yeast. J Cell Sci 108: 3285–3294
Lane ME, Saeur K, Wallace K, Jan JY, Lehner CF, Vaessin H (1996) Dacapo, a cyclin-dependent kinase inhibitor, stops cell proliferation during *Drosophila* development. Cell 87: 1225–1235
Lee MH, Reynisdottir I, Massague J (1995) Cloning of $p57^{KIP2}$, a cyclin-dependent kinase inhibitor with unique domain structure and tissue distribution. Genes Dev 9:639–649
Luo Y, Hurwitz J, Massague J (1995) Cell-cycle inhibition by independent CDK and PCNA binding domains in $p21^{Cip1}$. Nature 375: 159–161
Luo Y, Marx SO, Kiyokawa H, Koff A, Massague J, Marks A (1996) Rapamycin resistance tied to defective regulation of $p27^{Kip1}$. Mol Cell Biol 16: 6744–6751
Matsuoka S, Edwards MC, Bai C, Parker S, Zhang P, Baldini A, Harper JW, Elledge SJ (1995) $p57^{KIP2}$, a structurally distinct member of the $p21^{CIP1}$ Cdk inhibitor family, is a candidate tumor suppressor gene. Genes Dev 9: 650–662
Medema RH, Herrera RE, Lam F, Weinberg RA (1995) Growth suppression by p16ink4 requires functional retinoblastoma protein. Proc Natl Acad Sci USA 92: 6289–6293
Missero C, Di Cunto F, Kiyokawa H, Koff A, Dotto GP (1996) The absence of p21Cip1/WAF1 alters keratinocyte growth and differentiation and promotes ras-tumor progression. Genes Dev 10: 3065–3075
Moreno S, Nurse P (1994) Regulation of progression through the G1 phase of the cell cycle by the $rum1^{+}$ gene. Nature 367: 236–242
Nakayama K, Ishida N, Shirane M, Inomata A, Inoue T, Shishido N, Horii I, Loh DY, Nakayama KI (1996) Mice lacking p27Kip1 display increased body size, multiple organ hyperplasia, retinal dysplasia, and pituitary tumors. Cell 85: 707–720
Noda A, Ning Y, Venable SF, Pereira-Smith OM, Smith JR (1994) Cloning of senescent cell-derived inhibitors of DNA synthesis using an expression screen. Exp Cell Res 211: 90–98
Ohtusbo M, Roberts JM (1993) Cyclin-dependent regulation of G1 in mammalian fibroblasts. Science 259: 1908–1912
Pan ZQ, Reardon JT, Li L, Flores-Rozas H, Legerski R, Sancar A, Hurwitz J (1995) Inhibition of nucleotide excision repair by the cyclin-dependent kinase inhibitor p21. J Biol Chem 270: 22008–22016
Parker SB, Eichele G, Zhang P, Rawls A, Sands AT, Bradley A, Olson EN, Harper JW, Elledge SJ (1995) p53-independent expression of $p21^{Cip1}$ in muscle and other terminally differentiating cells. Science 267: 1024–1027

Polyak K, Kato JY, Soloman MJ, Sherr CJ, Massague J, Roberts JM, Koff A (1994a) $p27^{KIP1}$, a cyclin-Cdk inhibitor, links transforming growth factor-b and contact inhibition to cell cycle arrest. Genes Dev 8: 9–22

Polyak K, Lee MH, Erdjument-Bromage H, Koff A, Roberts JM, Tempst P, Massague J (1994b) Cloning of $p27^{KIP1}$, a cyclin-dependent kinase inhibitor and a potential mediator of extracellular antimitogenic signals. Cell 78: 59–66

Poon RYC, Toyshima H, Hunter T (1995) Redistribution of the CDK inhibitor p27 between different cyclin·CDK complexes in the mouse fibroblast cell cycle and in cells arrested with lovastatin or ultraviolet irradiation. Mol Biol Cell 6: 1197–1213

Quelle DE, Zindy F, Ashmun RA, Sherr CJ (1995a) Alternative reading frames of the INK4a tumor suppressor gene encode two unrelated proteins capable of inducing cell cycle arrest. Cell 83: 993–1000.

Quelle DE, Ashmun RA, Hannon GJ, Rehberger PA, Trono D, Richter KH, Walker C, Beach D, Sherr CJ, Serrano M (1995b) Cloning and characterization of murine p16INK4a and p15INK4b genes. Oncogene 11: 635–645

Reynisdottir I, Massague J (1997) The subcellular locations of $p15^{Ink4b}$ and $p27^{Kip1}$ coordinate their inhibitory interactions with cdk4 and cdk2. Genes Dev 11: 492–503

Reynisdottir I, Polyak K, Iavarone A, Massague J (1995) Kip/Cip and Ink4 Cdk inhibitors cooperate to induce cell cycle arrest in response to TGF-beta. Genes Dev 9: 1831–1845

Russo AA, Jeffrey PD, Patten AK, Massague J, Pavletich NP (1996) Crystal structure of the $p27^{Kip1}$ cyclin-dependent-kinase inhibitor bound to the cyclin A-Cdk2 complex. Nature 382: 325–331

Serrano M, Hannon GJ, Beach D (1993) A new regulatory motif in cell-cycle control causing specific inhibition of cyclinD/CDK4. Nature 366: 704–707

Serrano M, Lee H-W, Chin L, Cordon-Cardo C, Beach D, DePinho RA (1996) Role of the INK4a locus in tumor suppression and cell mortality. Cell 85: 27–37

Sherr CJ, Roberts JM (1995) Inhibitors of mammalian G1 cyclin-dependent kinases. Genes Dev 9: 1149–1163

Snow MHL (1977) Gastrulation in the mouse: growth and regionalization of the epiblast. J Embryol Exp Morphol 42: 293–303

Soos TJ, Kiyokawa H, Yan JS, Rubin MS, Giordano A, DeBlasio A, Bottega S, Wong B, Mendelsohn J, Koff A (1996) Formation of p27-CDK complexes during the human mitotic cell cycle. Cell Growth Diff 7: 135–146

Tam PPL, Snow MHL (1981) Proliferation and migration of primordial germ cells during compensatory growth in mouse embryos. J Embryol Exp Morphol 64: 133–147

Tam SW, Shay JW, Pagano M (1994) Differential expression and cell cycle regulation of the cyclin-dependent kinase 4 inhibitor $p16^{Ink4}$. Cancer Res 54: 5816–5820

Toyoshima H, Hunter T (1994) p27, a novel inhibitor of G1 cyclin-Cdk protein kinase activity, is related to p21. Cell 78: 67–74

Waga S, Hannon GJ, Beach D, Stillman B (1994) The p21 inhibitor of cyclin-dependent kinases controls DNA replication by interaction with PCNA Nature 369: 574–578

Xiong Y, Hannon GJ, Zhang H, Casso D, Kobayashi R, Beach D (1993a) p21 is a universal inhibitor of cyclin kinases. Nature 366: 701–704

Xiong Y, Zhang H, Beach D (1993b) Subunit rearrangement of the cyclin-dependent kinases is associated with cellular transformation. Genes & Dev. 7: 1572–1583

Yan Y, Frisen J, Lee MH, Massague J, Barbacid M (1997) Ablation of the CDK inhibitor $p57^{Kip2}$ results in increased apoptosis and delayed differentiation during mouse development. Genes Dev 11: 973–983

Zhang H, Xiong Y, Beach D (1993) Proliferating cell nuclear antigen and p21 are components of multiple cell cycle kinase complexes. Mol Biol Cell 4: 897–906

Zhang H, Hannon GJ, Beach D (1994) p21-containing cyclin kinases exist in both active and inactive states. Genes Dev 8:1750–1758

Zhang P, Liegeois NJ, Wong C, Finegold M, Hou H, Thompson JC, Silverman A, Harper JW, DePinho RA, Elledge SJ (1997) Altered cell differentiation and proliferation in mice lacking $p57^{Kip2}$ indicates a role in Beckwith-Wiedemann syndrome. Nature 387: 151–158

Regulation of the Cell Cycle by the Ubiquitin Pathway

J. Slingerland and M. Pagano[1]

1 Introduction

Ubiquitin-mediated proteolysis is a major mechanism of regulation of cellular processes in which speed, specificity and timing are critical. Degradation of key substrates by the ubiquitin pathway controls cell cycle progression, circadian rhythms, cell fate commitment in development, signal transduction pathways, and immune responses (see Hochstrasser 1995 for review).

The order and timing of cell cycle transitions are governed by the sequential activation and inactivation of a series of cyclin-dependent kinases (CDKs) (Multiple Authors 1996). The activity of CDKs is regulated by activating subunits, inhibitory subunits, and by phosphorylation/dephosphorylation events (see also more specific chapters in this book). Both positive regulators of CDKs, the cyclins, and negative regulators, the CDK inhibitors (CKIs), are subject to regulated degradation by the ubiquitin proteolytic pathway which dictates their timed elimination in response to mitogenic and anti-mitogenic stimuli.

2 The Ubiquitin-Mediated Proteolytic Pathway

Polyubiquitination of substrates targets them for degradation by the 26S proteasome, a multiprotein complex conserved from archaebacteria to humans (for a review see Rolfe et al. 1997). Ubiquitin is an evolutionarily highly conserved 76 amino acid polypeptide which is abundant in all eukaryotic cells. The initial step in the ubiquitin pathway is ATP-dependent and involves the linkage of ubiquitin to a ubiquitin-activating enzyme, or E1, in a high energy thioester bond. Ubiquitin is then transferred in a second thioester linkage to a ubiquitin conjugating enzyme (Ubc), or E2, which in turn catalyzes the transfer of ubiquitin to the substrate protein in a covalent bond. In some cases, substrate polyubiquitination requires another enzyme, the ubiquitin ligase, or E3. The ubiquitin ligase can participate in the hierarchic transfer of ubiquitin into the sub-

[1] Department of Pathology and Kaplan Cancer Center, MSB 548, New York University Medical Center, 550 First Avenue, New York, New York 10016, USA,
E-mail: paganm02@mcrcr.med.nyu.edu

strate, or can function as an adaptor to facilitate positioning and transfer of ubiquitin from the E2 directly onto the substrate (see Fig. 1). A number of E3s have been shown to bind physically to the substrate. Ubiquitination of the target substrate occurs through linkage of the α-carboxyl glycine of ubiquitin to a lysine ε-amino group on the protein substrate. The consecutive addition of ubiquitin moieties to a substrate generates a polyubiquitin chain. The nomenclature of E1, E2, and E3 came from the identification of ubiquitin enzymes in eluates from ubiquitin affinity columns (Hershko et al. 1983). Both E2 and E3 proteins exist as large families and it is thought that different combinations of E2s with different E3 proteins define substrate specificity. At least 13 different E2s have been identified in both *Saccharomyces cerevisiae* and humans. In contrast to the E2s, whose catalytic sites are well conserved among themselves, only some E3 ligases share a few motifs (e.g., the hect domain, the TPR motif, the F box, the WD40 repeat; reviewed in King et al. 1996a). In particular, the hect domain defines a class of E3s that the recent explosion of sequence information from the various organism-based genome efforts indicates to be remarkably large (reviewed by Jentsch 1992; Hochstrasser 1995; Rolfe et al. 1997).

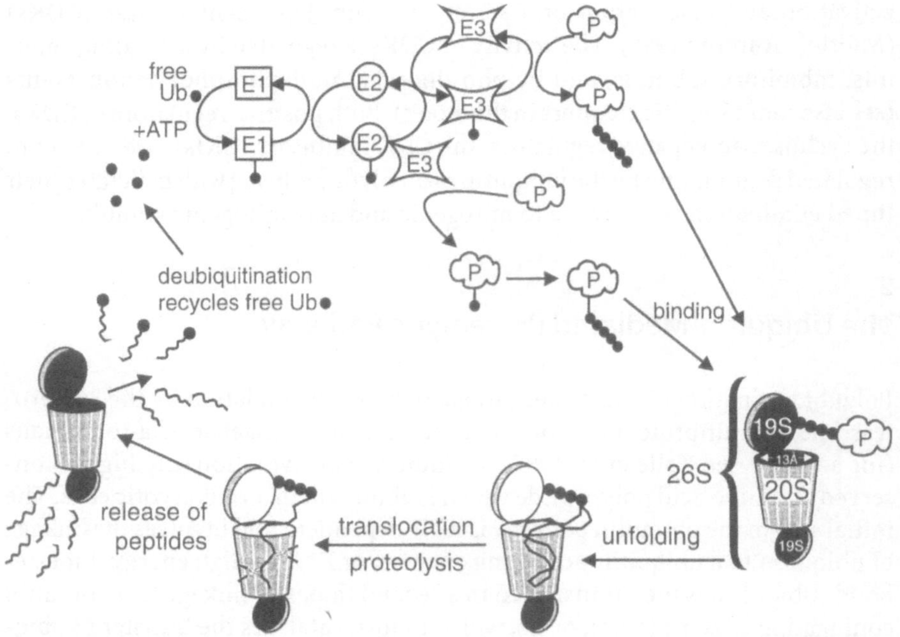

Fig. 1. The ubiquitin-proteasome pathway. Ubiquitin (Ub) is activated by an E1 enzyme in the presence of ATP. It is then transferred to an E2 enzyme (ubiquitin-conjugating enzyme) and finally to an E3 (ubiquitin ligase). This transfers the ubiquitin molecule to the target protein (P). In some cases, the E3 enzyme acts as an adaptor to facilitate positioning and transfer of ubiquitin from the E2 directly onto the substrate. The consecutive addition of ubiquitin moieties to a substrate generates a polyubiquitin chain. Proteins that are modified in this manner are recognized by the proteasome, a 26S multiprotein complex. The substrate P is unfolded and exposed for proteolytic cleavage within the proteasome. Peptides fragments are generated and ubiquitin is recycled by deubiquitinating enzymes. See also text for further details.

The 26S proteasome is composed of a catalytic 20S core of four heptameric rings of α and β subunits stacked into a hollow cylinder (reviewed by Rubin and Finley 1995). Two 19S subunits, also called PA700, contain 700-kDa proteasome activators and are found at the ends of the 20S cylinder (DeMartino et al. 1994; Ma et al. 1994; Peters et al. 1994). The polyubiquitin side chain both targets and tethers substrates to the S5 subunit of the PA700. ATP-dependent unfolding of the substrate allows its translocation through the 13 Å entrance of the 20S catalytic channel (Lowe et al. 1995; Rubin and Finley 1995). This latter translocation may be accompanied by partial disassembly of the ubiquitin chains. Peptidases on the inner surface of β subunits degrade the substrate, releasing ubiquitinated peptides. Ubiquitin is then recycled by the action of deubiquitinating enzymes on these fragments (see Fig. 1).

Another family of enzymes participating in the ubiquitin-proteasome pathway is represented by the deubiquitinating enzymes (dubs), or ubiquitin carboxyl-terminal hydrolases. Over 15 different dubs have been identified in *S. cerevisiae* (Hochstrasser 1995). These remove ubiquitin from substrate proteins and represent an important level of control. Deubiquitinating enzymes govern the rate of recycling of ubiquitin from end products cleaved from the 26S proteasome (Papa and Hochstrasser 1993). In addition, dubs may serve a more specific role by influencing the rate of recognition and degradation of specific substrates by "editing" the number of ubiquitin moieties in the polyubiquitin chain (Baker et al. 1992).

3 Regulation of CDK Activity by the Ubiquitin Pathway

Two ubiquitin-dependent proteolytic pathways that regulate cell division have been well characterized (Fig. 2). In *S. cerevisiae*, the initiation of DNA replication requires ubiquitin-mediated degradation of a CDK inhibitor, Sic1, in a pathway that involves the E2 enzyme Cdc34 and an E3 multiprotein complex which includes Cdc53, Cdc4, and Skp1. Recent studies reviewed herein show that this pathway targets phosphorylated substrates (see Sect. 3. 1). A second pathway that regulates chromosome segregation and mitotic exit by degrading anaphase inhibitors and mitotic cyclins involves a different E2 and a large molecular weight E3 complex called the anaphase promoting complex (APC) or cyclosome. This pathway targets substrates containing one or more destruction box motifs (see Sect. 3. 2).

3.1 G1/S Regulation of CDK Activity by the Ubiquitin Pathway

In *S. cerevisiae*, transition from the G1 to S phase requires the activity of B-type cyclins Clb5 and Clb6 in association with Cdc28 [the yeast CDK1; (Epstein and Cross 1992; Schwob and Nasmyth 1993; Schwob et al. 1994)]. The Clb/Cdc28 ac-

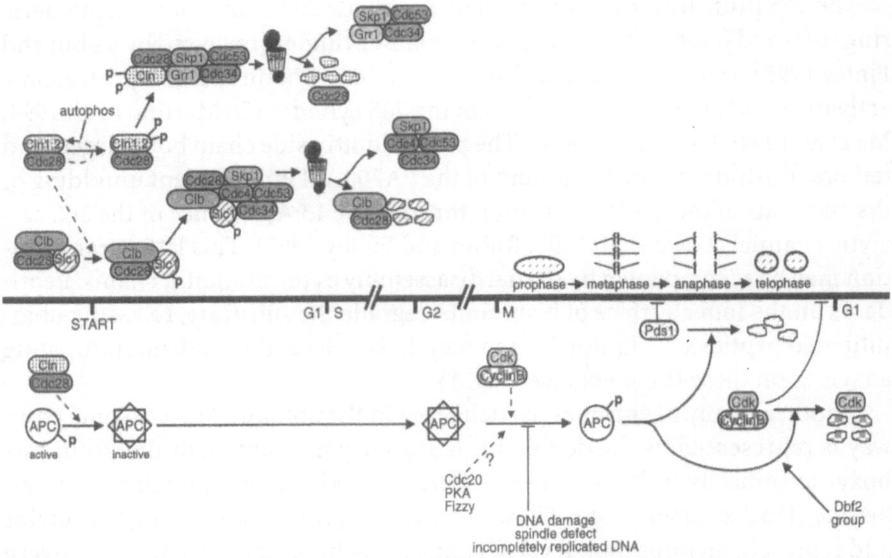

Fig. 2. Proteolytic paths involved in cell cycle control. The phosphorylation of both G1 cyclins, Clns 1 and 2, and of the CDK inhibitor, Sic1, triggers their recognition by multiprotein E2/E3 complexes. The E2, Cdc34, binds an ubiquitin ligase (E3) complex which includes Cdc53, Skp1 and Grr1, for the Clns, and Cdc53, Skp1 and Cdc4, in the case of Sic1. The anaphase promoting complex, or APC, is a multiprotein E3 complex which regulates the G2/M transition. The APC is regulated by proteins which may include cyclin B/CDK1, Cdc20 and PKA. DNA damage, spindle defects and incompletely replicated DNA inhibit APC activation. APC-mediated degradation of Pds1 is required for anaphase progression and APC-mediated degradation of cyclin B is required for the exit from mitosis. APC action on B-type cyclins is regulate by the Dbf2 group of protein kinases. The APC remains active until it is inactivated at START in a Cln/Cdc28-dependent manner. See text for details.

tivities are regulated by the CKI, Sic1 (Mendenhall 1993). *CDC34*, a gene required for the G1 to S phase transition, encodes a ubiquitin conjugating enzyme, namely Ubc3 (Goebl et al. 1994). Cdc34 is needed to generate S phase CDK activity through the degradation of Sic1 (Schwob et al. 1994). Sic1 is degraded at S phase in wild-type yeast cells and accumulates in *cdc34*^{ts} mutant cells (Schwob et al. 1994). Extracts from Cdc34-deficient yeast inhibit Clb5- and Clb6-dependent Cdc28 activities in vitro. The fact that *cdc34*^{ts} *sic1*Δ double mutants can replicate DNA at the non-permissive temperature suggested that ubiquitin mediated proteolysis of Sic1 by Cdc34 is required for S phase entrance (Schwob et al. 1994). Conditional alleles in three other genes have been identified that share the *cdc34*^{ts} phenotype: *CDC53* (Mathias et al. 1996), *CDC4* (Hareford and Hartwell 1974) and *SKP1* (Bai et al. 1996). Biochemical experiments showed that polyubiquitination of Sic1 is sustained by extracts from wild-type cells but not ts *cdc4* mutants, and that extracts from mutant cells are restored by the addition of exogenous Cdc4 protein (Verma et al. 1997b). In addition, purified recombinant Cdc4, Cdc53 and Skp1 in the presence of ubiquitin, Cdc34, ATP and E1 are sufficient for Sic1 polyubiquitination (Feldman et al.

1997; Skowyra et al. 1997). Thus, the complex formed by Cdc4, Cdc53, and Skp1 seems to play a role as an E3 ubiquitin ligase for Sic1. Cdc4 physically interacts with Sic1 (Skowyra et al. 1997) and acts as the catalytic subunit, forming a thioester bond with ubiquitin (P. Zhou and P. Howley, pers. comm.).

Skp1 and Cdc53 are parts of a ubiquitin ligase which binds and targets Cln2 for destruction by the proteasome (Willems et al. 1996). Thus, the E2/E3 complexes involved in the proteolysis of Sic1 and Clns share some components, namely, Cdc34, Cdc53 and Skp1. The substrate specificity and binding are defined by a third subunit (Skowyra et al. 1997): Cdc4 in the case of Sic1 (Bai et al. 1996; Feldman et al. 1997; Skowyra et al. 1997) and Grr1 in the case of Cln2 (Barral et al. 1995; Skowyra et al. 1997). Both Cdc4 and Grr1 interact with Skp1 via a highly conserved "F box" domain, so called because it was first identified in cyclin F, a protein interacting with human Skp1 (Bai et al. 1996; Ning and Johnston 1997). Interestingly, neither Cdc4 nor Grr1 appear to be essential in Cln3 degradation (D. Finley, pers. comm.), suggesting the existence of yet another member of this F box-containing protein group. Cdc34 and some of the subunits of the G1/S ubiquitin ligases are responsible not only for the polyubiquitination of Sic1 and Clns, but also for that of other substrates. For example, Cdc34 (Kornitzer et al. 1994), Skp1, Cdc53, and Cdc4 (D. Kornitzer, pers. comm.) are essential for the degradation of Gcn4, a transcriptional activator of amino acid biosynthesis. Furthermore, Cdc34 plays a role in the degradation of the cell cycle inhibitor Far1 (McKinney et al. 1993), and, together with Cdc4 and Cdc53, is in involved in the degradation Cdc6, a protein necessary for DNA replication (Piatti et al. 1996; Drury et al. 1997). Finally, Skp1 may potentially have a role in addition to those postulated in G1, since it has been found to be part of the kinetochore complex (Connelley and Hieter 1996; Stemmann and Lechner 1996).

Proteolysis at the G1 to S phase transition is important for the integration of CDK activity with environmental signals. Nutrient sensing in yeast and responsiveness to growth factors in mammalian cells occur in G1 and these signals modulate G1 cyclin transcription (reviewed in Sherr 1994). In turn, G1 cyclins (Clns) regulate levels of CKIs. The *S. cerevisiae* Cln/Cdc28 complexes play a critical function to phosphorylate the CDK inhibitor Sic1, thereby targeting it for degradation (Schneider et al. 1996; Skowyra et al. 1997). The periodicity of Cln/Cdc28 activation dictates the timing of Sic1 degradation. A *sic1* deletion can rescue the lethality of a triple *cln* mutant, suggesting that Sic1 inactivation is the only non-redundant function of Clns (Schneider et al. 1996; Tyers 1996). Sic1 has nine CDK phosphorylation consensus sites, and mutation of three of these sites eliminates ubiquitination and causes Sic1 stability (Verma et al. 1997a). Cln2 is also ubiquitinated in a Cln/Cdc28 dependent manner both in vitro and in vivo (Deshaies et al. 1995; Lanker et al. 1996; Willems et al. 1996). Cln/Cdc28 dependent phosphorylation of carboxyl residues on Cln2 mediates recognition by and binding to the Cln2-specific ubiquitin ligase (Willems et al.

1996; Skowyra et al. 1997). Thus, in the current model, the G1 ubiquitin ligases are constitutively active and phosphorylation of their substrates (G1 cyclins, Sic1 and probably others) may create novel epitopes for recognition by the ubiquitin machinery (the E2/E3 complex).

In fission yeast, the G1 to S transition appears regulated by a ubiquitin pathway homologous to that in *S. cerevisiae*. In this yeast, an F box-containing protein, Pop-1, plays a role analogous to that of Cdc4 in the ubiquitination of the fission yeast CKI Rum-1 and of Cdc18, the homologue of *S. cerevisiae* Cdc6 (Kominami and Toda 1997). In addition, a *Schizosaccharomyces pombe CDC53* homologue designated *puc3+* has been identified (T. Toda, pers. comm.).

Several components of the E2/E3 complexes regulating the timed degradation of G1 cyclins and CDK inhibitors appear to be conserved from yeast to humans. Homologues of the *CDC34* gene have been identified in vertebrates (Plon et al. 1993; Yew and Kirschner 1997) and, as in yeast, they seem to play a role in the proteolysis of CKIs [e.g., *Xenopus* Xic1 (Yew and Kirschner 1997) and mammalian p27 (Pagano et al. 1995)]. *SKP1* is also highly conserved through phylogeny in eukaryotes. Mammalian Skp1 binds to cyclins A and F (Zhang et al. 1995; Bai et al. 1996). While the association of Skp1 with cyclin F occurs via the F box in cyclin F, Skp1 binds to cyclin A through another F box-containing protein, Skp-2. Whether Skp1 and Skp2 regulate the degradation of cell cycle regulators in mammalian cells is not known.

CDC53 homologues (cullins) have been identified in *Caenorhabditis elegans* and in humans [cul 1-5; (Kipreos et al. 1996; Pause et al. 1997; Stankovic et al. 1997)]. One of the *C. elegans CDC53* homologues, *cul1*, appears to play an important inhibitory role in cell proliferation, as strongly suggested by a shortened G1 phase and the generalized hyperplasia observed in *cul1* null animals (Kipreos et al. 1996). It remains to be seen whether *CDC53* homologues carry out regulatory roles in the degradation of CKIs and cyclins in higher eukaryotes. The notion that E2/E3 complexes regulating G1 cyclins and CDK inhibitors may be conserved across phylogeny is supported by the high degree of conservation in the components of the mitotic ubiquitin system (see Sect. 3.2). Perhaps the most striking similarity between ubiquitin pathways in yeast and higher eukaryotes is the requirement for phosphorylation in the degradation of G1 regulatory proteins. An increasing body of evidence suggests that just as Sic1 phosphorylation targets the destruction of this inhibitor, the timed phosphorylation of p27 mediates its recognition by the ubiquitin proteasome (Sheaff et al. 1997; Vlach et al. 1997). Similarly, phosphorylation of cyclin E by CDK2 (Clurman et al. 1996; Won and Reed 1996) and of cyclin D1 by a still unknown kinase (Diehl et al. 1997) is required for their degradation.

3.2
Mitotic Regulation of CDK Activity by the Ubiquitin Pathway

The metaphase to anaphase transition requires the activity of an E3 complex called anaphase promoting complex (APC) or cyclosome. The APC catalyzes ubiquitination of destruction box (D box) -containing substrates. The D box motif is a stretch of nine amino acids (RxxLxxIxN) followed by a lysine-rich domain and was first identified in B-type cyclins. Deletion or mutation of the D box substantially stabilizes B-type cyclins (Glotzer et al. 1991; King et al. 1996b). In contrast to G1 substrates, phosphorylation of B type cyclins is not required for their degradation (Izumi and Maller 1991; Li et al. 1995). The B-type cyclins are necessary for entry into mitosis and have different names in different organisms: Clb1 to Clb4 in budding yeast, Cdc13 in fission yeast, and cyclin B1, B2 and B3 in higher eukaryotes. Destruction of B-type cyclins is required for the exit from telophase and entry into interphase of the next cell cycle (Draetta et al. 1989; Luca and Ruderman 1989; Murray et al. 1989; Luca et al. 1991), whereas the metaphase to anaphase transition does not require the degradation of B-type cyclins (Holloway et al. 1993; Yamano et al. 1996). Cyclin B mutants lacking the destruction box arrest the cell cycle in mitosis and maintain both cyclin B/CDK1 kinase and APC in their active states (Murray et al. 1989; Glotzer et al. 1991; Luca et al. 1991; Gallant and Nigg 1992). In *Xenopus*, destruction of cyclin B2 but not of cyclin B1 requires binding to CDK1 (Stewart et al. 1994; van der Velden and Lohka 1994). Importantly, the D box occurs in other APC substrates which may act as sister chromatid "glue" proteins and prevent premature chromatid segregation, thereby preventing entry into anaphase (see below). For this reason, in addition to the name "cyclosome", the ubiquitin ligase responsible for the polyubiquitination of D box-containing substrates has been designated also as APC. It appears that "chromosome glue" proteins are degraded before B-type cyclins, but it is not clear what targets the APC to different substrates at different times.

The APC is a multi-protein complex formed by at least eight subunits. The APC is highly conserved through evolution and subunit homologues have been identified in many species. Genetic screens for yeasts deficient in Clb proteolysis revealed *CDC16*, *CDC23*, *CDC27*, *CDC26* and *APC1* (an *Aspergillus BIME* homologue) to encode APC subunits (Lamb et al. 1994; Irniger et al. 1995; Zachariae et al. 1996). Cdc16 (Apc6), Cdc23 (Apc7), and Cdc27 (Apc3) mediate in vitro ubiquitination of Clbs in crude yeast extracts (Zachariae et al. 1996). *Xenopus* homologues of *BIME*, *CDC16*, *CDC23* and *CDC27* (King et al. 1995; Peters et al. 1996) as well as mammalian homologues of *CDC16* and *CDC27* have been identified (Tugendreich et al. 1995) and found to be involved in the degradation of B-type cyclins (see also King et al. 1996a and references therein). The subunits of yeast and *Xenopus* APCs do not share homology with other known E3s and do not appear to form a thioester bond with ubiquitin (King et al. 1995).

Thus, the APC may function as a scaffold to bring together the ubiquitin conjugating enzyme and D box substrates, rather than as a catalytic component.

At least two distinct E2s have been involved as partners of APC, Ubc4 (King et al. 1995) and Ubc10, also designated E2C and Ubc-X (Hershko et al. 1994; Aristarkhov et al. 1996; Yu et al. 1996). Ubc9 has also been implicated in the degradation of mitotic cyclins (Seufert et al. 1995). However, it is now known that Ubc9 conjugates ubiquitin-like molecules to protein substrates and that this post-translational modification does not target these substrates for degradation (M. Dasso, pers. comm.), suggesting that Ubc9 may have only an indirect role in cyclin degradation. Recently, UbcP4, an E2 sharing 60 % identity with Ubc10, has been cloned in the fission yeast (Osaka et al. 1997). In vivo depletion of UbcP4 inhibits both the G2/M and the metaphase/anaphase transitions and blocks degradation of the B-type cyclin Cdc13, suggesting that this E2 may be important for cyclin degradation during mitosis. It is not clear which one of these Ubcs works in vivo in concert with APC during mitosis, but it is clear that the activity of this E2 is not cell cycle regulated.

In contrast to its E2(s), the activity of the APC is cell cycle dependent, being activated at metaphase (Glotzer et al. 1991; Hershko et al. 1994; King et al. 1995; Sudakin et al. 1995; Aristarkhov et al. 1996; Yu et al. 1996). APC activity remains high in the G1 phase of the next cell cycle until cells pass the START, preventing the accumulation of B-type mitotic cyclins during G1 (Amon et al. 1994; Brandeis and Hunt 1996; Irniger and Nasmyth 1997). In yeast, APC inactivation by Cln/Cdc28 at START (Amon et al. 1994; Dirick et al. 1995) inhibits Clb proteolysis, allowing Clb accumulation. The APC may be kept inactive during S and G2 by Clb/Cdc28 complexes (Amon 1997).

Phosphorylation positively regulates the E3 activity of APC subunits in mitosis (King et al. 1995; Lahav-Baratz et al. 1995; Peters et al. 1996). CDK1 (Cdc28 in budding yeast) is a good candidate for the mitotic APC activating kinase (Irniger et al. 1995; Sudakin et al. 1995). Thus, on the one hand, during G2, B-type cyclins in complex with CDK1 appear to keep the APC inactive, thereby preventing their own premature degradation (Amon 1997). On the other hand, in mitosis, B-type cyclins activate their own destruction, allowing exit from mitosis. Phosphorylation of APC by CDK1 could be regulated, at least in part, by $p13^{Suc1}$. This protein seems necessary for cyclin degradation (Basi and Draetta 1995; Patra and Dunphy 1996) and associates at the same time with cyclin-CDK1 complexes (reviewed by Pines 1996) and with the APC (Sudakin et al. 1997), probably creating a bridge between these two complexes.

Activation of B-type cyclin/CDK1 complexes is not only required for the G2/M transition, but also prevents the replication of DNA more than once per cell cycle (reviewed in Stilman 1996). Low levels of cyclin B inhibit entry into mitosis and stimulate endoreduplication in megakaryocytes. Interestingly, in these cells, the low levels of cyclin B are due to lack of APC inactivation at the end of S phase (Zhang et al. 1998).

In addition to regulation by CDK1, biochemical and genetic data suggest roles for protein kinase A (Yamashita et al. 1996), protein-phosphatase-1 (Ishii et al. 1996), and Cdc20/Fizzy (Dawson et al. 1995; Sigrist et al. 1995) in APC regulation. Finally, in budding yeast, another group of proteins plays a role in APC function. A signaling pathway involving Tem-1, a ras-like GTPase, and Cdc5, Cdc15 and Dbf2 protein kinases controls exit from mitosis (Toyn and Johnston 1994). This pathway is required for proteolysis of B-type cyclins, but not for sister chromatid segregation at metaphase exit. The timing of Dbf2 protein kinase activation and the dephosphorylation of Dbf2 in late anaphase suggest that this regulator may be involved in Clb proteolysis either directly or indirectly. There is a lag between MPF activation and mitotic cyclin destruction which may be needed to permit completion of anaphase events (e.g., completion of spindle movements). The Dbf2 pathway may act to prevent destruction of B-type cyclins until spindle separation is complete.

Activation of CDK1 is required for chromosome segregation but degradation of B-type cyclins is not needed for anaphase progression (Holloway et al. 1993). Non-cyclin substrates, including *cut2+* in fission yeast and *PDS1* in budding yeast, inhibit anaphase until they are degraded by APC-mediated proteolysis (Funabiki et al. 1996a; Funabiki et al. 1996b; Yamamoto et al. 1996). It has been proposed that APC deficient yeast arrest in anaphase because of failure to degrade sister chromatid adhesion proteins (Holloway et al. 1993). Pds1 is a candidate chromatid adhesion molecule or may regulate adhesion by other proteins (Yamamoto et al. 1996). Its degradation at the metaphase-anaphase transition triggers chromosome segregation; it has a D box and is an APC substrate both in vivo and in vitro (Cohen-Fix et al. 1996). Finally, Cut2 of *S. pombe* has two D boxes whose mutation renders this protein highly stable and blocks chromosome segregation (Funabiki et al. 1996a; Funabiki et al. 1996b). The APC might control also the disassembly of the mitotic spindle through the ubiquitination of the budding yeast Ase1, a D box containing protein localized in the midzone of the mitotic spindle and required for spindle elongation in anaphase (Juang et al. 1997).

4
Perspectives

It is apparent that selective proteolysis by the ubiquitin pathway is a key mechanism by which the activity of cell cycle regulatory proteins are switched on and off. In some cases, these switches involve changes in the substrate (e.g., the phosphorylation of Clns and of Sic1), while in others, modifications of the ubiquitin ligase itself regulate degradation (e.g., activation of the APC by phosphorylation). Further complexity derives from changes in the composition of E3 complexes, as seen in G1 with F box proteins. Changes in the activities of specific deubiquitinating enzymes could theoretically afford a further level of control, but to date there is no evidence of a role for this type of regulation. Finally, changes in the intracellular localization of both enzymes and substrates and of substrate associated proteins could regulate substrate degradation.

We are only beginning to understand the extent of cell cycle regulation by proteolysis. A number of questions come to the fore. To what degree is there conservation of E2/E3 complexes in the G1 proteolytic paths between yeast and humans? What role, if any, do cullins play in the degradation of mammalian cell cycle regulators? What activates the APC and what targets the APC to different substrates and at different times? What keeps the APC active during early G1 and how does activation of Cln/Cdc28 in late G1 shut off APC-mediated proteolysis? What are the mechanistic links between G1/S and G2/M proteolytic pathways? What are the mechanisms of regulation of the APC in meiosis?

Finally, to what extent does deregulation of cell cycle regulators contribute to human cancer? Reduced levels of the CDK inhibitor p27 in aggressive colon cancers (Loda et al. 1997), in non-small cell lung cancer (Esposito et al. 1997), and possibly in breast cancers (Catzavelos et al. 1997; Porter et al. 1997; Tan et al. 1997) are due to an increased proteasome-mediated proteolysis of p27, although the mechanism of this abnormality is unknown. A frequent alteration in human cancers is the overexpression of G1 cyclins (reviewed in Hunter and Pines 1994; Sherr 1996). Cyclin protein levels are frequently elevated in the absence of gene overexpression (Buckley et al. 1993; Keyomarsi et al. 1995; Welker et al. 1996; Porter et al. 1997), and could derive from decreased cyclin proteolysis. By analogy to the results in yeast and *C. elegans*, human homologues of *cul1/CDC53* could conceivably play a role in limiting cell growth through involvement in the proteolysis of positive cell cycle regulators (e.g., cyclins). Given the negative growth regulatory role of Cul1, one could speculate that it acts as a tumor suppressor, whose loss or inactivation causes cyclin accumulation and stimulates uncontrolled cell proliferation. An understanding of the mechanisms underlying altered proteolysis of cell cycle regulators in human cancers may lead to the development of novel approaches in cancer therapy.

Acknowledgments. We thank M. Dasso, R. Verma, R. Deshaies, D. Finley, W. Harper, A. Hershko, P. Howley, D. Kornitzer, T. Toda, M. Tyers, and P. Zhou for communicating data before publication; and M. Tyers for critically reading the manuscript. We acknowledge Chris McDowell for artwork in Figs. 1 and 2. We apologize to colleagues whose work has not been cited, or cited indirectly through other articles, due to space limitations. J.M.S. is a clinician investigator supported by the Ontario Cancer Treatment and Research Foundation. M.P. is in part supported by the NIH grants CA66229-02 and CA57587.

References

Amon A (1997) Regulation of B-type cyclin proteolysis by CDC28-associated kinases in budding yeast. EMBO J 16:2693-2702

Amon A, Irniger S, Nasmyth K (1994) Closing the cell cycle circle in yeast: G2 cyclin proteolysis initiated at mitosis persists until the activation of G1 cyclins in the next cycle. Cell 77:1037-1050

Aristarkhov A, Eytan E, Moghe A, Adom A, Hershko A, Ruderman J (1996) E2-C, a cyclin-selective ubiquitin carrier protein required for the destruction of mitotic cyclins. Proc Natl Acad Sci USA 93:4294-4299

Bai C, Sen P, Hofman K, Ma L, Goebel M, Harper W, Elledge S (1996) Skp1 connects cell cycle regulators to the ubiquitin proteolysis machinery through a novel motif, the F-box. Cell 86:263-274

Baker R, Tobias J W, Varshavsky A (1992) Ubiquitin-specific proteases of *Saccharomyces cerevisiae*. Cloning of UBP2 and UBP3, and functional analysis of the UBP gene family. J Biol Chem 267:23364-23375

Barral Y, Jentsch S, Mann C (1995) G1 cyclin turnover and nutrient uptake are controlled by a common pathway in yeast. Genes Dev 9:399-409

Basi G, Draetta G (1995) p13Suc1 of *Schizosaccharomyces pombe* regulates two distinct forms of the mitotic cdc2 kinase. Mol Cell Biol 15:2028-2036

Brandeis M, Hunt T (1996) The proteolysis of mitotic cyclins in mammalian cells persists from the end of mitosis until the onset of S phase. EMBO J 15:5280-5289

Buckley M F, Sweeney K J E, Hamilton J A, Sini R L, Manning D L, Nicholson R I, deFazio A, Watts C K W, A. M E, Sutherland R L (1993) Expression and amplification of cyclin genes in human breast cancer. Oncogene 8:2127-2133

Catzavelos C, Bhattacharya N, Ung Y, Wilson J, Roncari L, Sandhu C, Shaw P, Yeger H, Morava-Protzner I, Kapusta L, Franssen E, Pritchard K, Slingerland J (1997) Decreased levels of the cell-cycle inhibitor $p27^{Kip1}$ protein: prognostic implications in primary breast cancer. Nature Med 3:227-230

Clurman B, Sheaff R, Thress K, Groudine M, Roberts J (1996) Turnover of cyclin E by the ubiquitin-proteasome pathway is regulated by CDK2 binding and cyclin phosphorylation. Genes Dev. 10:1979-1990

Cohen-Fix O, Peters J, Kirschne r M, Koshland D (1996) Anaphase initiation in *Saccharomyces cerevisiae* is controlled by the APC-dependent degradation of the anaphase inhibitor Pds1p. Genes Dev 10:3081-3093

Connelley C, Hieter P (1996) Budding yeast Skp1 is an evolutionarily conserved kinetochore protein required for cell cycle progression. Cell 86:275-285

Dawson I, Roth S, Artavanis-Tsakonas S (1995) The *Drosophila* cell cycle gene fizzy is required for normal degradation of cyclins A and B during mitosis and has homology to the CDC20 gene of *Saccharomyces cerevisiae*. J Cell Biol 129:725-737

DeMartino G N, Moomaw C R, Zagnitko O P, Proske R J, Ma C-P, Afendis J, Swaffield J C, Slaughter C A (1994) PA700, an ATP-dependent activator of the 20S proteasome, is an ATPase containing multiple members of a nucleotide binding family. J Biol Chem 269:20878-20884

Deshaies R J, Chau V, Kirschner M W (1995) Ubiquitination of the G1 cyclin Cln2p by a Cdc34p-dependent pathway. EMBO J 14:303-312

Diehl J A, Zindy F, Sherr C (1997) Inhibition of cyclin D1 phosphorylation on threonine 286 prevents its rapid degradation via the ubiquitin-proteasome pathway. Genes Dev. 11:957-972

Dirick L, Bohm T, Nasmith K (1995) Roles and regulation of Cln-Cdc28 kinases at the start of the cell cycle of *Saccaromiches Cerevisiae*. EMBO J 14:4803-4813

Draetta G, Luca F, Westendorf J, Brizuela L, Ruderman J, Beach D (1989) Cdc2 protein kinase is complexed with both cyclin A and B: evidence for proteolytic inactivation of MPF. Cell 56:829-838

Drury L, Perkins G, Diffley J (1997) The Cdc4/34/53 pathway targets Cdc6p for proteolysis in budding yeast. EMBO J 16:5566-5576

Epstein C, Cross F (1992) CLNB5: a novel B cyclin from budding yeast with a role in S phase. Genes Dev. 6:1695-1706

Esposito V, Baldi A, DeLuca A, Sgaramella G, Giordano G G, Caputi M, Baldi F, Pagano M, Giordano A (1997) Prognostic role of the cell cycle inhibitor p27 in non-small cell lung cancer. Cancer Res 57:3381-3385

Feldman R M, Correll C C, Kaplan K B, Deshaies R J (1997) A complex of Cdc4p, Skp1p, and Cdc53p/Cullin catalyzes ubiquitination of the phosphorylated CDK inhibitor Sic1p. Cell 91:221-230

Funabiki H, Kumada K, Yanagida M (1996a) Fission yeast Cut1 and Cut2 are essential for sister chromatid separation, concentrate along the metaphase spindle and form large complexes. EMBO J 15:6617-6628

Funabiki H, Yamano H, Kumada K, Nagao K, Hunt T, Yanagida M (1996b) Cut2 proteolysis required for sister-chromatid separation in fission yeast. Nature 381:438-441

Gallant P, Nigg E (1992) Cyclin B2 undergoes cell cycle dependent nuclear translocation and, when expressed as a non destructable mutant, causes mitotic arrest in HeLa cells. J Cell Biol 117:213-224

Glotzer M, Murray A, Kirschner M (1991) Cyclin is degraded by the ubiquitin pathway. Nature 349:132-138

Goebl M, Goetsch L, Byers B (1994) The Ubc3 (Cdc34) ubiquitin-conjugating enzyme Is ubiquitinated and phosphorylated in vivo. Mol Cell Biol 14:3022-3029

Hareford L, Hartwell L (1974) Sequential gene function in the initiation of Saccharomyces cerevisiae DNA synthesis. J Mol Biol 84:445-461

Hershko A, Heller H, Elias S, Ciechanover A (1983) Components of ubiquitin-protein ligase system. Resolution, affinity purification, and role in protein breakdown. J Biol Chem 258:8206-8214

Hershko A, Ganoth D, Sudakin V, Dahan A, Cohen L, Luca F, Ruderman J, Eytan E (1994) Components of a system that ligates cyclin to ubiquitin and their regulation by the protein kinase cdc2. J Biol Chem 269:4940-4946

Hochstrasser M (1995) Ubiquitin, proteasomes, and the regulation of intracellular protein degradation. Curr Op in Cell Biol. 7:215-223

Holloway S, Glotzer M, King R, Murray A (1993) Anaphase is initiated by proteolysis rather than by inactivation of maturation-promoting factor. Cell 73:1393-1402

Hunter T, Pines J (1994) Cyclins and Cancer II. Cyclin D and CDK inhibitors come of age. Cell 79:573-582

Irniger S, Nasmyth K (1997) The anaphase-promoting complex is required in G1 arrested yeast cells to inhibit B-type cyclin accumulation and to prevent uncontrolled entry into S-phase. J. Cell Sci. 110:1523-1531

Irniger S, Piatti S, Michaelis C, Nasmyth K (1995) Genes involved in sister chromatid separation are needed for B-type cyclin proteolysis in budding yeast. Cell 81:269-278

Ishii K, Kumada K, Toda T, Yanagida M (1996) Requirement for PP1 phosphatase and 20S cyclosome/APC for the onset of anaphase is lessened by the dosage increase of a novel gene sds23+. EMBO J 15:6629-6640

Izumi T, Maller J (1991) Phosphorylation of *Xenopus* cyclins B1 and B2 is not required for cell cycle transitions. Mol Cell Biol 11:3860-3867

Jentsch S (1992) The ubiquitin-conjugation system. Annu Rev Genet 26:179-207

Juang Y, Huang J, Peters J M, McLaughlin M, Tai C, Pellman D (1997) APC-mediated proteolysis of Ase1 and the morphogenesis of the mitotic spindle. Science 275:1311-1314

Keyomarsi K, Conte D Jr, Toyofuku W, Fox M P (1995) Deregulation of cyclin E in breast cancer. Oncogene 11:941-950

King R, Peters J, Tugendreich S, Rolfe M, Hieter P, Kirschner M (1995) A 20S complex containing Cdc27 and Cdc16 catalyzes the mitosis-specific conjugation of ubiquitin to cyclin B. Cell 81:279-288

King R, Deshaies R, Peters J, Kirschner M (1996a) How proteolysis drives the cell cycle. Science 274:1652-1659

King R, Glotzer M, Kirschner M (1996b) Mutagenic analysis of the destruction signal for mitotic cyclins and structural characterization of ubiquitinated intermediates. Mol Biol Cell 7:1343-1357

Kipreos E T, Lander L, Wing J, He W, Hedgecock E (1996) cul-1 is required for cell cycle exit in *C. elegans* and identifies a novel gene family. Cell 85:829-839

Kominami K, Toda T (1997) Fission yeast WD-repeat protein Pop-1 regulates genome ploidy through ubiquitin-proteasome-mediated degradation of the CDK inhibitor Rum-1 and the S-phase initiator Cdc18. Genes Dev 11:1548-1560

Kornitzer D, Raboy B, Kulka R, Fink G (1994) Regulated degradation of the transcription factor Gcn4. EMBO J 13:6021-6030

Lahav-Baratz S, Sudakin V, Ruderman J, Hershko A (1995) Reversible phosphorylation controls the activity of cyclosome-associated cyclin-ubiquitin ligase. Proc. Natl. Acad. Sci.USA 92:9303-9308

Lamb J, Michaud W, Sikorski R, Hieter P (1994) Cdc16p, Cdc23p and Cdc27p form a complex essential for mitosis. EMBO J 13:4321-4328

Lanker S, Valdivieso M, Wittenberg C (1996) Rapid degradation of the G1 cyclin Cln2 indudec by CDK-dependent phosphorylation. Science 271:1597-1601

Li J, Meyer A, Donoghue D (1995) Requirement for phosphorylation of cyclin B1 for *Xenopus* oocyte maturation. Mol Biol Cell 6:1111-1124

Loda M, Cukor B, Tam S, Lavin P, Fiorentino M, Draetta G, Jessup J, Pagano M (1997) Increased proteasome-dependent degradation of teh cyclin-dependent kinase inhibitor p27 inaggressive colorectal carcinomas. Nature Med 3:231-234

Lowe J, Stock D, Jap B, Zwickl P, Baumeister W, Huber R (1995) Crystal structure of the 20S proteasome from the archaeon *T. acidophilum* at a 3.4 Å resolution. Science 268:533-539

Luca FC, Ruderman J V (1989) Control of programmed cyclin destruction in a cell-free system. J Cell Biol 109:1895-1909

Luca FC, Shibuya E, Dohrmann C, Ruderman J (1991) Both cyclin AÆ60 and BÆ97 are stable and arrest cells in M-phase, but only cyclin B Æ97 turns on cyclin destruction. EMBO J 10:4311-4320

Ma C-P, Vu J H, Proske R J, Slaughter J A, DeMartino G N (1994) Identification, purification, and characterization of a high molecular weight, ATP-dependent activator (PA700) of the 20S proteasome. J Biol Chem 269:3539-3547

Mathias N, Johnson S, Winey M, Adams A, Goetsch L, Pringle J, Byers B, Goebl M (1996) Cdc53p acts in concert with cdc4 and cdc34 to control the G1 phase transition and identifies a conserved family of proteins. Mol Cell Biol 16:6634-6643

McKinney J, Chang F, Heintz N, Cross F R (1993) Negative regulation of FAR1 at the start of the yeast cell cycle. Genes Dev 7:833-843

Mendenhall M (1993) An inhibitor of $p34^{cdc28}$ protein kinase activity from *Saccharomyces cerevisiae*. Science 259:216-219

Multiple Authors (1996) Six reviews on the cell cycle. Science 274:1643-1677

Murray A W, Solomon M J, Kirschner M W (1989) The role of cyclin synthesis and degradation in the control of maturation promoting factor activity. Nature 339:280-286

Ning F, Johnston M (1997) Grr1 of Saccharomyces cerevisiae is connected to the ubiquitin proteolysis machinery through Skp1: coupling glucose sensing to gene expression and the cell cycle. EMBO J 16:5629-5638

Osaka F, Seino H, Seno T, Yamao F (1997) A ubiquitin-conjugating enzyme in fission yeast that is essential for the onset of anaphase in mitosis. Mol Cell Biol 17:3388-3397

Pagano M, Tam S W, Theodoras A M, Beer P, Delsal G, Chau V, Yew R, Draetta G, Rolfe M (1995) Role of the ubiquitin-proteasome pathway in regulating abundance of the cyclin-dependent kinase inhibitor p27. Science 269:682-685

Papa F, Hochstrasser M (1993) The yeast DOA4 gene encodes a deubiquitinating enzyme related to a product of the human tre-2 oncogene. Nature 366:313-319

Patra D, Dunphy W (1996) Xe-p9, a *Xenopus* Suc1.Cks homolog, has multiple essential roles in cell cycle control. Genes Dev 10:1503-1515

Pause A, Lee S, Worrel R, Chen D, Burgess W, Linehan M, Klausner R (1997) The von Hippel-Lindau tumor suppressor gene product forms a stable complex with human Cul2, a member of the Cdc53 family of proteins. Proc Natl Acad Sci U.S.A. 94:2156-2161
Peters J-M, Franke W W, Kleinschmidt J A (1994) Distinct 19S and 20S subcomplexes of the 26S proteasome and their distribution in the nucleus and cytoplasm. J Biol Chem 269:7709-7718
Peters J, King R, Hoog C, Kirschner M (1996) Identification of BIME as a subunit of the anaphase-promoting complex. Science 274:1199-1201
Piatti S, Bohm T, Cocker J, Diffley J, Nasmyth K (1996) Activation of S-phase-promoting CDKs in late G1 defines a "point of no return" after which Cdc6 synthesis cannot promote DNA replication in yeast. Genes Dev 10:1516-1531
Pines J (1996) Reaching a role for the Cks proteins. Curr Biol 11:1399-1402
Plon S, Leppig K, Do H, Groudine M (1993) Cloning of the human homolog of the *CDC34* cell cycle gene by complementation in yeast. Proc Natl Acad Sci USA 90:10484-10488
Porter P, Malone K, Heagerty P, Alexander G, Gatti L, Firpo E J, Daling J, Roberts J (1997) Expression of cell-cycle regulators p27Kip1 and Cyclin E, alone and in combination, correlate with survival in young breast cancer patients. Nature Med 3:222-225
Rolfe M, Chiu I, Pagano M (1997) The ubiquitin-mediated proteolytic pathway as a therapeutic area. J Mol Med 75:5-17
Rubin D, Finley D (1995) The proteasome: a protein-degrading organelle? Curr Biol 5:854-858
Schneider B, Ying Q, Futcher B (1996) Linkage of replication to start by the CDK inhibitor sic1. Science 272:560-562
Schwob E, Böhm T, Mendenhall M, Nasmyth K (1994) The B-type cyclin kinase inhibitor p40sic1 controls the G1 to S transition in *S. cerevisiae.* Cell 79:233-244
Schwob E, Nasmyth K (1993) CLB5 and CLB6, a new pair of B type cyclins involved in DNA replication in *Saccharomyces cerevisiae.* Genes Dev. 7:1160-1175
Seufert W, Futcher B, Jentsch S (1995) Role of a ubiquitin-conjugating enzyme in degradation of S- and M-phase cyclins. Nature 373:78-81
Sheaff R, Groudine M, Gordon M, Roberts J, Clurman B (1997) Cyclin E-CDK2 is a regulator of $p27^{Kip1}$. Genes Dev 11:1464-1478
Sherr C (1994) G1 phase progression: cycling on cue. Cell 79:551-555
Sherr C (1996) Cancer cell cycles. Science 274:1672-1677
Sigrist S, Jacobs H, Stratmann R, Lehner C (1995) Exit from mitosis is regulated by Drosophila fizzy and the sequential destruction of cyclins A, B and B3. EMBO J 14:4827-4838
Skowyra D, Craig K L, Tyers M, Elledge S J, Harper J W (1997) F-box proteins are receptors that recruit phosphorylated substrates to the SCF ubiquitin-ligase complex. Cell 91:209-219
Stankovic T, Byrd P J, Cooper P R, McConville C M, Munroe D J, Riley J H, Watts G D J, Ambrose H, McGuire G, Smith A D, Sutcliffe A, Mills T, Taylor A M R (1997) Construction of a transcription map around the gene for ataxia-telangiectasia; identification of at least four novel genes. Genomics 40:267-276
Stemmann O, Lechner J (1996) The *Saccharomyces cerevisiae* kinetochore contains a cyclin-CDK complexing homologue, as identified by in vitro reconstitution. EMBO J 15:3611-3620
Stewart E, Kobayashi H, Harrison D, Hunt T (1994) Destruction of *Xenopus* cyclins A and B2, but not B1, requires binding to $p34^{Cdc2}$. EMBO J 13:584-594
Stilman B (1996) Cell cycle control of DNA replication. Science 274:1659-1664
Sudakin V, Ganoth D, Dahan A, Heller H, Hershko J, Luca F, Ruderman J, Hershko A (1995) The cyclosome, a large complex containing cyclin selective ubiquitin ligase activity, targets cyclins for destruction at the end of mitosis. Mol Biol Cell 6:185-198
Sudakin V, Shteinberg M, Ganoth D, Hershko J, Hershko A (1997) Binding of activated Cyclosome to $p13^{Suc1}$. J Biol Chem 272:18051-1859
Tan P, Cady B, Wanner M, Worland P, Cukor B, Fiorentino M, Magi-Galluzzi C, Lavin P, Pagano M, Loda M (1997) The cell cycle inhibitor p27 is an independent prognostic marker in small (T1a,b) invasive breast carcinomas. Cancer Res 57:1259-1263
Toyn J, Johnston L (1994) The Dbf2 and Dbf20 protein kinases of budding yeast are activated after the metaphase to anaphase cell cycle transition. EMBO J 13:1103-1113

Tugendreich S, Tomkiel J, Earnshaw W, Hieter P (1995) Cdc27Hs colocalizes with Cdc16Hs to the centrosome and mitotic spindle and is essential for the metaphase to anaphase transition. Cell 81:261-268

Tyers M (1996) The cyclin-dependent kinase inhibitor p40^{SIC1} imposes the requirement for Cln G1 cyclin function at Start. Proc Natl Acad Sci USA 93:7772-7776

van der Velden H, Lohka M (1994) Cell cycle-regulated degradation of *Xenopus* cyclin B2 requires binding to p34^{cdc2}. Mol Biol Cell 5:713-724

Verma R, Annan R, Huddleston M, Carr S, Reynard G, Deshaies R J (1997a) Phosphorylation of Sic1p by G1 cyclin/CDK is required for its degradation and entry into S phase. Science 278:455-460

Verma R, Feldman R M R, Deshaies R J (1997b) SIC1 is ubiquitinated in vitro by a pathway that requires CDC4, CDC34, and cyclin/CDK activities. Mol Biol Cell 8:1427-1437

Vlach J, Hennecke S, Amati B (1997) Phosphorylation-dependent degradation of the cyclin-dependent kinase inhibitor p27^{Kip1}. EMBO J16:5334–5344

Welker M, Lukas J, Strauss M, Bartek J (1996) Enhanced protein stability: a novel mechanism of D-type cyclin over-abundance identified in human sarcoma cells. Oncogene 13:4191–4201

Willems A, Lanker S, Patton E, Craig K, Nason T, Mathias N, Kobayashi R, Wittenberg C, Tyers M (1996) Cdc53 targets phosphorylated G1 cyclins for degradation by the ubiquitin proteolytic pathway. Cell 86:453–463

Won K, Reed S (1996) Activation of cyclin E/CDK2 is coupled to site-specific autophosphorylation and ubiquitin-dependent degradation of cyclin E. EMBO J 15:4182–4193

Yamamoto A, Guacci V, Koshland D (1996) Pds1p, an inhibitor of anaphase in budding yeast, plays a critical role in the APC and checkpoint pathway(s). J Cell Biol 133:99–110

Yamano H, Gannon J, Hunt T (1996) The role of proteolysis in cell cycle progression in *Schizosaccharomyces pombe.* EMBO J 15:5266–5279

Yamashita Y, Nakaseko Y, Samejima I, Kumada K, Yamada H, Michaelson D, Yanagida M (1996) 20S cyclosome complex formation and proteolytic activity inhibited by the cAMP/PKA pathway. Nature 384:276–279

Yew R, Kirschner M W (1997) Proteolysis and DNA replication: the *CDC34* requirement in the *Xenopus* egg cell cycle. Science 277:1672–1675

Yu H, King R, Peters J, Kirschner M (1996) Identification of a novel ubiquitin-conjugating enzyme involved in mitotic cyclin degradation. Curr Biol 6:455–466

Zachariae W, Shin T, Galova M, Obermaier B, Nasmyth K (1996) Identification of subunits of the anaphase-promoting complex of *Saccharomyces cerevisiae.* Science 274:1201-1204

Zhang H, Kobayashi R, Galaktionov K, Beach D (1995) p19Skp-1 and p45Skp-2 are essential elements of the cyclin A-CDK2 S phase kinase. Cell 82:915–925

Zhang Y, Wang Z, Liu D, Pagano M, Ravid K (1997) Ubiquitin-dependent degradation of Cyclin B is accelerated in polyploid megakaryocytes. J Biol Chem In press

Regulation of the Cell Cycle by the Rb Tumor Suppressor Family

M.E. Ewen[1]

1 Introduction

The rare childhood cancer retinoblastoma is caused, at least in part, by inactivating mutations in the retinoblastoma gene. The study of the retinoblastoma protein (Rb) over the past decade has led to a significant advance in our molecular understanding of many forms of cancer which take the lives of thousands of people (of all ages) every year. Deregulated cell proliferation is a common feature of many cancers. Rb, though not essential for cell proliferation, plays a significant role in regulating cell cycle progression and perturbation of Rb, or its upstream regulators appear to have an important role in the development of the majority of human solid tumors. This review will describe how Rb and two of its relatives, p107 and p130, are thought to regulate cell proliferation and differentiation, and what is known about how loss of Rb function might contribute to tumor formation.

2 The Rb Pathway

An important function of Rb is to control progression through the G1 phase of the cell cycle and entry into S phase. The ability of Rb to restrain progression through G1 is influenced by its phosphorylation status. In its un- or hypophosphorylated form Rb is active with respect to its ability to control cell proliferation. Phosphorylation of Rb inactivates its cell cycle-constraining function. A number of kinases and phosphatases operating upstream of Rb dictate its functional state (active or inactive). It is thought that a major biochemical function of Rb is to modulate the activity of certain transcription factors, such as E2F, which control the expression of genes whose function influences progression through the cell cycle. The ability of Rb to regulate transcription is thought to be regulated by its phosphorylation status (dictated by upstream Rb effectors). Collectively, the factors working upstream and downstream of Rb, as well as Rb itself, define what is referred to as the Rb pathway. This pathway is disrupted in a large percentage of human cancers.

[1] The Dana-Farber Cancer Institute and the Harvard Medical School, 44 Binney Street, Boston, Massachusetts 02115, USA, E-mail: mark_ewen@macmailgw.dfci.harvard.edu

Two Rb homologues, p107 and p130, in generic terms, function in the same pathway in which Rb operates. A similar set of kinases influences the function of p107 and p130. These proteins, like Rb, influence the expression of a number of genes by interacting with transcription factors. That the Rb family of proteins share similar biochemical functions follows from the observation that a large part of their function is dictated through similar protein-protein interactions. Rb, p107 and p130 can form complexes with oncoproteins encoded by simian virus 40 (SV40), adenovirus and human papillomavirus (HPV), namely, T antigen, E1A and E7, respectively. The minimal T/E1A/E7 binding region for the Rb family is composed of a linear sequence: A domain-spacer-B domain (Ewen 1994). The A and B domains show the highest degree of homology between Rb family members. These two domains are critical components in the interaction with both upstream (e.g., D-type cyclins) and downstream (e.g., E2F) factors in the Rb pathway. Tumor-derived Rb mutants are defective for binding to E1A, T, E7, E2F and D cyclins. p107 and p130 also show a significant degree of homology over their entire sequence (Hannon et al. 1993; Li et al. 1993; Mayol et al. 1993). A notable difference between p107/p130 and Rb is that no naturally occurring mutations which disrupt the function of either p107 or p130 have been reported.

2.1
Downstream of Rb, p107 and p130

The most extensively studied downstream target of the Rb family of proteins is the E2F family of cellular transcription factors, E2F1 through E2F5 (see Chap. 9). The involvement of E2F in cell proliferation follows from the observation that a number of genes whose products are required for cell cycle progression and DNA replication have E2F DNA binding sites in their promoters. Furthermore, ectopic expression of E2F1 can drive quiescent cells into S phase (Johnson et al. 1993; Qin et al. 1994; Shan and Lee 1994; DeGregori et al. 1995b; Kowalik et al. 1995; Schwartz et al. 1995; Asono et al., 1996). Conversely, a dominant-negative E2F1 can inhibit E1A-induced DNA synthesis (Dobrowolski et al. 1994b) and dominant-negative DP1 (one of the heterodimeric partners of E2Fs) can cause G1 cell cycle arrest (Wu et al. 1996). In *Drosophila* E2F is required for S phase entry (Duronio et al. 1995). Together, these results suggest that E2F is both necessary and sufficient for G1 progression and S phase entry, depending on the particular cell system analyzed.

Several lines of evidence suggest that E2F operates downstream of Rb/p107/p130. Ectopic expression of E2F1 can override an Rb-induced G1 arrest (Zhu et al. 1993; Qin et al. 1995). Similarly, enforced expression of E2F1 can reverse the G1 arrest induced by ectopic expression of the cyclin-dependent kinase inhibitor (CDK) inhibitor $p16^{INK4a}$ (DeGregori et al. 1995b; Lukas et al. 1996b; see Sect. 2.2). The ability of Rb to control cell proliferation directly correlates with its ability to associate with and regulate E2F (Bagchi et al. 1991; Chittenden et al. 1991; Kaelin et al. 1991; Arroyo and Raychaudhuri 1992; Hiebert et al. 1992; Qin et al. 1992; Hiebert 1993). There is also in vivo evidence that E2Fs operate downstream of Rb. The cyclin E gene is thought to be regulated by E2Fs (DeGregori et al. 1995a; Ohtani et al. 1995; Botz et al. 1996; Geng et al. 1996), and in *Drosophila*, cyclin E is an E2F target (and is required for E2F-induced S phase entry; Duronio and O'Farrell 1995; Duronio et al. 1996). Loss of Rb function results in deregulated expression of cyclin E, suggesting that Rb-E2F complexes act to repress the expression of cyclin E, and that "free" E2F (resulting from phosphorylation of Rb) drives the expression of cyclin E (Herrera et al. 1996c).

Much of the current research in this field is directed at determining how the combinatorial action of Rb, p107 and p130 together with E2F1, E2F2, E2F3, E2F4 and E2F5 regulates cell cycle progression. Complexes containing E2F1, E2F2 and E2F3 bind to Rb and not p107 in vivo (Dyson et al. 1993; Lees et al. 1993). E2F5 binds preferentially to p130 (Hijmans et al. 1995). E2F4 has been shown to interact with Rb, p107 and p130 (Beijersbergen et al. 1994b; Ginsberg et al. 1994; Vairo et al. 1995; Moberg et al. 1996). The relative ability of various E2Fs to override the G1 arrest induced by Rb, p107 and p130 has complemented these studies. It has been noted, for example, that E2F1 overrides the G1 block induced by Rb more efficiently than that induced by p130, and the opposite result is found for E2F4 (Vairo et al. 1995). The results suggest a possible difference in the function of Rb-E2F1 and p130-E2F4 complexes. However, whether these overexpression experiments reflect in vivo biological function of these proteins remains to be determined. One difficulty with such experiments is the observation that ectopic expression of Rb causes a G1 arrest in Rb-deficient tumor cell lines but not Rb-positive cells, a phenomenon which is not fully understood. Similarly, p107 and p130 have been shown to cause a G1 arrest in tumor cells lacking functional Rb, but not in Rb-positive cells, with one exception (Claudio et al. 1994). Another difficulty stems from the observation that, in transfection experiments, some of the E2F family members and their heterodimeric DP partners do not localize to the nucleus unless coexpressed with certain Rb family members (Magae et al. 1996). Thus, many of the above-noted E2F "overrides" need to be reconsidered once we have a better understanding of the role of subcellular localization of E2Fs in the regulation of E2F-mediated transactivation during the cell cycle.

The temporal pattern of Rb, p107 and p130 complex formation with various E2Fs through the cell cycle has also been investigated. In quiescent (G0) cells, the major E2F species is p130-E2F4 (Cobrinik et al. 1993; Vairo et al. 1995), which may distinguish between G0 and G1 cells (Smith et al. 1996). The p130-E2F4 complex is thought to repress transcription of certain genes, i.e., *E2F1* and *cdc2*, and to contribute to the maintenance of a quiescent state (Johnson 1995; Tommasi and Pfeifer 1995; Smith et al. 1996). Importantly, the E2F sites in the E2F1 promoter were found to be important for cell cycle regulation of this gene when progressing from G0 to S phase but not in cycling cells (Neuman et al. 1994; Smith et al. 1996). Consistent with the role of Rb in controlling progression through G1, Rb-E2F complexes are found in mid- to late G1. These complexes persist into S and G2 (Shirodkar et al. 1992). In late G1 p107-E2F complexes can be found which contain cyclin E-CDK2 (Lees et al. 1992). In S phase p107-E2F complexes which contain cyclin A-CDK2 can be found (Bandara et al. 1991;Mudryj et al. 1991; Cao et al. 1992; Lees et al. 1992; Pagano et al. 1992; Shirodkar et al. 1992). p130-E2F can also form higher order cyclin E-CDK2 or cyclin A-CDK2 complexes when reconstituted in vitro (Cobrinik et al. 1993), and E2F-p130-CDK2 complexes have been purified from mammalian cell extracts (Shiyanov et al. 1996). The functional significance of the association of cyclin E/A-cdk2 with p107/p130 remains to be determined. By analogy to cyclin A-mediated transphosphorylation of DP1 in the E2F1-DP1 complex (Xu et al. 1994; Krek et al. 1994, 1995; Dynlacht et al. 1995; see also Chap. 9), these kinases may phosphorylate E2F and/or DP members when in complex with p107/p130. Such a mechanism has been proposed for the inactivation of c-Myc transactivation function by p107 (Beijersbergen et al. 1994a; Gu et al. 1994; Hoang et al. 1995).

Consistent with the notion that Rb regulates the activity of E2F, changes in expression of E2F target genes have been noted in *Rb*$^{-/-}$ mouse embryo fibroblasts (MEFs) (Almasan et al. 1995; Herrera et al. 1996c; Hurford et al. 1997). In an effort to understand the roles of Rb, p107 and p130 in E2F-dependent gene expression, MEFs derived from embryos lacking Rb family members, either alone or in combination, have been analyzed (Hurford et al. 1997). In the *Rb*$^{-/-}$ MEFs the expression of a particular subset of E2F-regulated genes was deregulated. By contrast, no changes in the expression of E2F target genes were noted in either *p107*$^{-/-}$ or *p130*$^{-/-}$ MEFs. However, in cells lacking both *p107* and *p130*, deregulated expression of E2F targets was observed. Interestingly, the particular genes deregulated in the *Rb*$^{-/-}$ MEFs and the *p107*$^{-/-}$*p130*$^{-/-}$ MEFs were completely different (Hurford et al. 1997). This measure of Rb family member function is consistent with the observation that *Rb*, *p107* or *p130* and *p107p130*-deficient mice each have a different phenotype. Disruption of the *Rb* gene in mice results in embryonic lethality with defects in the liver, central nervous system, ocular lens and erythropoiesis (Clarke et al. 1992; Jacks et al. 1992; Lee et al. 1992). Mice lacking either p107 or p130 are without obvious abnormalities (Cobrinik et al. 1996; Lee et al. 1996), but inactivation of both the *p107* and *p130* genes results in neonatal lethality, deregulated chondrocyte growth, defective endochondral bone development and shortened limbs (Cobrinik et al. 1996), i.e., a different spectrum than in *Rb*-deficient embryos. Though these are interesting correlations, there are no data suggesting that the particular spectrum of E2F targets deregulated in the "knockout" MEFs contributes to the particular phenotypes seen in the mice or embryos.

2.2 Upstream of Rb, p107 and p130

Progression through the cell cycle is regulated by the coordinate action of a number of CDKs. A major form of positive regulation of CDKs is through their association with and activation by cyclins. Another important form of CDK regulation, directly pertinent to the study of Rb/p107/p130, is through the action of CDK inhibitors (see Chap. 6). Traversal of mid- and late G1 phase is regulated by the three D-type cyclins, D1, D2 and D3, which activate CDK4 and CDK6, and by cyclin E, which activates CDK2. D cyclins are generally thought to function earlier in G1 than cyclin E (see Chap. 2).

Cyclins and/or cyclin-CDK complexes can interact with Rb/p107/p130. D-type cyclins can associate with Rb, p107 and p130 (Dowdy et al. 1993; Ewen et al. 1993a; Hannon et al. 1993). This association requires the intact T/E1A/E7 binding region (A region-spacer-B region) plus the C-terminus of the particular Rb family member. Ternary complexes between Rb, D-type cyclins and CDK4 or CDK6 have not been observed in mammalian cells (Kato et al. 1993). Phosphorylation of Rb by CDK4 prevents its association with D cyclins. p107 and p130, and to a significantly lesser extent Rb, can form stable complexes with both cyclin E- and cyclin A-cdk2, in a manner which is independent of the association of p107/p130 with viral oncoproteins (Ewen et al. 1992; Faha et al. 1992; Hannon et al. 1993; Li et al. 1993; Peeper et al. 1993). This interaction is mediated in part by a short sequence motif found in the spacer elements of p107 and p130 (Zhu et al. 1995; Adams et al. 1996; J. Chen et al. 1996).

The interaction between D-type cyclins and Rb has been studied in terms of its biochemical and biological significance. The cyclin D-dependent kinases can phosphorylate Rb, which leads to inactivation of its ability to cause a G1 arrest (Matsushime et al. 1992; Ewen et al. 1993a; Kato et al. 1993). Microinjection of neutralizing antibodies against cyclin D1 or antisense cyclin D1 arrest cells in G1 (Baldin et al. 1993; Quelle et al. 1993; Lukas et al. 1994, 1995a). Cells which do not express functional Rb, or which express viral oncoproteins (i.e., T, E1A or E7) that inactivate Rb do not arrest when the function of cyclin D1 is inhibited (Lukas et al. 1994, 1995a; Tam et al. 1994). Likewise ectopic expression of the CDK inhibitor $p16^{INK4a}$, which is specific for CDK4 and CDK6 (Serrano et al. 1993), causes a G1 arrest and/or growth inhibition in an Rb-dependent manner (Guan et al. 1994; Koh et al. 1995; Lukas et al. 1995b; Medema et al. 1995). Together, these results indicate that, in terms of cell cycle progression, a major target of cyclin D-CDK4/6 is Rb.

In contrast to cyclin D, cyclin E operates in both an Rb-dependent and -independent manner. Like D-type cyclins, ectopic expression of cyclin E can induce the phosphorylation and inactivation of Rb (Hinds et al. 1992). However, cyclin E function has been shown to be essential for G1 progression and S phase entry in fibroblasts that express T antigen and therefore lack functional Rb

(Ohtsubo et al. 1995; Hofmann and Livingston 1996). Cyclin E likely controls a different rate limiting step during G1 progression than D cyclin-dependent kinases (Resnitzky and Reed 1995). Consistent with this, enforced expression of the p21 family of CDK inhibitors, which can inhibit the action of both D cyclin- and cyclin E-dependent kinases, can cause a G1 arrest in cells lacking functional Rb (for review see Sherr and Roberts 1995). Thus, cyclin E-CDK2 must have other physiologically important substrates, e.g., p27, cyclin E itself, etc. (see Chap. 2).

The relative contribution of D-type cyclins and cyclin E to the inactivation of Rb is an ongoing line of research. Data suggesting that these cyclins cooperate to induce the phosphorylation and inactivation of Rb exist (in yeast; Hatakeyama et al. 1994). More recently, purified Rb phosphorylated by cyclin D-CDK4, but not by cyclin E-CDK2, has been shown to inactivate the growth suppressive function of Rb (Connell-Crowley et al. 1997). Serine 795 in Rb (located C-terminal to the minimal T/E1A/E7 binding region in exon 23) can be efficiently phosphorylated by cyclin D-CDK4 but not cyclin E-CDK2. Mutation of serine 795 to alanine prevented the inactivation of Rb by CDK4 in a microinjection assay (Connell-Crowley et al. 1997). In an independent study, serine 780 was efficiently phosphorylated in vitro by cyclin D1-CDK4, but not by either cyclin E-CDK2 or cyclin A-CDK2. Serine 780 was also phosphorylated in G1 in a cell cycle-dependent manner. In addition, Rb phosphorylated at serine 780 was incapable of binding to E2F1 in vivo (Kitagawa et al. 1996). Given the role of Rb-E2F complexes in repressing transcription and mediating a G1 arrest (see Sect. 3.1), these results suggest that cyclin D-CDK4-mediated phosphorylation of Rb inactivates its antiproliferative function. The functional role of un- versus hypophosphorylated Rb has also been investigated (S.A. Ezhevsky and S.F. Dowdy, pers. comm.). It was shown that hypophosphorylated Rb associates with E2F in vivo, and therefore is active. Two different experimental approaches were used to demonstrate that expression of $p15^{INK4b}$ and $p16^{INK4a}$ results in the accumulation of unphosphorylated Rb and prevents Rb hypophosphorylation. Clearly, more work needs to be done to address the biological significance of the various phosphorylations of Rb and how they are affected when the upstream regulators of Rb (cyclin D1, CDK4 and $p16^{INK4a}$) are deregulated/mutated in human cancers.

As is the case for Rb, p107 and p130, phosphorylation is cell cycle regulated (Beijersbergen et al. 1995; Mayol et al. 1995, 1996; Xiao et al. 1996). Also like Rb, the growth suppressive functions of both p107 and p130 are inactivated by D-type cyclin-mediated phosphorylation (Zhu et al. 1993; Beijersbergen et al. 1995; Claudio et al. 1996). By contrast to Rb, the growth suppressive function of p107 is not overridden by ectopic expression of cyclins E or A (Zhu et al. 1993; Beijersbergen et al. 1995). However, these cyclins have been reported to overcome the growth suppressive properties of p130 (Claudio et al. 1996). More work will be required to clarify the functional significance of the interaction of cyclin E/A-CDK2 with p107/p130. It is possible that the cyclin-CDK works downstream of p107/p130, by contrast to the situation with D-type cyclins (as discussed in Sect. 2.1).

3 Transcriptional Regulation by Rb, p107 and p130

There has been considerable evolution in our understanding of how the Rb family regulates transcription. These studies have also broadened our understanding of Rb biology. Summarized below are some of the more recent developments in this area of research. To gain a more historical perspective, see Ewen (1994) and Nevins (1992). For a fairly comprehensive list of Rb binding proteins see Taya (1997).

3.1 Transcriptional Repression

The region of E2F1 responsible for binding to Rb is embedded within its transactivation domain (Helin et al. 1992; Kaelin et al. 1992). Complex formation between Rb and E2F results in inhibition of E2F-mediated transactivation. More recently, it has been shown that E2F-mediated targeting of the Rb-E2F and p107-E2F complexes to an E2F site within a promoter actively represses transcription (Adnane et al. 1995; Bremner et al. 1995; Sellers et al. 1995; Weintraub et al. 1995; Zacksenhaus et al. 1996b). How the E2F complexes repress transcription is not clear. Several possibilities have been proposed. Once bound to DNA the Rb-E2F complex may inactivate surrounding transcription factors by blocking their interaction with the basal transcription machinery (Weintraub et al. 1995), an effect possibly mediated through interaction with TAFII250 (Adnane et al. 1995; Shao et al. 1995). Alternatively, the minimal T/E1A/E7 binding region of Rb may bind a second protein in addition to E2F which is involved in transcriptional repression (Sellers et al. 1995). A different twist on this same notion proposes that the A and B regions of Rb and p107 (minimal T/E1A/E7 binding region) interact to form a transcriptional repressor motif (Chow and Dean 1996; Chow et al. 1996; Starostik et al. 1996). In this model, the phosphorylation of specific sites in the C-terminus of Rb by G1 CDKs is proposed to modulate the interaction of the A and B regions and repression (Chow et al. 1996). Phosphorylation site mutants will be needed to help validate this hypothesis.

The potential importance of Rb-E2F repression is highlighted by the observation that inactivation of the *E2F1* gene in mice leads to hyperplasia in certain tissues (Field et al. 1996; Yamasaki et al. 1996). If E2F is the major downstream target for Rb function, then this result would suggest that transcriptional repression by Rb-E2F complexes is a mechanism whereby Rb acts as a tumor suppressor. Considerable effort will be needed to prove such a concept, however, many of the tools to approach this question are becoming available. It will be of interest to determine the phenotype of mouse germline mutations of other E2F family members. In addition, whether the phenotype of $Rb^{+/-}$ mice is altered by crosses with $E2F^{-/-}$ mice will be quite informative.

3.2 Interaction with Other Transcription Factors

Repression of c-fos transcription by Rb was the first demonstration that Rb can regulate transcription (Robbins et al. 1990). This was shown to occur via a retinoblastoma control element present in certain promoters and to be mediated by Sp1 (for reviews see Ewen 1994; Sellers and Kaelin 1996). More recently, a link has been made between Sp1 and E2F. Sp1 has been shown to bind to and act synergistically with E2F1, E2F2 and E2F3, but not E2F4 and E2F5, leading to activation of transcription (Karlseder et al. 1996; Lin et al. 1996). These experiments were performed using the thymidine kinase and the dihydrofolate reductase promoters which contain both E2F and Sp1 binding sites. The role of Rb in these two studies was not examined. It will be of interest to determine under what conditions Rb positively and negatively regulates Sp1-mediated transactivation (see for example Kim et al. 1992a) given that, as noted above, Rb-E2F complexes can repress Sp1 transactivation (Adnane et al. 1995).

hBRM and hBRG1 are mammalian homologues of the SNF2/SWI2 yeast transcriptional activator. SNF2/SWI2 do not bind specific DNA sequences but rather assist DNA binding of certain transcription factors by altering chromatin structure (Carlson and Laurent 1994; Peterson and Tamkun 1995). Rb, p107 and p130 can interact with hBRM/hBRG1 (Dunaief et al. 1994; Singh et al. 1995; Strober et al. 1996). The ability of hBRM to potentiate glucocorticoid receptor-mediated transactivation has been shown to be Rb-dependent (Singh et al. 1995). hBRM and hBRG1, like Rb, have been shown to induce flat cells (upon transfection into SAOS-2 cells) in a manner which depends on their ability to interact with Rb, p107 and p130 (Dunaief et al. 1994; Strober et al. 1996). In this context, it is important to note that Rb-induced flat cells and the induction of G1 cell cycle arrest can be genetically unlinked (W. Sellers and W. Kaelin, pers. comm; see below). Thus it will be of interest to determine if hBRM and hBRG1 induce flat cells in the absence of G1 arrest. Given the role of SNF2/SWI2 in chromatin remodeling, it is interesting to note that $Rb^{-/-}$ MEFs show increased histone H1 phosphorylation and relaxed chromatin structure (Herrera et al. 1996a).

Several lines of investigation suggest an intriguing link between Rb and c-Abl. The c-Abl proto-oncogene is a nonreceptor tyrosine kinase. The extreme C terminus of Rb has been reported to bind to and regulate the kinase activity of c-Abl (Welch and Wang 1993, 1995). Using an SAOS-2 cell (Rb-minus, p53-minus) growth suppression assay, both kinase-active and -inactive c-Abl were shown to overcome the growth suppressive function of ectopically expressed Rb (Welch and Wang 1995). In rodent fibroblasts, kinase-defective c-Abl can accelerate the transition from a quiescent state to S phase (Welch and Wang 1995). Another group has shown that wild type c-Abl negatively regulates cell growth (Sawyers et al. 1994) in a p53-dependent and Rb-independent manner (Goga et al. 1995). Others have reported that the cytostatic function of c-Abl requires both p53 and Rb (Wen et al. 1996). More work needs to be done to fully understand the upstream/downstream nature of c-Abl and Rb/p53 signaling and its relevance to: (1) Rb's role as a tumor suppressor; and (2) Rb's ability to control progression through G1 and entrance into S phase.

The ability of Rb to regulate cell cycle progression is well documented, while the potential role of Rb in growth regulation is largely unexplored. The activity of RNA polymerase I transcription factor UBF (upstream binding factor) can be blocked by Rb in vitro (Cavanaugh et al. 1995). Similarly, RNA polymerase III transcription has been reported to be inhibited by Rb, and $Rb^{-/-}$ MEFs show elevated levels of pol III activity (White et al. 1996). $Rb^{-/-}Rb^{+/+}$ chimeras develop normally (Maandag et al. 1994; Williams et al. 1994b), suggesting that Rb may not play a critical role in general intracellular growth control. However, the results do suggest the possibility that cell cycle regulation by Rb is linked to growth control through its effect on pol I and pol III transcription.

3.3 Transcription and Proliferation/Differentiation

The Id protein family (Id1, Id2 and Id3) are helix-loop-helix (HLH) proteins which lack a basic amino acid domain necessary for DNA binding. They are thought to act in a dominant negative manner by forming inactive complexes with basic HLH transcription factors (Benenzra et al. 1990). Their expression is thought to inhibit lineage-specific transcription. For example, Id1 levels are high in proliferating myoblasts where it is complexed with E2A proteins. During muscle differentiation Id1 levels fall, allowing formation of E12/E47-myoD heterodimers, thereby allowing for MyoD-specific transcription. Id2, but not Id1 or Id3, can associate with Rb, p107 and p130 (Iavarone et al. 1994; Lasorella et al. 1996). Ectopic expression of Id2 can reverse the G1 cell cycle arrest induced by expression of Rb, p107 or p130 in SAOS-2 cells (Iavarone et al. 1994; Lasorella et al. 1996). An intact T/E1A/E7 binding region appears to be required for Rb to interact with Id2 (Iavarone et al. 1994). Consistent with the data suggesting that Id2 binds to and inactivates Rb, U2OS cells (Rb-positive) constitutively expressing Id2 have many of the characteristics of Rb-minus tumor cell lines, including reduced levels of cyclin D1 and loss of cyclin D1-CDK4 complexes. Finally, these Id2 expressing cells are resistant to G1 arrest induced by ectopic expression of $p16^{INK4a}$ (Lasorella et al. 1996). Thus far, complexes between endogenous levels of Rb and Id2 have not been reported, and more work needs to be done to understand the regulation of Id2 expression in physiological settings.

It has been suggested that Rb has a role in terminal adipocyte differentiation. *Rb*$^{+/+}$, but not *Rb*$^{-/-}$ fibroblasts derived from lung buds can be induced to differentiate into adipocytes. Ectopic expression of Rb can restore the ability of *Rb*$^{-/-}$ fibroblasts to differentiate (Chen et al. 1996a). Adipocyte differentiation is known to involve CCAAT/enhancer-binding proteins (C/EBPs) (McKnight et al. 1989). Rb associates with C/EBPs only in differentiating cells. Rb can enhance transcription of a C/EBPβ-responsive promoter (Chen et al. 1996a) and stimulate C/EBP DNA binding (Chen et al. 1996a, 1996b), however, stable Rb-C/EBP-DNA complexes have not been demonstrated. The above observations suggest that Rb has a positive role in differentiation, which is distinct from its role in cell cycle regulation.

The interplay between MyoD and Rb during skeletal muscle differentiation also suggests an active role for Rb in differentiation. Mice lacking *Rb* form normal skeletal muscle when analyzed at the tissue level (Clarke et al. 1992; Jacks et al. 1992; Lee et al. 1992). However, in vitro analysis of myocyctes lacking *Rb* indicates that, unlike their wild-type counterparts, multinucleate myocytes do not permanently withdraw from the cell cycle (Schneider et al. 1994; Novitch et al. 1996). In addition, expression of "late" differentiation markers, i.e., myosin heavy chain, is attenuated in these cells (Novitch et al. 1996). Analysis of transgenic mice expressing low levels of Rb in an *Rb*$^{-/-}$ background have demonstrated similar results (Zacksenhaus et al. 1996a). *Rb*$^{-/-}$ myocytes accumulate in S and G2, but do not progress through mitosis (Novitch et al. 1996). By contrast, myocyte differentiation appears normal in *p107/p130*-deficient mice. Thus, Rb appears to have an important role in expression of muscle-specific genes as well as permanent withdrawal from the cell cycle during muscle differentiation.

Skeletal muscle differentiation is achieved in part by the activities of myogenic basic HLH transcription factors (Weintraub 1993; Lassar and Munsterberg 1994). These transcription factors are involved in activation of muscle-specific genes and irreversible cell cycle withdrawal leading to differentiation of skeletal muscles. MyoD, a member of the basic HLH family, can initiate both of these processes in several cell types (Davis et al. 1987). MyoD-induced expression of muscle-specific genes and withdrawal from the cell cycle are Rb-dependent processes (Gu et al. 1993; Novitch et al. 1996). It has been suggested that they are mediated by the interaction of Rb and MyoD (Gu et al. 1993; Schneider et al. 1994), although these results have not been reproduced by others. *Rb*$^{-/-}$ myocytes, but not wild-type cells, show elevated levels of p107 when induced to differentiate (Schneider et al. 1994), suggesting that p107 may compensate for the loss of *Rb*. Consistent with this notion, ectopic expression of p107, like Rb, can inhibit DNA synthesis in *Rb*$^{-/-}$ myocytes (Schneider et al. 1994). These results were extended by the suggestion that the role of Rb family members in muscle-specific gene expression and cell cycle withdrawal may involve different biochemical activities. Enforced expression of p107 or p130 was less effective than Rb in cooperating with MyoD to induce muscle-specific gene expression (Novitch et al. 1996). Interestingly, expression of p107 was as efficient as Rb in inhibiting DNA synthesis (Schneider et al. 1994), but was only approximately 40 % as efficient as Rb in facilitating MyoD-dependent transcription (Novitch et al. 1996). In addition, expression of p21 or $p16^{INK4a}$, which induces G1 arrest, does not augment MyoD-mediated transactivation (Novitch et al. 1996). Furthermore, G1 cyclin-mediated inhibition of muscle differentiation has been reported to be partially independent of Rb hyperphosphorylation (Skapek et al. 1996).

A genetic analysis of Rb has been undertaken which sheds further light on the possibility that the ability of Rb to regulate cell proliferation and differentiation are separable functions. Using a series of Rb mutants, it has been demonstrated that the ability of Rb to bind to E2F and repress transcription is strongly linked to its ability to cause G1 cell cycle arrest (W. Sellers and W. Kaelin, pers. comm.). The ability of Rb to induce a large/flat cell phenotype in SAOS-2 cells and its ability to cooperate with MyoD in transcription were strongly correlated and genetically separable from E2F binding. Importantly, the ability of Rb to cause G1 arrest could be dissociated from its ability to induce large/flat cells (W. Sellers and W. Kaelin, pers. comm.). Consistent with this, Rb, p107 and p130 were all equally capable of inducing G1 arrest in SAOS-2 cells, but only Rb, as opposed to less effective p107 and p130, was efficient in inducing a flat/large cell phenotype in these cells. Thus, the relative abilities of Rb, p107 and p130 to induce flat/large cells and to facilitate MyoD transcriptional activation parallel one another. Introduction of Rb into SAOS-2 cells (human osteosarcoma) induced the expression of bone alkaline phosphatase, consistent with the idea that the flat/large cells represent a differentiation and not a cell cycle phenotype. Interestingly, two naturally occurring partially penetrant Rb mutants failed to bind to E2F and repress transcription but were competent in inducing large/flat cells and in cooperating with MyoD. These results suggest the possibility that both the E2F binding and differentiation function of Rb contribute to its overall function as a tumor suppressor. (See Chap. 13 for a more details on cell cycle regulation and differentiation).

4
Signal Transduction and Rb

Both positive and negative growth signals act during the G1 phase of the cell cycle to influence cell proliferation. These extracellular signals act up to a point in late G1 termed "restriction point" (R) (Pardee 1974) after which the cells become largely refractory to serum factors influencing proliferation. Once cells have passed this point of serum dependence, they are committed to progress until mitosis (see Temin 1971; Pardee 1974, 1989; Zetterberg et al. 1995). That Rb might regulate progression through G1 was first suggested by the observations that SV40 T antigen targets the un- or hypophosphorylated forms of Rb (the forms present in G1; Ludlow et al. 1989, 1990) and that ectopic expression of Rb could cause G1 arrest in Rb-minus SAOS-2 cells (Goodrich et al. 1991). Importantly, in this setting Rb remains hypophosphorylated. In addition, *Rb*$^{-/-}$ MEFs have a shortened G1 interval (Lukas et al. 1995a; Herrera et al. 1996c). That the function of Rb might be influenced by extracellular signals was originally suggested by the observation that phosphorylation of Rb in G1 roughly coincides with the passage of cells from serum dependence to serum independence (Buchkovich et al. 1989; Chen et al. 1989; DeCaprio et al. 1989; Mihara et al. 1989). As discussed below, these observations have been significantly extended to place Rb as an important transducer of extracellular signals influencing cell proliferation.

4.1 TGF-β Signaling Pathways

How transforming growth factor-beta (TGF-β) induces G1 cell cycle arrest has received considerable attention. The precise mechanism of action appears to be cell type dependent. However, several lines of evidence suggest that Rb is a critical player. TGF-β treatment leads to the accumulation of unphosphorylated Rb (Laiho et al. 1990). The viral oncoproteins SV40 T antigen, adenovirus E1A, and HPV E7 can all overcome a TGF-β-induced G1 arrest (Laiho et al. 1990; Pietenpol et al. 1990). Regulation of CDK4 and CDK2 activity/expression, both of which are thought to be upstream regulators of Rb function during G1 (at least in part in the case of CDK2), are affected by TGF-β treatment (Ewen et al. 1993b; Geng and Weinberg 1993; Koff et al. 1993; Slingerland et al. 1994). Ectopic expression of E2F1, a downstream target of Rb, can override a TGF-β-induced G1 arrest (Schwartz et al. 1995). The CDK4/CDK6 specific inhibitor p15^{INK4b}, which like p16^{INK4a} functions in an Rb-dependent manner to cause G1 arrest, is induced by TGF-β (Hannon and Beach 1994). Finally, using *Rb*$^{-/-}$ and *Rb*$^{+/+}$ primary fibroblasts from matched littermates, growth inhibition caused by TGF-β was shown to depend on Rb, further strengthening the link between TGF-β action and Rb (Herrera et al. 1996b).

c-Myc is also thought to be a target of TGF-β action. TGF-β has been shown to downregulate c-Myc expression in an Rb-dependent manner (Pietenpol et al. 1990, 1991), although Rb-independent mechanisms also exist (Zentella et al. 1991). Consistent with the above- mentioned role of CDK2 activity in TGF-β action, a c-Myc-dependent step in formation of active cyclin E-CDK2 has been identified (Steiner et al. 1995). In addition, enforced expression of c-Myc can reverse TGF-β-induced (Alexandrow et al. 1995), p27-induced (Vlach et al. 1996) and an Rb-induced (Goodrich and Lee 1992) G1 arrest. How c-Myc overrides a TGF-β- and Rb-induced cell cycle arrest remains to be determined.

4.2 D-Type Cyclins, Ras, and Rb

Consistent with the idea that Rb phosphorylation is a mitogen-dependent process the synthesis of D-type cyclins, unlike that of other cyclins, is regulated by extracellular factors (Matsushime et al. 1991; Sherr 1993). Further studies have suggested that multiple regulatory pathways may contribute to the synthesis of cyclins D1, D2 and D3 (Ajchenbaum et al. 1992; Cocks et al. 1992; Won et al. 1992; Sewing et al. 1993; Winston and Pledger 1993; Lukas et al. 1995c; Brooks et al. 1996). Mice genetically lacking either *cyclin D1* or *D2*, which display tissue-specific developmental defects, have reinforced these data at the level of the whole animal (Fantl et al. 1995; Sicinski et al. 1995, 1996).

A role for Rb in signal transduction is suggested by the interplay with Ras. Ras, an inner plasma membrane-bound G protein, plays a significant role in receiving and transducing extracellular signals (for a review see Barbacid 1987). Ras proteins function as downstream mediators for several membrane-bound receptor and nonreceptor tyrosine kinases (reviewed in Egan and Weinberg 1993; Pronk and Bos 1994; Marshall 1995). Ras proteins affect a number of cellular processes, including cell cycle progression. Overexpression of (oncogenic) Ras is sufficient to drive fibroblasts from G0 into S phase in the absence of serum (Feramisco et al. 1984; Stacey and Kung 1984). Ras activity has also been shown to be required for progression from G0 to the G1/S transition (Mulcahy et al. 1985; Feig and Cooper 1988; Dobrowolski et al. 1994a; Okuda et al. 1994). There are data suggesting that the Ras/Raf/MEK/MAPK pathway converges on cyclin D1 expression/activity (Filmus et al. 1994; Herber et al. 1994; Albanese et al. 1995; Liu et al. 1995; Lavoie et al. 1996; Lukas et al. 1996a; Winston et al. 1996). Two lines of investigation have suggested that Ras operates upstream of Rb. First, ectopic expression of p16^{INK4a}, which functions in an Rb-dependent manner, blocks DNA synthesis induced by oncogenic Ras (Serrano et al. 1995). Also consistent with this being an Rb-dependent phenomenon, p16^{INK4a} was shown to inhibit Ras plus Myc-mediated transformation, but not Ras plus E1A-mediated transformation (with E1A, unlike c-Myc, directly binding to and functionally inactivating Rb; Serrano et al. 1995). Second, using *Rb*$^{-/-}$ MEFs, it has been demonstrated that G1 arrest induced by inactivation of Ras activity is Rb-dependent (Mittnacht et al. 1997; Peeper et al. 1997). Consistent with the idea that Rb might mediate G1 arrest, expression of dominant-negative Ras, which blocks Ras-dependent mitogenic signaling, results in the accumulation of unphosphorylated Rb (Peeper et al. 1997). Ectopic expression of cyclin D1 can override G1 arrest induced by Ras inactivation, which is consistent with the above data suggesting that this cyclin functions downstream of Ras (Peeper et al. 1997) and with the observation that adenovirus E1A can override the requirement for Ras activity during G1 progression and S phase entry (Stacey et al. 1994). Importantly, in contrast to the above noted results which apply to mid- to late G1, Ras function during the G0 to G1 transition is Rb-independent (Peeper et al. 1997).

4.3
v-Abl, c-Myc, and Rb

A series of experiments have suggested that Rb might be involved in v-Abl oncogenic signaling. c-Myc and Ras have been shown to be essential for v-Abl-mediated transformation (Sawyers et al. 1992 1995; Zou et al. 1997). By contrast to c-Abl discussed above, v-Abl is cytoplasmic and is thought to function in a different signal transduction pathway than c-Abl. Induction of c-Myc transcription by v-Abl has been shown to require the activities of CDK2 and CDK4. This was suggested by the observation that v-Abl expression leads to the activation of these kinases and that v-Abl-mediated activation of c-Myc transcription can be inhibited by ectopic expression of p21 (Zou et al. 1997). Consistent with a role for CDK4 in this pathway, cyclin D1 has been shown to act synergistically with v-Abl in transformation and to rescue a src-homology 2 (SH2) domain mutant of BCR-ABL for transformation (Afar et al. 1995). In both of these assays, cyclin D1 required an intact N-terminal LXCXE motif previously shown to be required for complex formation with Rb (Dowdy et al. 1993; Ewen et al. 1993a). This suggests that cyclin D1 might function in v-Abl signaling by an Rb-dependent pathway (Afar et al. 1995). Consistent with this notion, v-Abl-mediated induction of c-Myc is accompanied by the hyperphosphorylation of Rb, p107 and p130 and can be blocked by overexpression of Rb (Zou et al. 1997). On the basis of these latter results, Zou et al. suggested that cyclin D1-CDK4 phosphorylation of Rb and the consequent release of free E2F were responsible for the activation of c-Myc transcription. This suggestion was based on the previous observation that E2F can regulate c-Myc transcription (Hamel et al. 1992; Oswald et al. 1996) and the observation that v-Abl can activate c-Myc transcription through an E2F site (Wong et al. 1995). This wiring diagram will not likely be so simple given that the relationship between cyclin D1 and c-Myc is not straightforward (see Afar et al. 1995; Roussel et al. 1995 for further discussion).

4.4
Gene Knockouts and Signal Transduction

Due to the complexity of the various tissues and organs which make up a mammal, it is difficult to consider signal transduction pathways in the same light as discussed above. However, to some degree the overall importance of a particular molecule in signal transduction can be assessed. In terms of Rb, this is exemplified by the analysis of the rescue (generating live mice) of *Rb*-deficient cells in chimeric mice (Maandag et al. 1994; Williams et al. 1994b). In the various tissues analyzed, $Rb^{-/-}$ cells made up a significant percentage of the total population. How the chimeras rescue the embryonic lethality seen with $Rb^{-/-}$ embryos is not entirely clear. It was argued that in cells where Rb function is critical, $Rb^{-/-}$ cells were either selected against or rescued by *Rb*-positive cells. The observation that erythropoiesis was rescued in the chimeras suggests that Rb function in differentiation of erythrocytes is not cell autonomous (Maandag et al. 1994; Williams et al. 1994b), and there is the possibility that *Rb*-positive cells produce the necessary extracellular signals to allow differentiation of the *Rb*-minus cells. Here the involvement of Rb in signal transduction is such that the Rb-positive cells would be sending the signal. Consistent with this notion, Rb is thought to positively regulate the expression of secreted factors such as TGF-β2 (Kim et al. 1992b) and IL6 (Chen et al. 1996b). On the other hand, $Rb^{-/-}$ chimeras also exhibit retinal abnormalities, which suggest a cell autonomous role for Rb in the development of the retina. Ectopic mitosis and cellular degeneration were noted in these mice at 16.5 to 18.5 days gestation in the inner half of the retina, i.e., at a time when Rb expression is normally confined to this region (Szekely et al. 1992; Maandag et al. 1994; Williams et al. 1994b). In this setting, Rb function may be required to receive, rather then send, extracellular signals influencing proliferation and/or differentiation.

The results from the *E2F1*-deficient mice also suggest a critical role for the Rb family in signal transduction. Several cell types of these knockout mice show a high proliferation rate (Field et al. 1996; Yamasaki et al. 1996). Three models were proposed to explain the findings. In a cell autonomous role for E2F, it was proposed that Rb-E2F1 complexes play a critical role in repressing the expression of nuclear genes which promote cell proliferation (Field et al. 1996; Weinberg 1996; Yamasaki et al. 1996). Loss of this repression would lead to uncontrolled proliferation. In a non cell autonomous model for E2F function, it was argued that hyperplastic/neoplastic growth may be due to defects in the cellular environment of *E2F1*-deficient cells (Weinberg 1996). In this model it might be argued that the primary role of E2F1 is to repress the expression of secreted factors which function in an autocrine/paracrine fashion to suppress cell growth. Loss of E2F1 function in this setting would also lead to deregulated growth. The E2F-responsive genes identified to date are nuclear and involved in DNA synthesis rather than cell signaling, an observation arguing in favor of a cell autonomous role for E2F1. However, given the above arguments for the role of Rb in non cell autonomous cell proliferation and differentiation, E2F1 may have a role in extracellular signaling, assuming this family of transcription factors represents the critical downstream target for Rb function. The third model stems from the observation that tumor formation can result from an escape from apoptosis and that deregulated expression of E2F1 can induce apoptosis (see Sect. 5.2). Loss of *E2F1* may result in reduced apoptosis resulting in tumor development (Field et al. 1996; Yamasaki et al. 1996).

5
Cancer

The Rb/cyclin D1/CDK4/p16^{INK4a} pathway is disrupted in a high percentage of human tumors (Hall and Peters 1996; Sherr 1996). Rb is inactivated in tumors by either inactivating mutations of the gene or expression of HPV E7 (in cervical cancer). Cyclin D1 is overexpressed and/or the gene is amplified. *CDK4* is found either amplified or, in one case, bearing a mutation which renders the protein insensitive to binding and regulation by p16^{INK4a}. p16^{INK4a} is inactivated by either mutation or methylation of the gene, which inhibits its expression. Importantly, disruption of different components of this pathway should not necessarily be thought of as functionally equivalent. This follows from several lines of investigation. For example, when analyzing human cancers, mutations in p16^{INK4a} in familial melanoma do not predispose individuals to retinoblastoma. Moreover, when studying mouse models, loss of *Rb* has a different outcome than loss of *p16^{INK4a}* (Serrano et al. 1996; see also Chap. 12).

The data reviewed above indicate that Rb plays a critical role in regulation of cell proliferation, differentiation and signal transduction. The *Rb* knockout mice indicate that Rb plays a role in murine development and suppression of tumor formation (in the heterozygotes). However, these results do not explain how or why loss of Rb function contributes to tumor formation. A major clue to answering this question has come from a line of investigations which have gone on simultaneously with the study of Rb, and involved the study of the p53 tumor suppressor.

5.1 p53 Function and Cancer

Like elements of the Rb pathway, which is disrupted in a high percentage of human cancer, *p53* mutations are found in more than 50 % of human cancers (Hollstein et al. 1991; Levine et al. 1991). The most studied biochemical function of p53 is its transcriptional activation function. Among p53's most studied biological functions are its ability to cause G1 cell cycle arrest, suppress transformation, induce apoptosis, and prevent genomic instability. An analysis of various p53 mutants suggests that these four functions of p53 may involve separate activities of the protein. For example, there are reports that induction of apoptosis and suppression of transformation may, in part, be independent of p53 transcriptional function (reviewed in Levine 1997). Transcription- dependent and -independent functions of p53 have been shown to synergize to elicit a full apoptotic response as a function of cell type and context (X. Chen et al. 1996; Polyak et al. 1996). In terms of cancer, p53 appears to suppress tumor growth induced by oncogenic events by inducing apoptosis (Symonds et al. 1994). With respect to the clinical relevance of this observation, it has been demonstrated that many chemotherapeutic agents work by inducing apoptosis. Using an animal model, it has been demonstrated that following either gamma radiation or adriamycin, a high apoptotic index and tumor regression correlated with a p53 wild-type status (Lowe et al. 1994).

5.2
p53 and Rb and the Decision Between G1 Arrest and Apoptosis

An emerging theme in p53 research is that G1 arrest and apoptosis, though partially genetically separable, are two alternative biological outcomes of p53 action. The cellular decision between these two fates appears to be a function of the particular signals which impinge upon p53 (for reviews see Ko and Prives 1996; Levine 1997). The G1 arrest function of p53 is mediated, in part, by one of its transcriptional targets, the CDK inhibitor p21 (Dulic et al. 1994; Brugarolas et al. 1995; Deng et al. 1995). p21 can arrest cells in both an Rb-dependent and -independent manner (Sherr and Roberts 1995), and can protect cells from apoptosis (X. Chen et al. 1996; Polyak et al. 1996; Wang and Walsh 1996). How p53 induces apoptosis is less clear and likely involves several pathways (Levine 1997).

Given that p21 can inhibit the kinases responsible for Rb phosphorylation, it would appear that Rb functions downstream of p53 to mediate G1 arrest. In terms of apoptosis, deregulation of the Rb pathway appears to be one of the signals that can trigger p53-dependent apoptosis, thus placing Rb upstream of p53. Put another way, p53 appears to monitor whether the Rb pathway controlling progression through G1 is intact. Several lines of investigation support this hypothesis. The first was provided by the observation that p53 induces apoptosis caused by expression of adenovirus E1A, which binds to and inactivates Rb, p107 and p130 as well as other proteins such as p300/CBP (Debbas and White 1993; Lowe and Ruley 1993). Deregulated expression of E2F1, a major downstream target of Rb, can induce apoptosis in a p53-dependent manner (Qin et al. 1994; Shan and Lee 1994; Wu and Levine 1994). Consistent with these data, T antigen binding to both Rb family members and p53 contributes to tumor formation in animal models (Chen et al. 1992; Saenz Robles et al. 1994). Expression of HPV E7 in either the lens or photoreceptors was shown to induce p53-dependent apoptosis (Howes et al. 1994; Pan and Griep 1994). Importantly, in these two studies inactivation of p53 resulted in tumor development. By studying the lenses of $Rb^{-/-}$ and $Rb^{-/-}p53^{-/-}$ murine embryos, it was demonstrated that p53-dependent apoptosis was produced by loss of *Rb* (Morgenbesser et al. 1994). These studies were extended by demonstrating that germline mutations in *Rb* and *p53* cooperate in tumorigenesis (Williams et al. 1994a).

Elucidating the mechanism(s) by which "deregulation of the Rb pathway" elicits p53-dependent and -independent apoptosis is one of the challenges in the field. A clue to this question may be provided by the analysis of c-Myc-mediated apoptosis. E2F1 shares with c-Myc the ability to drive cells from quiescence into S phase and to induce apoptosis in the absence of serum growth factors (see Chaps. 9, 10). c-Myc-mediated apoptosis, like that mediated by E2F1, is p53 dependent (Hermeking and Eick 1994; Wagner et al. 1994). It has been demonstrated that c-Myc-induced apoptosis can be inhibited by Ras signaling through phosphatidylinositol-3-kinase (PI3K) and Akt/PKB (Kauffmann-Zeh et al. 1997; Kennedy et al. 1997). In this context, Ras is thought to function upstream of PI3K (Rodriguez-Viciana et al. 1994). In this report the p53 dependence of the observed phenomena was not investigated. The Akt/PKB proto-oncogene is likely a target of PI3K (Burgering and Coffer 1995; Franke et al. 1995). E2F1 and c-Myc-mediated apoptosis may be similar given that there is some data suggesting that c-Myc expression is controlled, at least in part, by E2F1. Thus, it will be of interest to determine if Ras signaling is involved in E2F1-dependent apoptosis. Three other possibilities (which are not completely unrelated) have been suggested on the basis of our current understanding (Macleod et al. 1996). First, E2F1 may regulate the expression of genes involved in inducing apoptosis, possibly in cooperation with p53. A second possibility is that p53 is induced by deregulation of Rb-E2F-mediated transcriptional repression. A third possibility is that exhaustion of growth and survival factors (perhaps operating through Ras/PI3K/Akt) induces p53 when cell proliferation/growth is deregulated by loss of Rb function.

The clinical implications of the above findings are exemplified by the observation that expression of E1A can sensitize cells to p53-dependent apoptosis induced by chemotherapeutic agents (Lowe et al. 1993). Using $Rb^{-/-}$ and $Rb^{+/+}$ MEFs it has been demonstrated that p53 is induced in response to chemotherapeutic agents, however, the $Rb^{-/-}$ MEFs undergo apoptosis and the $Rb^{+/+}$ cells arrest in G1 under these conditions (Almasan et al. 1995). Importantly, as the above results suggest, Rb has a protective role against apoptosis (Haas-Kogan et al. 1995; Macleod et al. 1996). Thus, it appears that disruption of the Rb and p53 pathways can cooperate in transformation, as originally suggested by the observation that three classes of DNA tumor viruses, adenoviruses, polyomaviruses and human papillomaviruses, all target these two tumor suppressors. In addition, the status of these two pathways appears to dictate the effectiveness of chemotherapy.

5.3
p107 and p130 in Human Cancer?

There is no direct evidence that loss of *p107* or *p130* is involved in either human or mouse tumor formation. This does not necessarily imply that these Rb family proteins do not contribute to human cancer. As discussed earlier, the upstream factors influencing Rb function, i.e., D-type cyclins and p16^{INK4a}, also influence the function of p107 and p130. Given that these upstream components are frequently affected in human cancer, it is likely that the functions of p107 and p130 are deregulated in human cancers. A similar argument can be made for E2F. No E2F mutations have been reported in human cancer, however, the activity of E2F is thought to be deregulated in human cancer because Rb, cyclin D1, CDK4 and p16^{INK4a} are targets in human cancer. Furthermore, p107 and Rb have been shown to have overlapping functions in vivo (Lee et al. 1996); and p107 and p130 also have shared functions in vivo (Cobrinik et al. 1996). The observation that p107 and p130 can compensate for loss of Rb, in various assays, also argues in favor of a role for p107 and p130 in human cancer. In addition, it may be the compensation for Rb loss by p107/p130 that allows Rb loss to contribute to tumor formation.

A more direct experimental approach has suggested the involvement of functional inactivation of p107 and p130 in transformation. It has been argued on the basis of a genetic analysis of SV40 T antigen that its transforming ability is achieved, at least in part, through the LXCXE motif, which mediates binding to the Rb family of proteins (DeCaprio et al. 1988; Dyson 1989, 1990; Ewen et al. 1989, 1991; Ewen 1994). Primary MEFs derived from $Rb^{-/-}$ embryos have been used to study transformation mediated by SV40 T antigen. The LXCXE motif is required for T antigen-mediated transformation of $Rb^{-/-}$ MEFs, suggesting that inactivation of p107 and p130 by T antigen may contribute to the transforming mechanism of this oncoprotein (Christensen and Imperiale 1995; Zalvide and DeCaprio 1995). Importantly, how and whether these results impact on the development of human cancer remains to be determined. The results have been extended by demonstrating that T antigen alters the phosphorylation state of p107 and p130 (Stubdal et al. 1996) and stability of p130 (H. Stubdal and J.A. DeCaprio, pers. comm.), which is in contrast to the result found with Rb (Ludlow et al. 1989; Ludlow et al. 1990). Both the LXCXE motif and the first 82 amino acids of T antigen were required for this effect on p107 and p130 (Stubdal et al. 1996). The N-terminal 82 amino acids of T antigen have homology to the J domain of the DnaJ family of molecular chaperone proteins. The J domain homology region of T antigen has been shown to be required for altering the phosphorylation state of p107 and p130 and for p130 degradation (H. Stubdal and J.A. DeCaprio, pers. comm.). In terms of biology, the J domain homology region of T antigen was shown to confer a growth advantage in wild type MEFs, but was not required in MEFs derived from $p107^{-/-}p130^{-/-}$ embryos (J. Zalvide and J.A. DeCaprio, pers. comm.). In independent studies, cell lines expressing HPV E7 were shown to express lower levels of all three Rb family members, suggesting that E7 also targets p107 and p130 (Boyer et al. 1996; Jones and Munger 1997). Together these results suggest that functional inactivation of p107 and p130 contributes to T antigen- and E7-mediated neoplastic growth. Why loss of Rb, and not p107/p130, is selected for during tumor development remains a question in the field.

Acknowledgments I thank Drs. David Cobrinik, James DeCaprio, Steve Dowdy, Nick Dyson, Sergei Ezhevsky, Robert Hurford, Bill Kaelin, Myung-Ho Lee, Bill Sellers, Hilde Stubdal and Juan Zalvide for communicating data before publication. I thank Drs. Peter Adams, Francesco Hofmann, Mohamed Ladha, Christine McMahon, Bill Sellers and Hilde Stubdal for critical review of this chapter.

References

Adams PD, Sellers WR, Sharma SK, Wu AD, Nalin CM, Kaelin WG Jr (1996) Identification of a cyclin-cdk2 recognition motif present in substrates of p21-like cyclin-dependent kinase inhibitors. Mol Cell Biol 16: 6623–6633

Adnane J, Shao Z, Robbins PD (1995) The retinoblastoma susceptibility gene product represses transcription when directly bound to the promoter. J Biol Chem 270: 8837–8843

Afar DEH, McLaughlin J, Sherr CJ, Witte ON, Roussel MF (1995) Signaling by ABL oncogenes through cyclin D1. Proc Natl Acad Sci USA 92: 9540–9544

Ajchenbaum F, Ando K, DeCaprio JA, Griffin JD (1992) Independent regulation of human D-type cyclin gene expression during G1 phase in primary human T lymphocytes. J Cell Biol 268: 4113–4119

Albanese C, Johnson J, Watanabe G, Eklund N, Vu D, Arnold A, Pestell RG (1995) Transforming p21ras mutants and c-Ets-2 activate the cyclin D1 promoter through distinguishable regions. J Biol Chem 270: 23589–23597

Alexandrow MG, Kawabata M, Aakre M, Moses HL (1995) Overexpression of the c-Myc oncoprotein blocks the growth-inhibitory response but is required for the mitogenic effects of transforming growth factor β1. Proc Natl Acad Sci USA 92: 3239–3243

Almasan A, Yin Y, Kelly RE, Lee EY, Bradley A, Li W, Bertino JR, Wahl GM (1995) Deficiency of retinoblastoma protein leads to inappropriate S-phase entry, activation of E2F-responsive gene, and apoptosis. Proc Natl Acad Sci USA 92: 5436–5440

Arroyo M, Raychaudhuri P (1992) Retinoblastoma-repression of E2F-dependent transcription depends on the ability of the retinoblastoma protein to interact with E2F and is abrogated by the adenovirus E1A oncoprotein. Nucleic Acids Res 20: 5947–5954

Asono M, Nevins JR, Wharton RP (1996) Ectopic E2F expression induces S phase and apoptosis in *Drosophila* imaginal discs. Genes Dev 10: 1422–1432

Bagchi S, Weinmann R, Raychaudhuri P (1991) The retinoblastoma protein copurifies with E2F-I, an E1A-regulated inhibitor of the transcription factor E2F. Cell 65: 1063–1072

Baldin V, Lukas J, Marcote MJ, Pagano M, Draetta G (1993) Cyclin D1 is a nuclear protein required for cell cycle progression in G1. Genes Dev 7: 812–821

Bandara LR, Adamczewski JP, Hunt T, La Thangue NB (1991) Cyclin A and the retinoblastoma gene product complex with a common transcription factor. Nature 352: 249–251

Barbacid M (1987) ras Genes. Annu Rev Biochem 56: 779–827

Beijersbergen RL, Hijman EM, Zhu L, Bernards L (1994a) Interaction of c-Myc and the pRb-related protein p107 results in inhibition of c-Myc mediated transactivation. EMBO J 13: 4080–4086

Beijersbergen RL, Kerkhoven RM, Zhu L, Carlee L, Voorhoeve PM, Bernards R (1994b) E2F-4, a new member of the E2F gene family, has oncogenic activity and associates with p107 in vivo. Genes Dev 8: 2680–2690

Beijersbergen RL, Carlee L, Kerkhoven RM, Bernards R (1995) Regulation of the retinoblastoma protein-related p107 by G1 cyclin complexes. Genes Dev 9: 1340–1353

Benenzra R, Davis RL, Lockshon D, Turner DL, Weintraub H (1990) The protein Id: a negative regulator of helix-loop-helix DNA binding proteins. Cell 61: 49–59

Botz J, Zerfass-Thome K, Spitkovsky D, Delius H, Vogt B, Eilers M, Hatzigeorgiou A, Jansen-Durr P (1996) Cell cycle regulation of the murine cyclin E gene depends on an E2F binding site in the promoter. Mol Cell Biol 16: 3401–3409

Boyer SN, Wazer DE, Band V (1996) E7 protein of human papilloma virus-16 induces degradation of retinoblastoma protein through the ubiquitin-proteasome pathway. Cancer Res 56: 4620–4624

Bremner R, Cohen BL, Sopta M, Hamel PA, Ingles CJ, Gallie BL, Phillips RA (1995) Direct transcriptional repression by pRB and its reversal by specific cyclins. Mol Cell Biol 15: 3256–3265

Brooks AR, Shiffman D, Chan CS, Brooks EE, Milner PG (1996) Functional analysis of the human cyclin D2 and cyclin D3 promoters. J Biol Chem 271: 9090–9099

Brugarolas J, Chandrasekaran C, Gordon JI, Beach D, Jacks T, Hannon GJ (1995) Radiation-induced cell cycle arrest by p21 deficiency. Nature 377: 552–557

Buchkovich K, Duffy LA, Harlow E (1989) The retinoblastoma protein is phosphorylated during specific phases of the cell cycle. Cell 58: 1097–1105

Burgering BMT, Coffer PJ (1995) Protein kinase B (c-Akt) in phosohatidylinositol-3-OH kinase signal transduction. Nature 376: 599–602

Cao L, Faha B, Dembski M, Tsai L-H, Harlow E, Dyson N (1992) Independent binding of the retinoblastoma protein and p107 to the transcription factor E2F. Nature 355: 176–179

Carlson M, Laurent BC (1994) The SNF/SWI family of global transcriptional activators. Curr Opin Cell Biol 6: 396–402

Cavanaugh AH, Hempel WM, Taylor LJ, Rogalsky V, Todorov G, Rothblum LI (1995) Activity of RNA polymerase I transcription factor UBF blocked by *Rb* gene product. Nature 374: 177–180

Chen J, Saha P, Kornbluth S, Dynlacht BD, Dutta A (1996) Cyclin-binding motifs are essential to the function of p21CIP1. Mol Cell Biol 16: 4673–4682

Chen J, Tobin GJ, Pipas JM, Van Dyke T (1992) T-antigen mutant activities in vivo: roles of p53 and pRB binding in tumorigensis of the choroid plexus. Oncogene 7: 1167–1175

Chen P, Scully P, Shew J, Wang JYJ, Lee W (1989) Phosphorylation of the retinoblastoma gene product is modulated during the cell cycle and cellular differentiation. Cell 58: 1193–1198

Chen P-L, Riley DJ, Chen Y, Lee W-H (1996a) Retinoblastoma protein positively regulates terminal adipocyte differentiation through direct interaction with C/EBPs. Genes Dev 10: 2794–2804

Chen P-L, Riley DJ, Chen-Kiang S, Lee W-H (1996b) Retinoblastoma protein directly interacts with and activates the transcription factor NF-IL6. Proc Natl Acad Sci USA 93: 465–469

Chen X, Ko LJ, Jayaraman L, Prives C (1996) p53 levels, functional domains, and DNA damage determine the extent of the apoptotic response of tumor cells. Genes Dev 10: 2438–2451

Chittenden T, Livingston DM, Kaelin WG Jr (1991) The T/E1A-binding domain of the retinoblastoma product can interact selectively with a sequence-specific DNA-binding protein. Cell 65: 1073–1082

Chow KNB, Dean DC (1996) Domains A and B in the Rb pocket interact to form a transcriptional repressor motif. Mol Cell Biol 16: 4862–4868

Chow KNB, Starostik P, Dean DC (1996) The Rb family contains a conserved cyclin-dependent kinase-regulated transcriptional repressor motif. Mol Cell Biol 16: 7173–7181

Christensen JB, Imperiale MJ (1995) Inactivation of the retinoblastoma susceptibility protein is not sufficient for the transformation function of the conserved region 2-like domain of simian virus 40 large T antigen. J Virol 65: 3945–3948

Clarke AR, Maandag ER, van Roon M, van der Lugt NMT, van der Valk M, Hooper ML, Berns A, Riele H (1992) Requirement for a functional *Rb-1* gene in murine development. Nature 359: 328–330

Claudio PP, Howard CM, Baldi A, De Luca A, Fu Y, Condorelli G, Sun Y, Colburn N, Calabretta B, Giordano A (1994) p130/pRb2 has growth suppressive properties similar to yet distinctive from those of retinoblastoma family members pRb and p107. Cancer Res 54: 5556–5560

Claudio PP, DeLuca A, Howard CM, Baldi A, Firpo EJ, Paggi MG, Giordano A (1996) Functional analysis of pRb2/p130 interaction with cyclins. Cancer Res 56: 2003–2008

Cobrinik D, Whyte P, Peeper DS, Jacks T, Weinberg RA (1993) Cell cycle-specific association of E2F with the p130 E1A-binding protein. Genes Dev 7: 2392–2404

Cobrinik D, Lee M-H, Hannon G, Mulligan G, Bronson RT, Dyson N, Harlow E, Beach D, Weinberg RA, Jacks T (1996) Shared role of the pRB-related p130 and p107 proteins in limb development. Genes Dev 10: 1633–1644

Cocks BG, Vairo G, Bodrug SE, Hamilton JA (1992) Suppression of growth factor-induced CYL1 cyclin gene expression by antiproliferative agents. J Biol Chem 267: 12307–12309

Connell-Crowley L, Harper JW, Goodrich DW (1997) Cyclin Dl/cdk4 regulates retinoblastoma protein-mediated cell cycle arrest by site-specific phosphorylation. Mol Biol Cell 8: 287-301

Davis RL, Weintraub H, Lassar AB (1987) Expression of a single transfected cDNA converts fibroblasts to myoblasts. Cell 51: 987–1000

Debbas M, White E (1993) Wild-type p53 mediates apoptosis by E1A, which is inhibited by E1B. Genes Dev 7: 546–554

DeCaprio JA, Ludlow JW, Figge J, Shew J-Y, Huang C-M, Lee W-H, Marsilio E, Paucha E, Livingston DM (1988) SV40 large T antigen forms a specific complex with the product of the retinoblastoma susceptibility gene. Cell 54: 275–283

DeCaprio JA, Ludlow JW, Lynch D, Furukawa Y, Griffin J, Piwnica-Worms H, Huang C-M, Livingston DM (1989) The product of the retinoblastoma susceptibility gene has properties of a cell cycle regulatory element. Cell 58: 1085–1095

DeGregori J, Kowalik T, Nevins JR (1995a) Cellular targets for activation by E2F1 transcription factor induces DNA synthesis- and G1/S-regulatory genes. Mol Cell Biol 15: 4215–4224

DeGregori J, Leone G, Ohtani K, Miron A, Nevins JR (1995b) E2F-1 accumulation bypasses a G1 arrest resulting from the inhibition of G1 cyclin-dependent kinase activity. Genes Dev 9: 2873–2887

Deng C, Zhang P, Harper JW, Elledge SJ, Leder P (1995) Mice lacking $p21^{CIP1/WAF1}$ undergo normal development, but are defective in G1 checkpoint control. Cell 82: 675–684

Dobrowolski S, Harter M, Stacey DW (1994a) Cellular ras activity is required for passage through multiple points of the G0/G1 phase in Balb/c 3T3 cells. Mol Cell Biol 14: 5441–5449

Dobrowolski SF, Stacey DW, Harter ML, Stine JT, Hiebert SW (1994b) An E2F dominant negative mutant blocks E1A induced cell cycle progression. Oncogene 9: 2605–2612

Dowdy SF, Hinds PW, Louie K, Reed SI, Arnold A, Weinberg RA (1993) Physical interaction of the retinoblastoma protein with human D cyclins. Cell 73: 499–511

Dulic V, Kaufmann WK, Wilson SJ, Tisty TD, Lees E, Harper JW, Elledge SJ, Reed SI (1994) p53-dependent inhibition of cyclin-dependent kinase activities in human fibroblasts during radiation-induced G1 arrest. Cell 76: 1013–1023

Dunaief JL, Stober BE, Guha S, Khavari PA, Alin K, Luban J, Begemann M, Crabtree GR, Goff SP (1994) The retinoblastoma protein and BRG1 form a complex and cooperate to induce cell cycle arrest. Cell 79: 119–130

Duronio RJ, O'Farrell PH (1995) Developmental control of the G1 to S transition in *Drosophila*: cyclin E is a limiting downstream target of E2F. Genes Dev 9: 1456–1468

Duronio RJ, O'Farrell PH, Xie JE, Brook A, Dyson N (1995) The transcription factor E2F is required for S phase entry during *Drosophila* embryogenesis. Genes Dev 9: 1445–1455

Duronio RJ, Brook A, Dyson N, O'Farrell PH (1996) E2F-induced S phase requires cyclin E. Genes Dev 10: 2505–2513

Dynlacht BD, Flores O, Lees JA, Harlow E (1995) Differential regulation of E2F trans-activation by cyclin/cdk2 complexes. Genes Dev 8: 1772–1786

Dyson N, Buchkovich K, Whyte P, Harlow E (1989) The cellular 107K protein that binds to adenovirus E1A also associates with the large T antigens of SV40 and JC virus. Cell 58: 249–255

Dyson N, Bernards R, Friend S H, Gooding LR, Hassell JA, Major E O, Pipas JM, Van Dyke T, Harlow E (1990) Large T antigens of many polyomaviruses are able to form complexes with the retinoblastoma protein. J Virol 64: 1353–1356

Dyson N, Dembski M, Fattaey A, Ngwu C, Ewen M, Helin K (1993) Analysis of p107-associated proteins: p107 associates with a form of E2F that differs from pRb-associated E2F. J Virol 67: 7641–7647

Egan SE, Weinberg RA (1993) The pathway to signal achievement. Nature 365: 781–783

Ewen ME (1994) The cell cycle and the retinoblastoma protein family. Cancer Metastasis Rev 13: 45–66

Ewen ME, Ludlow JW, Marsilio E, DeCaprio JA, Millikan RC, Cheng S, Paucha E, Livingston DM (1989) An N-terminal transformation-governing sequence of SV40 large T antigen contributes to the binding of both p110RB and a second cellular protein, p120. Cell 58: 257–267

Ewen ME, Xing Y, Lawrence JB, Livingston DM (1991) Molecular cloning, chromosomal mapping, and expression of the cDNA for p107, a retinoblastoma gene product-related protein. Cell 66: 1155–1164

Ewen ME, Faha B, Harlow E, Livingston DM (1992) Interaction of p107 with cyclin A independent of complex formation with viral oncoproteins. Science 255: 85–87

Ewen ME, Sluss HK, Sherr CJ, Matsushime H, Kato J-Y, Livingston DM (1993a) Functional interactions of the retinoblastoma protein with mammalian D-type cyclins. Cell 73: 487–497

Ewen ME, Sluss HK, Whitehouse LL, Livingston DM (1993b) TGF-β inhibition of cdk4 synthesis is linked to cell cycle arrest. Cell 74: 1009–1020
Faha B, Ewen M, Tsai L, Livingston DM, Harlow E (1992) Interaction between human cyclin A and adenovirus E1A-associated p107 protein. Science 255: 87–90
Fantl V, Stamp G, Andrews A, Rosewell I, Dickson C (1995) Mice lacking cyclin D1 are small and show defects in eye and mammary gland development. Genes Dev 9: 2364–2372
Feig LA, Cooper GM (1988) Inhibition of NIH 3T3 cell proliferation by a mutant ras protein with preferential affinity for GDP. Mol Cell Biol 8: 3235–3243
Feramisco JR, Gross M, Kamata T, Rosenberg M, Sweet RW (1984) Microinjection of the oncogene form of the human H-ras (T-24) protein results in rapid proliferation of quiescent cells. Cell 38: 109–117
Field S, Tsai F-Y, Kuo F, Zubiaga AM, Kaelin WG Jr, Livingston DM, Orkin SH, Greenberg ME (1996) E2F-1 functions in mice to promote apoptosis and suppress proliferation. Cell 85: 549–561
Filmus J, Robles AI, Shi W, Wong MJ, Colombo LL, Conti CJ (1994) Induction of cyclin D1 overexpression by activated ras. Oncogene 9: 3627–3633
Franke TF, Yang S-I, Chan TO, Datta K, Kazlauskas A, Morrison DK, Kaplan DR, Tsichlis PN (1995) The protein kinase encoded by the Akt proto-oncogene is a target of the PDGF-activated phosphatidylinositol 3-kinase. Cell 81: 727–736
Geng Y, Weinberg RA (1993) Transforming growth factor β effects on expression of G1 cyclins and cyclin-dependent protein kinases. Proc Natl Acad Sci USA 90: 10315–10319
Geng Y, Eaton EN, Picon M, Roberts JM, Lundberg AS, Gifford A, Sardet C, Weinberg RA (1996) Regulation of cyclin E transcription by E2Fs and retinoblastoma protein. Oncogene 16: 2402–2407
Ginsberg D, Vairo G, Chittenden T, Xiao Z-X, Xu G, Wydner KL, DeCaprio JA, Lawrence JB, Livingston DM (1994) E2F-4, a new member of the E2F transcription factor family, interacts with p107. Genes Dev 8: 2665–2679
Goga A, Liu X, Hambuch TM, Senechal K, Major E, Berk AJ, Witte ON, Sawyers CL (1995) p53-dependent growth suppression by the c-Abl nuclear tyrosine kinase. Oncogene 11: 791–799
Goodrich DW, Lee W-H (1992) Abrogation by c-myc of G1 phase arrest induced by RB protein but not by p53. Nature 360: 177–179
Goodrich DW, Wang NP, Qian Y-W, Lee EY-HP, Lee W-H (1991) The retinoblastoma gene product regulates progression through the G1 phase of the cell cycle. Cell 67: 293–302
Gu W, Schneider JW, Condorelli G, Kaushal S, Mahdavi V, Nadal-Ginard B (1993) Interaction of myogenic factors and the retinoblastoma protein mediates muscle cell commitment and differentiation. Cell 72: 309–324
Gu W, Bhatia K, Magrath IT, Dang CV, Dalla-Favera R (1994) Binding and suppression of the Myc transcriptional activation by p107. Science 264: 251–254
Guan K-L, Jenkins CW, Li Y, Nickols MA, Wu X, O'Keefe CL, Matera AG, Xiong Y (1994) Growth suppression by p18, a $p16^{INK4/MTS1}$- and $p14^{INK4B/MTS2}$-related CDK6 inhibitor, correlates with wild-type pRb function. Genes Dev 8: 2939–2952
Haas-Kogan DA, Kogan SC, Levi D, Dazin P, T'Ang A, Fung Y-KT, Israel MA (1995) Inhibition of apoptosis by the retinoblastoma gene product. EMBO J 14: 461–472
Hall M, Peters G (1996) Genetic alterations of cyclins, cyclin-dependent kinases, and Cdk inhibitors in human cancer. Adv Cancer Res 68: 67–108
Hamel PA, Gill RM, Phillips RA, Gallie BL (1992) Transcriptional repression of the E2-containing promoters EIIaE, c-myc, and RB1 by the product of the RB1 gene. Mol Cell Biol 12: 3431–3438
Hannon GJ, Beach D (1994) $p15^{INK4B}$ is a potential effector of TGF-β-induced cell cycle arrest. Nature 371: 257–261
Hannon GJ, Demetrick D, Beach D (1993) Isolation of the Rb-related p130 through its interaction with CDK2 and cyclins. Genes Dev 7: 2378–2391
Hatakeyama M, Brill JA, Fink GR, Weinberg RA (1994) Collaboration of G1 cyclins in the functional inactivation of the retinoblastoma protein. Genes Dev 8: 1759–1771
Helin K, Lees JA, Vidal M, Dyson N, Harlow E, Fattaey A (1992) A cDNA encoding a pRB-binding protein with properties of the transcription factor E2F. Cell 70: 337–350

Herber B, Truss M, Beato M, Muller R (1994) Inducible regulatory elements in the human cyclin D1 promoter. Oncogene 9: 1295–1304
Hermeking H, Eick D (1994) Mediation of c-Myc-induced apoptosis by p53. Science 265: 2091–2093
Herrera RE, Chen F, Weinberg RA (1996a) Increased histone H1 phosphorylation and relaxed chromatin structure in *Rb*-deficient fibroblasts. Proc Natl Acad Sci USA 93: 11510–11515
Herrera RE, Makela TP, Weinberg RA (1996b) TGF-β-induced growth inhibition of primary fibroblasts requires the retinoblastoma protein. Mol Biol Cell 7: 1335–1342
Herrera RE, Sah VP, Williams BO, Makela TP, Weinberg RA, Jacks T (1996c) Altered cell cycle kinetics, gene expression, and G1 restriction point regulation in Rb-deficient fibroblasts. Mol Cell Biol 16: 2402–2407
Hiebert SW (1993) Regions of the retinoblastoma gene product required for its interaction with the E2F transcription factor are necessary for E2 promoter repression and pRb-mediated growth suppression. Mol Cell Biol 13: 3384–3391
Hiebert SW, Chellappan SP, Horowitz JM, Nevins JR (1992) The interaction of RB with E2F coincides with an inhibition of the transcriptional activity of E2F. Genes Dev 6: 177–185
Hijmans EM, Voorhoeve PM, Beijersbergen RL, van't Veer LJ, Bernards R (1995) E2F-5, a new E2F family member that interacts with p130 in vivo. Mol Cell Biol 15: 3082–3089
Hinds PW, Mittnacht S, Dulic V, Arnold A, Reed SI, Weinberg RA (1992) Regulation of retinoblastoma protein functions by ectopic expression of human cyclins. Cell 70: 993–1006
Hoang AT, Lutterbach B, Lewis BC, Yano T, Chou T-Y, Barrett JF, Raffeld M, Hann SR, Dang CV (1995) A link between increased transforming activity of lymphoma-derived MYC mutant alleles, their defective regulation by p107, and altered phosphorylation of the c-Myc transactivation domain. Mol Cell Biol 15: 4031–4042
Hofmann F, Livingston DM (1996) Differential effects of cdk2 and cdk3 on the control of pRb and E2F function during G1 exit. Genes Dev 10: 851–861
Hollstein M, Sidransky D, Vogelstein B, Harris CC (1991) p53 mutations in human cancers. Science 253: 49–53
Howes KA, Ransom N, Papermaster DS, Lasudry JGH, Albert DM, Windle JJ (1994) Apoptosis or retinoblastoma: alternative fates of photoreceptors expressing the HPV-16 E7 gene in the presence or absence of p53. Genes Dev 8: 1300–1310
Hurford RKJ, Cobrinik D, Lee M-H, Dyson N (1997) pRB and p107/p130 are required for the regulated expression of different sets of E2F responsive genes. Genes Dev 11: 1447–1463
Iavarone A, Garg P, Lasorella A, Hsu J, Israel MA (1994) The helix-loop-helix protein Id-2 enhances cell proliferation and binds the retinoblastoma protein. Genes Dev 8: 1270–1284
Jacks T, Fazeli A, Schmitt EM, Bronson RT, Goodell MA, Weinberg RA (1992) Effects of an Rb mutation in the mouse. Nature 359: 295–300
Johnson DG (1995) Regulation of *E2F-1* gene expression by p130 (Rb2) and D-type cyclin kinase activity. Oncogene 11: 1685–1692
Johnson DG, Schwarz JK, Cress WD, Nevins JR (1993) Expression of transcription factor E2F1 induced quiescent cells to enter S phase. Nature 365: 349–352
Jones DL, Munger K (1997) Analysis of the p53-mediated G1 growth arrest pathway in cell expressing the human papillomavirus type 16 E7 oncoprotein. J Virol 71: 2905–2912
Kaelin WG Jr, Pallas DC, DeCaprio JA, Kaye FJ, Livingston DM (1991) Identification of cellular proteins that can interact specifically with the T/E1A-binding region of the retinoblastoma gene product. Cell 64: 521–532
Kaelin WG Jr, Krek W, Sellers WR, DeCaprio JA, Ajchenbaum F, Fuchs CS, Chittenden T, Li Y, Farnham PJ, Blanar MA, Livingston DM, Flemington EK (1992) Expression cloning of a cDNA encoding a retinoblastoma-binding protein with E2F-like properties. Cell 70: 351–364
Karlseder J, Rotheneder H, Winterberger E (1996) Interaction of Sp1 with the growth- and cell cycle-regulated transcription factor E2F. Mol Cell Biol 16: 1659–1667
Kato J-Y, Matsushime H, Hiebert SW, Ewen ME, Sherr CJ (1993) Direct binding of cyclin D to the retinoblastoma gene product (pRb) and pRb phosphorylation by the cyclin D-dependent kinase cdk4. Genes Dev 7: 331–342

Kauffmann-Zeh A, Rodriguez-Viciana P, Ulrich E, Gilbert C, Coffer P, Downward J, Evan G (1997) Suppression of c-Myc-induced apoptosis by Ras signaling through PI(3)K and PKB. Nature 385: 544–548

Kennedy SG, Wagner AJ, Conzen SD, Jordan J, Bellacosa A, Tsichlis PN, Hay N (1997) The PI 3-kinase/Akt signaling pathway delivers an ani-apoptotic signal. Genes Dev 11: 701–713

Kim S-J, Onwuta US, Lee YI, Li R, Botchan MR, Robbins PD (1992a) The retinoblastoma gene product regulates Sp-1 mediated transcription. Mol Cell Biol 12: 2455–2463

Kim S-J, Wagner S, Liu F, O'Reilly MA, Robbins PD, Green MR (1992b) Retinoblastoma gene product activates expression of the human TGF-β2 gene through transcription factor ATF-2. Nature 358: 331–334

Kitagawa M, Higashi H, Jung HK, Suzuki-Takahashi I, Ikeda M, Tamai K, Kato J, Segawa K, Yoshida E, Nishimura S, Taya Y (1996) The consensus motif for phosphorylation by cyclin D1-cdk4 is different from that for phosphorylation by cyclin A/E-cdk2. EMBO J 15: 7060–7069

Ko LJ, Prives C (1996) p53: puzzle and paradigm. Genes Dev 10: 1054–1072

Koff A, Ohtsuki M, Polyak K, Roberts JM, Massagué J (1993) Negative regulation of G1 in mammalian cells: inhibition of cyclin E-dependent kinase by TGF-β. Science 260: 536–539

Koh J, Enders GH, Dynlacht BD, Harlow E (1995) Tumour-derived p16 alleles encoding proteins defective in cell-cycle regulation. Nature 375: 506–510

Kowalik TF, DeGregori J, Schwartz JK, Nevins JR (1995) E2F1 overexpression in quiescent fibroblasts leads to induction of cellular DNA synthesis and apoptosis. J Virol 69: 2491–2500

Krek W, Ewen ME, Shirodkar S, Arany Z, Kaelin WG Jr, Livingston DM (1994) Negative regulation of the growth-promoting transcription factor E2F-1 by a stably bound cyclin A-dependent protein kinase. Cell 78: 161–172

Krek W, Xu G, Livingston DM (1995) Cyclin A-kinase regulation of E2F-1 DNA binding function underlies suppression of an S phase checkpoint. Cell 83: 1149–1158

Laiho M, DeCaprio JA, Ludlow JW, Livingston DM, Massagué J (1990) Growth inhibition by TGF-β linked to suppression of retinoblastoma protein phosphorylation. Cell 62: 175–185

Lasorella A, Iavarone A, Israel MA (1996) Id2 specifically alters regulation of the cell cycle by tumor suppressor proteins. Mol Cell Biol 16: 2570–2578

Lassar AB, Munsterberg A (1994) Wiring diagrams: regulatory circuits and the control of skeletal myogenesis. Curr Opin Cell Biol 6: 432–442

Lavoie JN, L'Allemain G, Brunet A, Muller R, Pouyssegur J (1996) Cyclin D1 expression is regulated positively by the p42/p44^{MAPK} and negatively by the p38/HOG^{MAPK} pathway. J Biol Chem 271: 20608–20616

Lee EY-HP, Chang C-Y, Hu N, Wang Y-CJ, Lai C-C, Herrup K, Lee W-H, Bradley A (1992) Mice deficient for Rb are nonviable and show defects in neurogenesis and haematopoiesis. Nature 359: 288–294

Lee M-H, Williams BO, Mulligan G, Mukai S, Bronson RT, Dyson N, Harlow E, Jacks T (1996) Targeted disruption of p107: functional overlap between p107 and Rb. Genes Dev 10: 1621–1632

Lees E, Faha B, Dulic V, Reed SI, Harlow E (1992) Cyclin E/cdk2 and cyclin A/cdk2 kinases associate with p107 and E2F in a temporally distinct manner. Genes Dev 6: 1874–1885

Lees JA, Saito M, Vidal M, Valentine M, Look T, Harlow E, Dyson N, Helin K (1993) The retinoblastoma protein binds to a family of E2F transcription factors. Mol Cell Biol 13: 7813–7825

Levine AJ (1997) p53, the cellular gatekeeper for growth and division. Cell 88: 323–331

Levine AJ, Momand J, Finlay CA (1991) The p53 tumor suppressor gene. Nature 351: 453–456

Li Y, Graham C, Lacy S, Duncan AMV, Whyte P (1993) The adenovirus E1A-associated 130-kD protein is encoded by a member of the retinoblastoma gene family and physically interacts with cyclins A and E. Genes Dev 7: 2366–2377

Lin S-Y, Black AR, Kostic D, Pajovic S, Hoover CN, Azizkhan JC (1996) Cell cycle-regulated association of E2F1 and Sp1 is related to their functional interaction. Mol Cell Biol 16: 1668–1675

Liu J-J, Chao J-R, Jiang M-C, Ng S-Y, Yen JJ-Y, Yang-Yen HF (1995) Ras transformation results in an elevated level of cyclin D1 and acceleration of G1 progression in NIH 3T3 cells. Mol Cell Biol 15: 3654–3663

Lowe SW, Ruley HE (1993) Stabilization of the p53 tumor suppressor is induced by adenovirus 5 E1A and accompanies apoptosis. Genes Dev 7: 535–545

Lowe SW, Ruley HE, Jacks T, Housman DE (1993) p53-dependent apoptosis modulates the cytoxicity of anticancer agents. Cell 74: 957–967

Lowe SW, Bodis S, McClatchey A, Remington L, Ruley HE, Fisher DE, Housman DE, Jacks T (1994) p53 status and the efficacy of cancer therapy in vivo. Science 266: 807–810

Ludlow JW, DeCaprio JA, Huang C, Lee W, Paucha E, Livingston DM (1989) SV40 Large T antigen binds preferentially to an underphosphorylated member of the retinoblastoma susceptibility gene product family. Cell 56: 57–65

Ludlow JW, Shon J, Pipas JM, Livingston DM, DeCaprio JA (1990) The retinoblastoma susceptibility gene product undergoes cell cycle-dependent dephosphorylation and binding to and release from SV40 large T. Cell 60: 387–396

Lukas J, Muller H, Bartkova J, Spitkovsky D, Kjerulff AA, Jansen-Durr P, Strauss M, Bartek J (1994) DNA tumor virus oncoproteins and retinoblastoma gene mutations share the ability to relieve the cell's requirement for cyclin D1 function in G1. J Cell Biol 125: 625–638

Lukas J, Bartkova J, Rohde M, Strauss M, Bartek J (1995a) Cyclin D1 is dispensable for G1 control in retinoblastoma gene-deficient cells independently of cdk4 activity. Mol Cell Biol 15: 2600–2611

Lukas J, Parry D, Aagaard L, Mann DJ, Barkova J, Strauss M, Peters G, Bartek J (1995b) Retinoblastoma-protein-depedent cell-cycle inhibition by tumour suppressor p16. Nature 375: 503–506

Lukas J, Bartkova J, Welcker M, Peterson OW, Peters G, Strauss M, Bartek J (1995c) Cyclin D2 is a moderately oscillating nucleoprotein required for G1 phase progression in specific cell types. Oncogene 10: 2125–2134

Lukas J, Bartkova J, Bartek J (1996a) Convergence of mitogenic signalling cascades from diverse classes of receptors at the cyclin D-cyclin-dependent kinase-pRb-controlled G1 checkpoint. Mol Cell Biol 16: 6917–6925

Lukas J, Petersen BO, Holm K, Bartek J, Helin K (1996b) Deregulated expression of E2F family members induces S-phase entry and overcomes $p16^{INK4A}$-mediated growth suppression. Mol Cell Biol 16: 1047–1057

Maandag ECR, van der Valk M, Vlaar M, Feltkamp C, O'Brien J, van Roon M, van der Lugt N, Berns A, te Riele H (1994) Developmental rescue of an embryonic-lethal mutation in the retinoblastoma gene in chimeric mice. EMBO J 13: 4260–4268

Macleod KF, Hu Y, Jacks T (1996) Loss of Rb activates both p53-dependent and independent cell death pathways in the developing mouse nervous system. EMBO J 15: 6178–6188

Magae J, Wu C-L, Illenye S, Harlow E, Heintz NH (1996) Nuclear localization of DP and E2F transcription factors by heterodimeric partners and retinoblastoma protein family members. J Cell Sci 109: 1717–1726

Marshall CJ (1995) Specificity of receptor tyrosine kinase signaling: transient versus sustained extracellular signal-regulated kinase activation. Cell 80: 179–185

Matsushime H, Roussel MF, Ashmun RA, Sherr CJ (1991) Colony-stimulating factor 1 regulates novel cyclins during the G1 phase of the cell cycle. Cell 65: 701–713

Matsushime H, Ewen ME, Strom DK, Kato J-Y, Hanks SK, Roussel MF, Sherr CJ (1992) Identification and properties of an atypical catalytic subunit ($p34^{PSK-J3}$/cdk4) for mammalian D type cyclins. Cell 71: 323–334

Mayol X, Grana X, Baldi A, Sang N, Hu Q, Giordano A (1993) Cloning of a new member of the retinoblastoma gene family (pRb2) which binds to the E1A transforming domain. Oncogene 8: 2561–2566

Mayol X, Garriga J, Grana X (1995) Cell cycle-dependent phosphorylation of the retinoblastoma-related protein p130. Oncogene 11: 801–808

Mayol X, Garriga J, Grana X (1996) G1 cyclin/CDK-independent phosphorylation and accumulation of p130 during the transition from G1 to G0 lead to its association with E2F-4. Oncogene 13: 237–246

McKnight SL, Lane MD, Gluecksohn-Waelsch S (1989) Is CCAAT/enhancer-binding protein a central regulator of energy metabolism? Genes Dev. 3: 2021–2024

Medema RH, Herrera RE, Lam F, Weinberg RA (1995) Growth suppression by p16ink4 requires functional retinoblastoma protein. Proc Natl Acad Sci USA 92: 6289–6293

Mihara K, Cao X, Yen A, Chandler S, Driscoll B, Murphree AL, T'Ang A, Fung Y (1989) Cell cycle-dependent regulation of phosphorylation of the human retinoblastoma gene product. Science 246: 1300–1303

Mittnacht S, Paterson H, Olson MF, Marshall CJ (1997) Ras signalling is required for inactivation of tumour suppressor pRb cell-cycle control protein. Curr Biol 7: 219–221

Moberg K, Starz MA, Lees JA (1996) E2F-4 switches from p130 to p107 and pRB in response to cell cycle entry. Mol Cell Biol 16: 1436–1449

Morgenbesser SD, Williams BO, Jacks T, DePinho RA (1994) *p53*-dependent apoptosis produced by *Rb*-deficiency in the developing mouse lens. Nature 371: 72–74

Mudryj M, Devoto SH, Hiebert SW, Hunter T, Pines J, Nevins JR (1991) Cell cycle regulation of the E2F transcription factor involves an interaction with cyclin A. Cell 65: 1243–1253

Mulcahy LS, Smith MR, Stacey DW (1985) Requirement for ras proto-oncogene function during serum-stimulated growth of NIH3T3 cells. Nature 313: 241–243

Neuman E, Flemington EK, Sellers WR, Kaelin WG Jr (1994) Transcription of the E2F-1 gene is rendered cell cycle dependent by E2F DNA-binding sites within its promoter. Mol Cell Biol 14: 6607–6615

Nevins JR (1992) E2F: A link between the Rb tumor supressor protein and viral oncoproteins. Science 258: 424–429

Novitch BG, Mulligan GJ, Jacks T, Lassar AB (1996) Skeletal muscle cells lacking the retinoblastoma protein display defects in muscle gene expression and accumulate in S and G2 phases of the cell cycle. J Cell Biol 135: 441–456

Ohtani K, DeGregori J, Nevins JR (1995) Regulation of the cyclin E gene by transcription factor E2F1. Proc Natl Acad Sci USA 92: 12146–12150

Ohtsubo M, Theodoras AM, Schumacher J, Roberts JM, Pagano M (1995) Human cyclin E, a nuclear protein essential for the G1 to S phase transition. Mol Cell Biol 15: 2612–2624

Okuda K, Ernst TJ, Griffin JD (1994) Inhibition of p21ras activation block proliferation but not differentiation of interleukin-3-dependent myeloid cells. J Biol Chem 269: 24602–24607

Oswald F, Dobner T, Lipp M (1996) The E2F transcription factor activates a replication-dependent human H2A gene in early S phase of the cell cycle. Mol Cell Biol 16: 1889–1895

Pagano M, Draetta G, Jansen-Durr P (1992) Association of cdk2 kinase with the transcription factor E2F during S phase. Science 255: 1144–1147

Pan H, Griep AE (1994) Altered cell cycle regulation in the lens of HPV-16 E6 or E7 transgenic mice: implications for tumor suppressor gene function in development. Genes Dev 8: 1285–1299

Pardee AB (1974) A restriction point for control of normal animal cell proliferation. Proc Natl Acad Sci USA 71: 1286–1290

Pardee AB (1989) G1 events and regulation of cell proliferation. Science 246: 603–608

Peeper DS, Parker LL, Ewen ME, Toebes M, Frederick FL, Xu M, Zantema A, van der Eb AJ, Pinwica-Worms H (1993) A- and B-type cyclins differentially modulate substrate specificity of cyclin-CDK complexes. EMBO J 12: 1947–1954

Peeper DS, Upton TM, Ladha MH, Neuman E, Zalvide J, Bernards R, DeCaprio JA, Ewen ME (1997) Ras signalling linked to the cell-cycle machinery by the retinoblastoma protein. Nature 386: 177–181

Peterson CL, Tamkun JW (1995) The SWI-SNF complex: a chromatin remodeling machine? Trends Biochem Sci 20: 143–146

Pietenpol JA, Stein RW, Moran E, Yaciuk P, Schlegel R, Lyons RM, Pittelkow MR, Munger K, Howley PM, Moses HL (1990) TGF-β inhibition of c-myc transcription and growth in keratinocytes is abrogated by viral transforming proteins with pRB binding domains. Cell 61: 777–785

Pietenpol JA, Münger K, Howley PM, Stein RW, Moses HL (1991) Factor-binding element in the human c-myc promoter involved in transcriptional regulation by transforming growth factor β1 and by the retinoblastoma gene product. Proc Natl Acad Sci USA 88: 10227–10231

Polyak K, Waldman T, He T-C, Kinzler KW, Vogelstein B (1996) Genetic determinants of p53-induced apoptosis. Genes Dev 10: 1945–1952

Pronk GJ, Bos JL (1994) The role of p21ras in receptor tyrosine kinase signalling. Biochim Biophys Acta 1198: 131–147

Qin X-Q, Chittenden T, Livingston DM, Kaelin WG Jr (1992) Identification of a growth suppression domain within the retinoblastoma gene product. Genes Dev 6: 953–964

Qin X-Q, Livingston DM, Kaelin WG, Adams PD (1994) Deregulated transcription factor E2F-1 expression leads to S-phase entry and p53-mediated apoptosis. Proc Natl Acad Sci USA 91: 10918–10922

Qin X-Q, Livingston DM, Ewen M, Sellers WR, Arany Z, Kaelin WG Jr (1995) The transcription factor E2F-1 is a downstream target of Rb action. Mol Cell Biol 15: 742–755

Quelle DE, Ashmun RA, Shurtleff SA, Kato J, Bar-Sagi D, Roussel MF, Sherr CJ (1993) Overexpression of mouse D-type cyclins accelerates G_1 phase in rodent fibroblasts. Genes Dev 7: 1559–1571

Resnitzky D, Reed SI (1995) Different roles for cyclins D1 and E in regulation of the G1-to-S transition. Mol Cell Biol 15: 3463–3469

Robbins PD, Horowitz JM, Mulligan RC (1990) Negative regulation of human c-fos expression by the retinoblastoma gene product. Nature 346: 668–671

Rodriguez-Viciana P, Warne PH, Dhand R, Vanhaesebroeck B, Gout I, Fry MJ, Waterfield MD, Downward J (1994) Phoshatidylinositol-3-OH kinase as a direct target of Ras. Nature 370: 527–532

Roussel MF, Theodoras AM, Pagano M, Sherr CJ (1995) Rescue of defective mitogenic signaling by D-type cyclins. Proc Natl Acad Sci USA 92: 6837–6841

Saenz Robles MT, Symonds H, Chen J, Van Dyke T (1994) Induction versus progression of brain tumor development: differential functions for the pRB- and p53-targeting domains of simian virus 40 T antigen. Mol Cell Biol 14: 2686–2698

Sawyers CL, Callahan W, Witte ON (1992) Dominant negative MYC blocks transformation by ABL oncogenes. Cell 70: 901–910

Sawyers CL, McLaughlin J, Goga A, Havlik M, Witte O (1994) The nuclear tyrosine kinase c-Abl negatively regulates cell growth. Cell 77: 121–131

Sawyers CL, McLaughlin J, Witte ON (1995) Genetic requirement for ras in the transformation of fibroblasts and hematopoietic cells by the Bcr-Abl oncogene. J Exp Med 181: 307–313

Schneider JW, Gu W, Zhu L, Mahdavi V, Nadal-Ginard B (1994) Reversal of terminal differentiation mediated by p107 in $Rb^{-/-}$ muscle cells. Science 264: 1467–1471

Schwartz JK, Bassing CH, Kovesdi I, Datto MB, Blazing M, George S, Wang X-F, Nevins JR (1995) Expression of the E2F1 transcription factor overcomes type β transforming growth factor-mediated growth suppression. Proc Natl Acad Sci USA 92: 483–487

Sellers WR, Kaelin WG (1996) RB as a modulator of transcription. Biochim Biophys Acta 1288: M1–M5

Sellers WR, Rodgers JW, Kaelin WG Jr (1995) A potent transrepression domain in the retinoblastoma protein induces a cell cycle arrest when bound to E2F sites. Proc Natl Acad Sci USA 92: 11544–11548

Serrano M, Hannon GJ, Beach D (1993) A new regulatory motif in cell-cycle control causing specific inhibition of cyclin D/cdk4. Nature 366: 704–707

Serrano M, Gomez-Lohoz E, DePinho RA, Beach D, Bar-Sagi D (1995) Inhibition of Ras-induced proliferation and transformation by $p16^{INK4}$. Science 267: 249–252

Serrano M, Lee H-W, Chin L, Cordon-Cardo C, Beach D, DePinho RA (1996) Role of the INK4a locus in tumor suppression and cell mortality. Cell 85: 27–37

Sewing A, Burger C, Brusselback S, Schalk C, Lucibello FC, Muller R (1993) Human cyclin D1 encodes a labile nuclear protein whose synthesis is directly induced by growth factors and suppressed by cyclic AMP. J Cell Sci 104: 545–555

Shan B, Lee W-H (1994) Deregulated expression of E2F-1 induces S-phase entry and leads to apoptosis. Mol Cell Biol 14: 8166–8173

Shao Z, Ruppert S, Robbins PD (1995) The retinoblastoma-susceptibility gene product binds directly to the human TATA-binding protein-associated factor TAFII250. Proc Natl Acad Sci USA 92: 3115–3119

Sherr CJ (1993) Mammalian G1 cyclins. Cell 73: 1059–1065

Sherr CJ (1996) Cancer cell cycles. Science 274: 1672–1677

Sherr CJ, Roberts JM (1995) Inhibitors of mammalian G1 cyclin-dependent kinases. Genes Dev 9: 1149–1163

Shirodkar S, Ewen M, DeCaprio JA, Morgan J, Livingston DM, Chittenden T (1992) The transcription factor E2F interacts with the retinoblastoma product and a p107-cyclin A complex in a cell cycle-regulated manner. Cell 68: 157–166

Shiyanov P, Bagchi S, Adami G, Kokontis J, Hay N, Arroyo M, Morozov A, Raychaudhuri P (1996) p21 disrupts the interaction between cdk2 and the E2F-p130 complex. Mol Cell Biol 16: 737–744

Sicinski P, Donaher JL, Parker SB, Li T, Fazeli A, Gardner H, Haslam SZ, Bronson RT, Elledge S, Weinberg RA (1995) Cyclin D1 provides a link between development and oncogenesis in the retina and breast. Cell 82: 621–630

Sicinski P, Donaher JL, Geng Y, Parker SB, Gardner H, Park MY, Robker RL, Richards JS, McGinnis LK, Biggers JD, Eppig JJ, Bronson RT, Elledge SJ, Weinberg RA (1996) Cyclin D2 is an FSH-responsive gene involved in gonadal cell proliferation and oncogenesis. Nature 384: 470–474

Singh P, Coe J, Hong W (1995) A role for the retinoblastoma protein in potentiating transcriptional activation of the glucocorticoid receptor. Nature 374: 562–565

Skapek SX, Rhee J, Kim PS, Novitch BG, Lassar AB (1996) Cyclin-mediated inhibition of muscle gene expression via a mechanism that is independent of pRB hyperphosphorylation. Mol Cell Biol 16: 7043–7053

Slingerland JM, Hengst L, Pan C-H, Alexander D, Stampfer MR, Reed SI (1994) A novel inhibitor of cyclin-cdk activity detected in transforming growth factor β-arrested epithelial cells. Mol Cell Biol 14: 3683–3694

Smith EJ, Leone G, DeGregori J, Jakoi L, Nevins JR (1996) The accumulation of an E2F-p130 transcriptional repressor distinguishes a G0 cell state from a G1 phase state. Mol Cell Biol 16: 6965–6976

Stacey DW, Kung HF (1984) Transformation of NIH 3T3 cells by microinjection of Ha-ras protein. Nature 310: 508–511

Stacey DW, Dobrowolski SF, Piotrkowski A, Harter ML (1994) The adenovirus E1A protein overrides the requirement for cellular ras in initiating DNA synthesis. EMBO J 13: 6107–6114

Starostik P, Chow KNB, Dean DC (1996) Transcriptional repression and growth suppression by the p107 pocket protein. Mol Cell Biol 16: 3606–3614

Steiner P, Philipp A, Lukas J, Godden-Kent D, Pagano M, Mittnacht S, Bartek J, Eilers M (1995) Identification of a Myc-dependent step during the formation of active G1 cyclin-cdk complexes. EMBO J 14: 4814–4826

Strober BE, Dunaief JL, Guha S, Goff SP (1996) Functional interactions between hBRM/hBRG1 transcriptional activators and the pRB family of proteins. Mol Cell Biol 16: 1575–1583

Stubdal H, Zalvide J, DeCaprio JA (1996) Simian virus 40 large T antigen alters the phosphorylation state of the RB-related proteins p130 and p107. J Virol 70: 2781–2788

Symonds H, Krall L, Remington L, Saenz-Robies M, Lowe S, Jacks T, Van Dyke T (1994) p53-dependent apoptosis suppresses tumor growth and progression in vivo. Cell 78: 703–711

Szekely L, Jiang WQ, Jakus-Bulic F, Rosen A, Ringerts N, Klein G, Wiman KG (1992) Cell type and differentiation dependent heterogeneity in retinoblastoma protein expression in SCID mouse fetuses. Cell Growth Diff 3: 149–156

Tam SW, Theodoras AM, Shay JW, Draetta GF, Pagano M (1994) Differential expression and regulation of cyclin D1 protein in normal and tumor human cells: association with Cdk4 is required for cyclin D1 function in G1 progression. Oncogene 9: 2663–2674

Taya Y (1997) RB kinases and RB-binding proteins: new points of view. Trends Biochem Sci 22: 14–17

Temin H (1971) Stimulation by serum of multiplication of stationary chicken cells. J Cell Physiol 78: 161–170

Tommasi S, Pfeifer GP (1995) In vivo structure of the human cdc2 promoter: release of a p130-E2F-4 complex from sequences immediately upstream of the transcription initiation site coincides with induction of cdc2 expression. Mol Cell Biol 15: 6901–6913

Vairo G, Livingston DM, Ginsberg D (1995) Functional interaction between E2F-4 and p130: evidence for distinct mechanisms underlying growth suppression by different retinoblastoma protein family members. Genes Dev 9: 869–881

Vlach J, Hennecke S, Alevizopoulos K, Conti D, Amati B (1996) Growth arrest by the cyclin-dependent kinase inhibitor p27^{Kip1} is abrogated by c-Myc. EMBO J 15: 6595–6604
Wagner AJ, Kokontis JM, Hay N (1994) Myc-mediated apoptosis requires wild-type p53 in a manner independent of cell cycle arrest and the ability of p53 to induce p21waf1/cip1. Genes Dev 8: 2817–2830
Wang J, Walsh K (1996) Resistance to apoptosis conferred by cdk inhibitors during myocyte differentiation. Science 273: 359–361
Weinberg RA (1996) E2F and cell proliferation: a world turned upside down. Cell 85: 457–459
Weintraub H (1993) The MyoD family and myogenesis: reduncancy, networks, and thresholds. Cell 75: 1241–1244
Weintraub SJ, Chow KN, Luo RX, Zhang SH, He S, Dean DC (1995) Mechanism of active transcriptional repression by the retinoblastoma protein. Nature 375: 812–815
Welch PJ, Wang JYJ (1993) A C-terminal protein-binding domain in the retinoblastoma protein regulates nuclear c-Abl tyrosine kinase in the cell cycle. Cell 75: 779–790
Welch PJ, Wang JYJ (1995) Abrogation of retinoblastoma protein function by c-Abl through tyrosine kinase-dependent and -independent mechanisms. Mol Cell Biol 15: 5542–5551
Wen ST, Jackson PK, Van Etten RA (1996) The cytostatic function of c-Abl is controlled by multiple nuclear localization signals and requires the p53 and Rb tumor suppressor gene products. EMBO J 15: 1583–1595
White RJ, Trouche D, Klaus M, Jackson SP, Kouzarides T (1996) Repression of RNA polymerase III transcription by the retinoblastoma protein. Nature 382: 88–90
Williams BO, Remington L, Albert DM, Mukai S, Bronson RT, Jacks T (1994a) Cooperative tumorigenic effects of germline mutations in *Rb* and *p53*. Nature Genet 7: 480–484
Williams BO, Schmitt EM, Remington L, Bronson RT, Albert DM, Weinberg RA, Jacks T (1994b) Extensive contribution of *Rb*-deficient cells to adult chimeric mice with limited histopathological consequences. EMBO J 13: 4251–4259
Winston JT, Pledger WJ (1993) Growth factor regulation of cyclin D1 mRNA expression through protein synthesis-dependent and independent mechanisms. Mol Biol Cell 4: 133–1144
Winston JT, Coats SR, Wang Y-Z, Pledger WJ (1996) Regulation of the cell cycle machinery by oncogenic Ras. Oncogene 12: 127–134
Won K-A, Xiong Y, Beach D, Gilman MZ (1992) Growth-regulated expression of D-type cyclin genes in human diploid fibroblasts. Proc Natl Acad Sci USA 89: 9910–9914
Wong KK, Zou X, Merrell KT, Patel AJ, Marcu KB, Chellappan S, Calame K (1995) v-Abl activates c-myc transcription through the E2F site. Mol Cell Biol 15: 6535–6544
Wu C-L, Classon M, Dyson N, Harlow E (1996) Expression of dominant-negative mutant DP-1 blocks cell cycle progression in G1. Mol Cell Biol 16: 3698–3706
Wu X, Levine AJ (1994) p53 and E2F-1 cooperate to mediate apoptosis. Proc Natl Acad Sci USA 91: 3602–3606
Xiao Z-X, Ginsberg D, Ewen M, Livingston DM (1996) Regulation of the retinoblastoma protein-related protein p107 by G1 cyclin-associated kinases. Proc Natl Acad Sci USA 93: 4633–4637
Xu M, Sheppard KA, Peng CY, Yee AS, Piwinica-Worms H (1994) Cyclin A/CDK2 binds directly to E2F-1 and inhibits the DNA-binding activity of E2F-1/DP-1 by phosphorylation. Mol Cell Biol 14: 8420–8431
Yamasaki L, Jacks T, Bronson R, Goillot E, Harlow E, Dyson NJ (1996) Tumor induction and tissue atrophy in mice lacking E2F-1. Cell 85: 537–548
Zacksenhaus E, Jiang Z, Chung D, Marth JD, Phillips RA, Gallie BL (1996a) pRb controls proliferation, differentiation, and death of skeletal muscle cells and other lineages during embryogenesis. Genes Dev 10: 3051–3064
Zacksenhaus E, Jiang Z, Phillips RA, Gallie B (1996b) Dual mechanisms of repression of E2F1 activity by the retinoblastoma gene product. EMBO J 15: 5917–5927
Zalvide J, DeCaprio JA (1995) Role of pRb-related proteins in simian virus 40 large-T-antigen-mediated transformation. Mol Cell Biol 15: 5800–5810
Zentella A, Weis FMB, Ralph DA, Laiho M, Massagué J (1991) Early gene responses to transforming growth factor-β in cells lacking growth suppressive RB function. Mol Cell Biol 11: 4952–4958

Zetterberg A, Larsson O, Wiman KG (1995) What is the restriction point? Curr Opin Cell Biol 7: 835–842

Zhu L, van der Heuvel S, Helin K, Fattaey A, Ewen M, Livingston D, Dyson N, Harlow E (1993) Inhibition of cell proliferation by p107, a relative of the retinoblastoma protein. Genes Dev 7: 1111–1125

Zhu L, Harlow E, Dynlacht BD (1995) p107 uses a $p21^{CIP1}$-related domain to bind cyclin/cdk2 and regulate interaction with E2F. Genes Dev 9: 1740–1752

Zou X, Rudchenko S, Wong K-k, Calame K (1997) Induction of c-*myc* transcription by the v-Abl tyrosine kinase requires Ras, Raf1, and cyclin-dependent kinases. Genes Dev 11: 654–662

Control of Cell Proliferation by Myc Proteins

A. Bürgin[1], C. Bouchard and M. Eilers

1 Introduction

myc genes were discovered as the transforming genes of four chicken retroviruses; soon cellular homologues were identified in a number of organisms ranging from birds and mammals to fish and sea urchins. The most recent addition to the list of organisms in which myc genes have been found is *Drosophila melanogaster*, opening up the possibility of a genetic analysis of Myc function (Gallant et al., 1996). Furthermore, genes encoding members of the Myc network of proteins have been identified in *Caenorhabditis elegans*, demonstrating an unexpected degree of evolutionary conservation (M.Cole, pers.comm.). No *myc* genes are present in *Saccharomyces cerevisiae* and they appear to be absent from *Schizosaccharomyces pombe.*

Interest in *myc* genes and the function of their encoded proteins was fuelled by their widespread involvement in both human and animal tumours. To review the evidence that implicates in the genesis of a variety of tumours is beyond what is attempted here (see for example Henriksson and Lüscher, 1996). However, it has become clear that Myc proteins are central regulators of cell proliferation even in non-tumorigenic cells, which will be the focus of this review. More recently, Myc proteins have also been implicated in apoptosis, the genetic control of programmed cell death. Finally, Myc proteins have been implicated in the control of cellular differentiation; the possible relationship of these roles to cell proliferation will be discussed.

2 Proteins of the Myc Network

To date, five *myc* genes have been described: *c-, N-, L-, B- and s-myc*. B-Myc differs structurally from the other proteins encoded, which are structurally related to another. Most of the work that links *myc* genes to cell proliferation has been performed with *c*-and *N-myc*; the available data suggest that *L-myc* has similar biological properties. Little information is available on the function of s-Myc proteins.

[1] Institut für Molekularbiologie und Tumorforschung (IMT), Universität Marburg, Emil-Mannkopff-Straße, 35033 Marburg, Germany

Myc proteins are highly unstable nuclear phosphoproteins that belong to the helix-loop-helix/leucine zipper family of transcription factors (for a review see Henriksson and Lüscher, 1996). Closely related proteins are the transcription factors USF, TFE-3 and TFE-B. All of these proteins share a related DNA-binding/dimerization motif and recognize the same sequence on DNA, CAC(A/G)TG, although the preference for flanking nucleotides varies (Blackwell et al., 1993; Blackwell and Weintraub, 1990). Strikingly, these proteins have widely divergent biological functions and the precise mechanism by which a cell discriminates between them is not fully understood (Desbarats et al., 1996). There are consensus nuclear localization motifs in Myc (Dang and Lee, 1988). At the amino-terminus, Myc carries a potent transcriptional activation domain (Desbarats et al., 1996; Kato et al., 1990). The integrity of both DNA-binding and dimerization and activation domains is critical for transformation by Myc (Stone et al., 1987).

In vivo, Myc does not form homodimers but heterodimerizes with a partner protein termed Max (Blackwood et al., 1991, 1992; Prendergast et al., 1991). Max is also a DNA-binding helix-loop-helix/leucine zipper protein and recognizes the same sequence on DNA as Myc. However, there are several notable differences: first, Max lacks a transcriptional activation domain and appears to be more or less inert in its effect on transcription (Kato et al., 1992); second, Max readily forms homodimers; and third, Max is a stable protein and its expression levels are only minutely affected by mitogenic stimulation (Blackwood et al., 1992).

Heterodimerization with Max is a prerequisite for all biological activities of Myc (Amati et al., 1993a, b). The inverse, however, is not true as Max has other partner proteins than Myc. To date, five have been described in the literature and designated Mad-1, Mxi-1(= Mad-2), Mad-3, Mad-4 and Mnt (Ayer et al., 1993; Hurlin et al., 1997; Hurlin et al., 1995; Zervos et al., 1993). All these proteins are related to each other and are also helix-loop-helix/leucine zipper proteins. In contrast to Max, they are potent repressors of transcription (Ayer et al., 1995; Schreiber Agus et al., 1995). While this property seems constitutive for Mad 1-4, the ability of Mnt to repress transcription is cell type dependent and may be regulated (Hurlin et al., 1997). Repression requires the integrity of an amino-terminal helical domain and is mediated by recruitment of a complex that contains two protein termed sin-3 and N-CoR. The complex has histone deacetylase activity which is required for repression and cell cycle arrest by Mad proteins (e.g. Alland et al., 1997; Laherty et al., 1997; Sommer et al., 1997). Together, the data strongly suggest that repression by Mad proteins occurs by modification of chromatin structure.

Several genes are known to date that are regulated by the Myc/Max/Mad family of proteins. These are for example genes which have been demonstrated to be required for cell proliferation, like the genes encoding ornithine decarboxylase (Bello-Fernandez et al., 1993), prothymosin-α (Eilers et al., 1991) and

CAD, an enzyme involved in nucleotide biogenesis (Miltenberger et al., 1995). One might conclude that one function of Myc is to stimulate expression of genes that prepare a cell for initiation of DNA synthesis. The eukaryotic translation initiation factor eIF-4E may be another example of such a gene (Jones et al., 1996). The only known direct target gene product that is part of the cell cycle machinery is the phosphatase cdc25A, and its potential role will be discussed below (Galaktionov et al., 1996).

Potentially the most enigmatic function of Myc is its ability to repress transcription in vivo (Penn et al., 1990). Several genes are known that are strongly repressed in Myc-transformed cells and the identity of the genes suggests that repression may contribute significantly to the phenotype of Myc-transformed cells. For example, several integrin genes are repressed in Myc-transformed cells (Inghirami et al., 1990). Moreover, expression of the c/EBP-α protein, a transcription factor that induces adipocyte differentiation is repressed by Myc (Freytag and Geddes, 1992). Finally, Myc negatively autoregulates its own expression (Penn et al., 1990) and negatively affects expression of cyclin D1 in some cellular contexts (Philipp et al., 1994); this will be discussed below. Several proteins have been implicated as potential candidates for gene repression by Myc; however, the molecular mechanisms of repression remains a conundrum (Roy et al., 1993; Shrivastava et al., 1993).

3
Cell Proliferation

A large number of observations demonstrate that Myc proteins are central regulators of cell proliferation in mammalian and potentially also other vertebrate cells. These observations are only briefly summarized here (for a detailed review see Henriksson and Lüscher, 1996).

First, expression of myc genes is closely linked to cellular proliferation in a vast number of experimental systems (e.g.Kelly et al., 1983; Kelly and Siebenlist, 1986). In contrast, *max* genes are expressed at a fairly constant level during mitogenic stimulation and in differentiating systems; thus, Max/Max homodimers are thought to be the predominant species of Myc network proteins in resting cells (Blackwood et al., 1992).

Conversely, expression of Mad genes is induced during cellular differentiation (Ayer and Eisenman, 1993; Larsson et al., 1994; Zervos et al., 1993). As a result, Mad displaces Myc from complex formation with Max during cellular differentiation. This switch has been demonstrated during the phorbol ester-induced differentiation of U937 myeloid cells (Ayer and Eisenman, 1993).

Second, expression of either constitutive or inducible alleles of Myc is sufficient to induce cell proliferation in tissue culture even in the absence of external growth factors (Eilers et al., 1991; Keath et al., 1984). Thus, Balb/c-3T3 or RAT1 fibroblasts have been engineered that express a hormone-inducible allele of myc (MycER: Eilers et al., 1989); in these cells, activation of Myc induces multiple rounds of S-phase and mitosis which appear to be limited only because cells undergo apoptosis. Similar observations have been made in transgenic mice that express a c-myc gene under the control of the potent IgH enhances: the mice display a characteristic preneoplastic phenotype which is characterised by an increased rate of proliferation and a larger pre-B cell compartment (Langdon et al., 1986). Together, these experiments document the potent mitogenic potential of Myc.

Myc function is not only sufficient, but is also required for cell cycle progression in culture and for growth of an organism in vivo. Experiments with antisense oligonucleotides demonstrated that inhibition of c-myc expression arrested proliferation in the G1 phase of the cell cycle (e.g.Heikkila et al., 1987). More recently, knockout mice have been generated in which the c-myc gene has been deleted; this is embryonically lethal (Davis et al., 1993). Conversely, ectopic expression of Mad proteins via microinjection blocks CSF-1-induced proliferation, further indicating a requirement for Myc function during normal G1 progression (Roussel et al., 1996).

While a role for Myc proteins in controlling cell proliferation is clearly established from these and related studies, it is an open question as to whether this is the sole role for Myc in cell metabolism. This concerns mainly two issues discussed below:

First, activation of Myc in quiescent, serum-deprived cells induces not only cell proliferation, but also apoptosis (Evan et al., 1992). Induction of apoptosis and cell proliferation by Myc are at least to some degree independent of each other: for example, loss of p53 or ectopic expression of Bcl-2 diminishes the apoptotic functions of Myc without affecting its ability to induce cell proliferation (Hermeking and Eick, 1994; Wagner et al., 1993, 1994). However, these lesions may affect the recognition process that recognises an inherent conflict when Myc is activated in low serum process or the execution of the death programme. The pro-apoptotic functions of Myc may still be an indirect consequence of its mitogenic properties. Alternatively, some of Myc's target genes may directly be part of pro-apoptotic pathways, and Myc activates two essentially independent genetic programmes as an organism's safeguard against tumorigenesis when single mutations at a Myc gene locus occurs.

One approach to address this issue was to analyze whether Myc-induced apoptosis was restricted to a specific phase of the cell cycle using RAT1 fibroblasts that harbour a hormone-inducible MycER protein (Harrington et al., 1994). The rationale behind these experiments was that a conflict should not exist if Myc was activated once cells were committed to enter the cell cycle. More precisely, as for example cyclin-dependent kinases act at specific points of the cell cycle, Myc should not induce apoptosis if activated at a point at which this kinase is active anyway. No such dependency was found and Myc was equally active at inducing apoptosis in cells in G2 as it was during the G1 phase.

A second approach was to inject inhibitors of cyclin-dependent kinases (CDKs) (Rudolph et al., 1996). Under conditions where these inhibitors effectively blocked cell cycle progression in response to activation of Myc, they did not affect Myc-induced apoptosis; thus, the requirement for CDK activity seemed to be much lower for apoptosis than for cell cycle progression.

However, antisense oligonucleotides directed against cdc25A effectively blocked Myc-induce apoptosis in a closely related experimental system, arguing for a causal role of this phosphatase in Myc-induced apoptosis (Galaktionov et al., 1996). At first glance, this appears to be in conflict with the microinjection studies described above as both treatments should lead to inhibition of CDK activity. Several possibilities exist to reconcile the conflict. First, it has not been demonstrated that blocking cdc25A expression blocks apoptosis via its effect on cyclin E/cdk2 kinase activity. Indeed this would be surprising as Myc-induced apoptosis occurs under low serum conditions. However, cdc25A needs to be phosphorylated by the growth factor-dependent kinase raf to be active; thus it should be largely inactive as a phosphatase in serum-deprived cells (Galaktionov et al., 1995). Therefore, cdc25A might have a more specific role in apoptotic signalling; this appears well possible as cdc25 is used as a checkpoint in yeast. Second, block by deprivation of cdc25A activity and overexpression of inhibitors induces different molecular changes in the state of cyclin-dependent kinases, and it is possible that these send out different signals.

A causal role in Myc-induced apoptosis has also been suggested for ornithine-decarboxylase (ODC), as inhibition of ODC retards Myc-induced apoptosis in myeloid cells (Packham and Cleveland, 1994). The suggestion is that oxidation of excessive polyamines generated upon induction of Myc might directly lead to the generation of reactive oxygen species. Therefore, it is possible that Myc-induced apoptosis results from different genetic lesions or from the activation of different target genes in different tissues. Alternatively, activation of both ornithine decarboxylase and cdc25A are necessary but not sufficient events that contribute to induction of apoptosis by Myc.

The second issue is whether the block Myc exerts on cell differentiation in a number of experimental systems is an indirect consequence of its mitogenic properties. Ectopic expression of Myc has been reported to block erythroid differentiation of murine erythroleukemia cells (Coppola and Cole, 1986), muscle cells (Miner and Wold, 1991), myeloid cells (Larsson et al., 1988) and 3T3-L1 preadipocytes (Freytag, 1988). In these cells, differentiation is coupled to withdrawal from the cell cycle, suggesting that both processes may be intimately linked.

One of the clearest examples for non-cell cycle related activity of Myc comes from studies of adipocyte differentiation in L1 cells. In these cells, ectopic expression of Myc suppresses differentiation and inhibits expression of c/EBP-α, a transcription factor that induces differentiation (Freytag and Geddes, 1992). Re-expression of c/EBP-α restores the differentiation programme, suggesting that inhibition of c/EBP-α expression is causal in this process. Further work implicates the start site of the c/EBP-α gene as the target for repression by Myc, suggesting that Myc directly interacts with factors recognizing the "initiator" of the c/EBP-α gene (Li et al., 1994). Thus, some of the repressive effects Myc exerts on transcription may contribute to inhibition of differentiation.

4
Upstream of Myc: Control of Myc Function

The function of Myc proteins is controlled at multiple levels. As mentioned above, expression of *c-myc* closely reflects the proliferative status of the cell in many experimental systems. However, which signal transduction pathways mediate this induction is still a matter of some debate. Studies with a mutant CSF-1 receptor showed that induction of *c-myc* expression is mediated by a signalling pathway distinct from that leading to *c-fos* induction (Roussel et al., 1991). Further studies showed that induction of *c-myc* expression in response to stimulation of cells by PDGF is mediated by a src-dependent, but ras-independent signalling pathway. In contrast, *c-fos* was shown to be mediated by ras-dependent signalling pathway (Barone and Courtneidge, 1995).

The Src-dependent pathway has not been molecularly defined yet and the idea that Ras does not affect *c-myc* expression has been challenged by recent experiments; for example, activation of a hormone-inducible allele of the protein kinase raf strongly activates expression of the endogenous *c-myc* gene (Kerkhoff and Rapp, submitted). Furthermore, the *c-myc* promoter harbours a binding site for the transcription factor *c-ets*, which can mediate a strong response to Raf and Ras (Langer et al., 1992; Roussel et al., 1994). Ets proteins are phosphorylated by MAP kinases and this is responsible for their activation (M.Mc Mahon., pers. comm.). Therefore, it seems likely, that Ras proteins are upstream activators of *c-myc* expression in at least some cellular contexts.

While expression of *max* is largely constitutive, expression of *mad* genes is controlled by pathways leading to cellular differentiation; for example, expression of *mad-1* mRNA is induced in U937 cells by addition of TPA, which induces cellular differentiation (Ayer and Eisenman, 1993; Larsson et al., 1994). The recent cloning of the *mad-1* promoter may yield insights into this regulation (B.Pulverer, pers.comm.).

Second, transactivation by Myc in a controlled process. There are several independent pieces of evidence for this notion. Sequences of *myc* alleles isolated from Burkitt Lymphomas revealed a strong selection for specific point mutation in the transactivating domain of Myc (Bhatia et al., 1993; Brennscheidt et al., 1994). This observation is open to several interpretations; one obvious possibility is the suggestion that the transactivation domain is not „optimal" and restrained by either intramolecular or intermolecular associations. Further indirect evidence for such association has been obtained for Myc BoxII, a highly conserved sequence motif in the transactivation domain of Myc; chimeric proteins that harbour this domain as a GAL4-chimera have been found to compete with transformation by both Myc and E1A, suggesting that they may bind and titrate a limiting factor (Brough et al., 1995).

Several candidate proteins have been identified that associate with the transactivation domain of Myc. One is p107, which will be discussed below (Beijersbergen et al., 1994; Gu et al., 1994). Another is bin-1, a protein found in two-hybrid screening with a fragment of the activation domain of Myc (Sakamuro et al., 1996). Bin-1 is a nuclear protein that associates with myc in vitro. Ectopic expression of Bin-1 inhibits Myc-induced transformation; tumour-derived mutants of Myc are more resistant to inhibition of transformation by Bin-1. Thus, bin-1 is a candidate for a negative regulator for transactivation by Myc; however it is also possible that bin-1 is an effector protein that acts downstream of Myc. Bin-1 has homology to amphiphysin and to Rvs-167, a protein involved in the response to nitrogen starvation in yeast. Mutations in Rvs-167 fail to arrest in response to nitrogen starvation, suggesting that both bin-1 ad *rvs-167* are involved in negative growth regulatory pathways (David et al., 1994; Munn et al., 1995; Revardel et al., 1995).

5
Cyclin-dependent kinases as effectors of Myc function

Stimulation of cell proliferation by Myc might occur by different mechanisms. Some of the known target genes are essential for cell proliferation without being directly implicated in the control of cell cycle progression. Thus, Myc might stimulate proliferation in an indirect manner by providing gene products that are required for cell growth.

A number of observations suggest a more intimate link between the control of cell cycle progression and Myc function and suggest that Myc directly controls the activity of the cyclin E/cdk2 kinase in mammalian cells. Initially, it was found that activation of conditional alleles of *myc* in RAT1 fibroblasts rapidly induced the activity of cyclin E/cdk2 kinase. In resting cells, cyclin E protein complexed with cdk2 was detected, yet with little associated kinase activity (Steiner et al., 1995). Induction of Myc triggered a change in the molecular weight of cyclin E/cdk2 complexes and a concomitant activation of kinase activity.

Recent work has led to a more precise definition of the molecular events involved. It was found that activation of Myc induces a loss of p27 from cyclin E and cyclin A/cdk2 complexes and that this is required for activation (Müller et al., 1997). Loss of p27 appears to occur in two steps: first, p27 is lost from cdk2 complexes and accumulates transiently in a non-cdk2 bound form. This is facilitated by phosphorylation of p27 by cdk2 kinase, providing a possible explanation as to why p27 is specifically lost from cdk2 not cdk4 complexes upon activation of Myc. Soon afterwards p27 is degraded in a reaction that is sensitive to inhibitors of proteasome function.

Indeed, Myc-transformed cells are resistant to moderately elevated p27 than normal fibroblasts; when RAT1 fibroblasts are infected with retroviruses that express p27, they cease to proliferate and show a decreased cyclin E/cdk2 kinase activity (Vlach et al., 1996). The mechanisms by which Myc affects p27 function has not been elucidated; however, the bulk of p27 is not degraded in these cells.

To evaluate the role of cdk2 activation for the mitogenic properties of Myc, microinjection studies were performed in which inhibitors of cyclin-dependent kinases were injected into RAT1-MycER cells (Rudolph et al., 1996). Expression of cyclin A protein was used as a marker for cell cycle progression. These studies demonstrated that activation of cyclin E/cdk2 kinase was both necessary and sufficient to account for the activation of cyclin A transcription by Myc, providing firm support for the idea that this kinase mediates at least some mitogenic functions of the protein.

These data are compatible with both a direct and indirect effect of Myc on cyclin E/cdk2 kinase activity. However, experiments in which exponentially growing RAT1-MycER cells were separated according to cell size by centrifugal elutriation demonstrated that the cell cycle of Myc-transformed cells differs strongly from that of their non-transformed counterparts (Pusch et al., 1997). In these studies, cyclin E/cdk2 kinase activity was found to be low in early G1 and high in the late G1/S-phase of non-transformed cells, similar to what has been found in other cell types (e.g.Koff et al., 1992). In contrast, Myc-transformed cells had an almost completely upregulated cyclin E/cdk2 kinase activity throughout the entire cell cycle, strongly suggesting that the effect Myc exerts on cyclin E/cdk2 kinase activity can be unlinked from its general mitogenic properties.

So how does Myc act on this cyclin E/cdk2 kinase? Three models appear to be compatible with the available data. First, the human *cdc25A* gene contains an E-box element within the first intron and has been reported to be upregulated by Myc in rodent fibroblasts (Galaktionov et al., 1996). These findings provide a direct link between Myc and the cell cycle machinery. Stimulation of *cdc25A* expression may increase cyclin E/cdk2 kinase activity and indirectly phosphorylation of p27; as phosphorylation triggers destruction of p27, it may well contribute to or even cause the loss of p27 from cyclin E/cdk2 complexes and the activation that is observed.

There appear to be three observations that argue against a sole role for cdc25A in *myc*-induced activation of cyclin E/cdk2 kinase activity. For one, Myc upregulates cyclin E/cdk2 kinase activity in at least one cell line, RAT1 fibroblasts, without significant changes in the amount of cdc25A protein (Perez-Roger et al., 1997). Furthermore, phosphorylation of p27 is not sufficient to release it from cyclin E/cdk2 complexes in vitro (Müller et al., 1997), yet it is dissociated from cyclin E/cdk2 complexes and degraded in vivo upon induction of Myc. Finally, cdc25A cannot substitute for Myc in its ability to overcome a p27-mediated block to cell cycle progression (Vlach et al., 1996); in addition, it cannot overcome a block to proliferation imposed by dominant-negative Myc while proteins binding to p27 can (Berns et al., 1997).

In the second model, Myc has been suggested to direct the synthesis of proteins that bind to and ultimately degrade p27 (Steiner et al., 1995; Vlach et al., 1996). This suggestion directly reflects the observations made in vivo by comparison of Myc-transformed cells with their normal counterparts and is thus easily compatible with the available data; however, it suffers from the fact that no candidate proteins have been suggested that might account for the Myc-induced effects, and the proteins involved in degradation of p27 have not been identified. Thus, the suggestion remains somewhat speculative at the present time.

In the third model, several reports have found that there is an increase in cyclin E mRNA and protein upon induction of Myc (Jansen-Dürr et al., 1993; Perez-Roger et al., 1997). While this is not nearly sufficient to account for the increase in cyclin E/cdk2 kinase activity, it might generate a small pool of free cyclin E/cdk2 complexes that then phosphorylates multiple molecules of p27, therefore triggering them for destruction and amplifying the initial signal (Perez-Roger et al., 1997). This is a possibility, although the underlying assumptions about on-and off-rates of p27 for cyclin E/cdk2 complexes have not been tested. Moreover, if phosphorylation does not strongly reduce the affinity of p27 for cdk2, this model by itself is not sufficient to account for, but may well contribute to the in vivo observations. The critical test will be to identify the element in the cyclin E gene that responds to activation by Myc and determine whether Myc proteins interact directly with this element.

6
The enigma: Myc and D-Type Cyclin-Dependent Kinases

Data from a number of experimental systems show that *c-myc* and D-type cyclins are members of a common signalling pathway. Most striking are results obtained from studying a mutant allele of the CSF-1 receptor: this mutant, which harbours a mutation at residue 809, fails to induce expression of *c-myc* in response to addition of ligand (Roussel et al., 1991). Constitutive expression of *c-myc* rescues the mitogenic signalling defect. Strikingly, ectopic expression of cyclin D1 does the same and rescues mitogenic signalling; under these conditions, expression of *c-myc* is restored by cyclin D1 (Roussel et al., 1995).

Similar results have been obtained using mutants alleles of *bcr-abl*: again, mutants in the SH2 domain of *bcr-abl* that are deficient in inducing expression of *c-myc* are rescued by ectopic expression of either *c-myc* or cyclin D1 (Afar et al., 1995). Thus, two different experimental systems yield the same conclusion:

the function of D-type cyclins and *c-myc* are tightly linked and may form a common signalling pathway. The critical questions is whether both can be linked into one linear pathway with a clear-cut upstream/downstream relationship and what the molecular links between both classes of proteins may be. For this, several models have been suggested:

6.1
Rescue by Transcriptional Control: Do Myc and D-Type Cyclins Rescue Each Others Expression?

The simplest model would be that either D-type cyclins act upstream of *c-myc* to induce its expression or that Myc acts upstream to induce expression of cyclin D1 or another D-type cyclin. A number of studies have addressed both questions.

Ectopic expression of cyclin D1 has been reported to induce expression of *c-myc* via an E2F-site in the *c-myc* promoter in transient transfection experiments (Oswald et al., 1994). However, *c-myc* is an immediate early gene in many cell types; that is, its expression can be induced by growth factors without the need to synthesize intermediate proteins (Kelly et al., 1983). Thus, a general role of D-type cyclins (which by themselves are synthesized in response to a mitogenic signal) upstream of *c-myc* expression is unlikely.

Conversely, a number of studies have addressed the potential role of Myc/Max complexes in the control of cyclin D1 expression. Here, the results are somewhat controversial: Studies with the CSF-1 receptor mutant have indicated that ectopic expression of *c-myc* rescues induction of cyclin D1 mRNA in the background of the mutant CSF-1 receptor (Roussel et al., 1995). A number of other studies found either no effect of *myc* on the expression of D-type cyclins or a moderate to strong repression of cyclin D1, which was shown to be mediated by a Max-independent pathway (Jansen-Dürr et al., 1993; Lovec et al., 1994; Philipp et al., 1994). In a more detailed study, *myc* was found to repress expression of cyclin D1 strongly in a pRb-deficient cellular background, but only marginally if cells contained functional retinoblastoma protein, suggesting that expression of cyclin D1 may be induced by two different signalling pathways, of which only one is blocked by Myc (Marhin et al., 1996). Studies with the cyclin D1 promoter mapped the minimal growth factor-responsive region to a fragment that lacks Myc/Max target sites and have implicated AP-1 and Ets-proteins in the response to growth factors (Herber et al., 1994). Taken together, the data make it somewhat unlikely that Myc function is generally required for induction of cyclin D1 expression by growth factors and suggest that other links between both proteins must exist.

6.2
Rescue by Relief of Pocket Protein-Dependent Repression?

D-type cyclins form complexes with cdk4 and cdk6 which phosphorylate the pocket proteins pRb, p107 and p130 (Sherr, 1994). p107 has been shown to form complexes with both c-Myc and N-Myc proteins in vivo and inhibits transcriptional activation by both proteins (Beijersbergen et al., 1994; Gu et al., 1994). Therefore, D-type cyclins may act as upstream positive regulators of transcriptional activation by Myc. Alternatively, binding of Myc to p107 may sequester p107 and relieve its inhibitory function on E2F-dependent transcription: however, no evidence for this latter view has been presented.

Inhibition of D-type kinases by ectopic expression of p16 inhibits transcriptional activation by Myc in an NIH3T3 cell background in which pocket proteins are functional, but not in HeLa cells, in which pocket proteins are sequestered by the papillomavirus E7 protein (Haas et al., 1997). This suggests that D-type cyclins may indeed act upstream of transactivation by Myc. Ectopic expression of p16 also inhibits malignant transformation and cell cycle entry by *c-myc* (Haas et al., 1997; Rudolph et al., 1996) Transactivation by Myc is also inhibited by p107 and this inhibition is relieved by co-expression of cyclin D1 and cdk4. A mutant allele of Myc that is partly resistant to inhibition by p107 in transactivation, transforms primary cells in co-operation with EJ-ras in a p16-independent manner, suggesting that inhibition by p16 is at least partly mediated by factors that bind to the amino terminus of Myc (Haas et al., 1997).

Taken together, the data suggest a model in which D-type cyclins act at least in part upstream of Myc by controlling transcriptional activation by Myc via p107. Such a model would also be easily compatible with the observed repression of cyclin D1 expression by Myc; this would appear as a negative feedback control, similar to the repression of the endogenous *c-myc* gene by ectopic expression of *c-myc*.

6.3
Rescue by Functional Complementation

Finally, Myc and D-type cyclins may control a common molecular pathway and thus complement each other's functions. Again, evidence is beginning to emerge that supports this view from a number of different approaches.

D-type cyclins control the activity of the E2F-family of transcription factors via phosphorylation of pocket proteins; a block to this pathway can be overcome by ectopic expression of cyclin E and expression of cyclin E is regulated by an E2F site in its promoter (Botz et al., 1996; Geng et al., 1996). Together, the data show that the rate-limiting function of E2F proteins during G1 progression is the control of cyclin E expression.

As discussed before, activation of Myc strongly induces cyclin E/cdk2 kinase activity in a number of experimental settings. Similar to E2F proteins, ectopic expression of Myc also renders cells somewhat more resistant to ectopic expression of p16, although at higher doses p16 inhibits both cell cycle progression and transformation by Myc (Rudolph et al., 1996; Alevizopoulos et al, 1997). In p16-expression cells that proliferate due to the presence of Myc, pocket proteins remain hypophosphorylated and associated with E2F complexes, suggesting that cyclin D1/cdk4 activity is partially bypassed and not restored by Myc. This is consistent with previous reports that Myc is only a poor activator of cyclin D/cdk4 kinase activity (Steiner et al., 1995). However, *myc*-rescued cells have strongly elevated levels of cyclin E/cdk2 kinase activity, and ectopic expression of cyclin E can substitute for Myc in overcoming a cell cycle block to p16. Thus, the data suggest that Myc and D-type cyclins may synergize in the activation of cyclin E/cdk2 kinase.

If so, one might expect that defects in Myc function could be complemented by ectopic expression of cyclin E. This has been observed in some experimental systems, but not always (Berns et al., 1997; Vlach et al., 1996). For example, arrest of cells by p27 is partially overcome by Myc, but not by cyclin E: under these conditions, Myc restores cyclin E/cdk2 kinase activity, suggesting that it acts not primarily on cyclin E synthesis but on p27 metabolism, cdc25A expression or both (Vlach et al., 1996). This could also be the explanation why cyclin E does not restore growth to CSF-1 mutant receptor cells (Roussel et al., 1995).

The cell cycle arrest imposed by dominant negative alleles of Myc in U20S cells is rescued by ectopic expression of either cyclins D1 or cyclin E (Berns et al., 1997). Strikingly, cyclin D1 still rescues the block when co-expressed with dominant negative alleles of cdk4 that inhibit phosphorylation of p107. Under these conditions, Myc would be expected to be largely inactive. However, one might expect that inhibitors such as p27 are bound by the excess of cyclin D/cdk4 complexes and this may relieve the requirement for Myc function. This is consistent with transformation data that show that cdk4 transforms rat embryo fibroblasts primarily by sequestering inhibitors as its catalytic activity is dispensable for transformation (Haas et al., 1997).

7 Summary

Taken together, the available data appear to be consistent with a model in which Myc proteins function downstream of D-type cyclins and synergize with E2F proteins in the activation of the cyclin E/cdk2 kinase. This view of Myc proteins appears strikingly similar to established models for the E2F/DP family of proteins.

However, it should be noted that there are clear differences and several predictions of such a model that have been critically tested for E2F proteins are still untested for Myc in this model. First, it appears that at least some target genes of Myc implicated in this process are still unknown; second, clear data from knockout cells that link p107 to Myc function are missing; and third, we are not aware of studies of tumour samples that clarify whether mutations in myc genes relieve the requirement for mutations in the cyclin D/p16 pathway.

Acknowledgements I apologise to the colleagues whose work I did not quote due to space constraints. Work from my lab was carried out at the ZMBH, University of Heidelberg, and supported by grants from the Deutsche Forschungsgemeinschaft, BMBF, Human Science Frontier Organization and BASF.

References

Afar, D. E. H., McLaughlin, J., Sherr, C. J., Witte, O. N., Roussel, M. F. (1995). Signaling by ABL oncogenes through cyclin D1. Proc Natl Acad Sci USA*92*, 9540–9544.

Alevizopoulos, K., Vlach, J., Hennecke, S., and Amati, B. (1997). Cyclin E and c-Myc promote cell proliferation in the presence of p16(INK4a) and of hypophosphorylated retinoblastoma family proteins EMBO J *16*, 5322–33.

Alland, L., Muhle, R., Hou, H., Potes, J., Chin, L., Schreiber-Agus, N., Pinho, R. D. (1997). Role for N-CoR and histone deacetylase in Sin3-mediated transcriptional repression. Nature 387, 49–55.

Amati, B., Brooks, M. W., Levy, N., Littlewood, T. D., Evan, G. I., Land, H. (1993). Oncogenic activity of the c-Myc protein requires dimerization with Max. Cell *72*, 233–245.

Amati, B., Littlewood, T. D., Evan, G. I., Land, H. (1993). The c-Myc protein induces cell cycle progression and apoptosis through dimerization with Max. EMBO J. *13*, 5083–5087.

Ayer, D. E., Eisenman, R. N. (1993). A switch from Myc: Max to Mad: Max heterocomplexes accompanies monocyte/macrophage differentation. Genes Dev. *7*, 2110–2119.

Ayer, D. E., Kretzner, L., Eisenman, R. N. (1993). Mad: a heterodimeric partner for Max that antagonizes Myc transcriptional activity. Cell *72*, 211–222.

Ayer, D. E., Lawrence, Q. A., Eisenman, R. N. (1995). Mad-Max transcriptional repression is mediated by ternary complex formation with mammalian homologs of yeast repressor Sin3. Cell *80*, 767–776.

Barone, M. V., Courtneidge, S. A. (1995). Myc but not Fos rescue of PDGF signalling block caused by kinase-inactive Src. Nature *387*, 509–512.

Beijersbergen, R. L., Hijmans, E. M., Zhu, L., Bernards, R. (1994). Interaction of c-Myc with the pRb-related protein p107 results in inhibition of c-Myc-mediated transactivation. EMBO J. *13*, 4080–4086.

Bello-Fernandez, C., Packham, G., Cleveland, J. L. (1993). The ornithine decarboxylase gene is a transcriptional target of c-MYC. Proc Natl Acad Sci USA*90*, 7804–7808.

Berns, K., Hijmans, E. M., and Bernards, R. (1997). Repression of c-Myc responsive genes in cycling cells causes G1 arrest through reduction of cyclin E/CDK2 kinase activity. Oncogene *15*, 1347–56.

Bhatia, K., Huppi, K., Spangler, G., Siwarski, D., Iyer, R., Magrath, I. (1993). Point mutations in the c-Myc transactivation domain are common in Burkitt's lymphoma and mouse plasmacytomas. Nat Genet 5, 56–61.

Blackwell, T. K., Huang, J., Ma, A., Kretzner, L., Alt, F. W., Eisenman, R. N., Weintraub, H. (1993). Binding of Myc proteins to canonical and noncanonical DNA sequences. Mol. Cell. Biol. *13*, 5216–5224.

Blackwell, T. K., Weintraub, H. (1990). Differences and similarities in DNA-binding preferenceesz of MyoD and E2A protein complexes revealed by binding site selection. Science *250*, 1104–1110.

Blackwood, E. M., Lüscher, B., Eisenman, R. N. (1992). Myc and Max associate in vivo. Genes Dev. *6*, 71–80.

Blackwood, E. M., Lüscher, B., Kretzner, L., Eisenman, R. N. (1991). The Myc: Max protein complex and cell growth regulation. In Symposia on Quantitative Biology (Cold Spring Harbor: David Beach Bruce Stillman James D. Watson), pp. 109–117.

Botz, J., Zerfass Thome, K., Spitkovsky, D., Delius, H., Vogt, B., Eilers, M., Hatzigeorgiou, A., Jansen-Dürr, P. (1996). Cell cycle regulation of the murine cyclin E gene depends on an E2F binding site in the promoter. Mol. Cell. Biol. *16*, 3401–3409.

Brennscheidt, U., Eick, D., Kunzmann, R., Martens, U., Kiehntopf, M., Mertelsmann, R., Herrmann, F. (1994). Burkitt-like mutations in the c-myc gene locus in prolymphocytic leukemia. Leukemia 8, 897-902 Issn: 0887–6924.

Brough, D. E., Hofmann, T. J., Ellwood, K. B., Townley, R. A., Cole, M. D. (1995). An essential domain of the c-myc protein interacts with a nuclear factor that is also required for E1A-mediated transformation. Mol. Cell Biol. *15*, 1536–1544.

Coppola, J. A., Cole, M. D. (1986). Constitutive c-myc oncogene expression blocks mouse erythroleukaemia cell differentiation but not commitment. Nature *320*, 760–3.

Dang, C. V., Lee, W. M. F. (1988). Identification of the human c-myc protein nuclear translocation signal. Mol. Cell. Biol. *8*, 4048–4054.

David, C., Solimena, M., De Camilli, P. (1994). Autoimmunity in stiff-Man syndrome with breast cancer is targeted to the C-terminal region of human amphiphysin, a protein similar to the yeast proteins, Rvs167 and Rvs161. FEBS Lett *351*, 73-9 Issn: 0014–5793.

Davis, A. C., Wims, M., Spotts, G. D., Hann, S. R., Bradley, A. (1993). A null c-myc mutation causes lethality before 10.5 days of gestation in homozygotes and reduced fertility in heterozygous female mice. Genes Dev. 7, 671–682.

Desbarats, L., Gaubatz, S., Eilers, M. (1996). Discrimination between different E-box binding proteins at an endogenous target gene of Myc. Genes Dev. *10*, 447–460.

Eilers, M., Picard, D., Yamamoto, K., Bishop, J. M. (1989). Chimaeras between the MYC oncoprotein and steroid receptors cause hormone-dependent transformation of cells. Nature *340*, 66–68.

Eilers, M., Schirm, S., Bishop, J. M. (1991). The MYC protein activates transcription of the alpha-prothymosin gene. EMBO J. *10*, 133–141.

Evan, G. I., Wyllie, A. H., Gilbert, C. S., Littlewood, T. D., Land, H., Brooks, M., Waters, C. M., Penn, L. Z., and Hancock, D. C. (1992). Induction of apoptosis in fibroblasts by c-myc protein. Cell *69*, 119–128.

Freytag, S. O. (1988). Enforced expression of the c-myc oncogene inhibits cell differentiation by precluding entry into a distinct predifferentiation state in G0/G1. Mol. Cell. Biol. 8, 1614–1624.

Freytag, S. O., Geddes, T. J. (1992). Reciprocal regulation of adipogenesis by Myc and C/EBP alpha. Science *256*, 379–382.

Galaktionov, K., Chen, X., Beach, D. (1996). Cdc25 cell-cycle phosphatase as a target of c-myc. Nature *382*, 511–517.

Galaktionov, K., Jessus, C., Beach, D. (1995). Raf1 interaction with cdc25 phosphatase ties mitogenic signal transduction to cell cycle activation. Genes & Dev. *9*, 1046–1058.

Gallant, P., Shiio, Y., Cheng, P. F., Parkhurst, S. M., Eisenman, R. N. (1996). Myc and Max homologs in Drosophila. Science *274*, 1523-7 Issn: 0036–8075.

Geng, Y., Eaton, E. N., Picon, M., Roberts, J. M., Lundberg, A. S., Gifford, A., Sardet, C., Weinberg, R. A. (1996). Regulation of cyclin E transcription by E2Fs and retinoblastoma protein. Oncogene *12*, 1173-80 Issn: 0950–9232.

Gu, W., Bhatia, K., Magrath, I. T., Dang, C. V., Dalla-Favera, R. (1994). Binding and suppression of the Myc transcriptional activation domain by p107. Science *264*, 251–254.

Haas, K., Staller, P., Geisen, C., Bartek, J., M.Eilers, Möröy, T. (1997). Oncogenic activity of cdk4 and Myc depends on p16ink4A: evidence for cyclin D1/cdk4 and p16ink4A as upstream regulators of Myc. Oncogene (in press).

Harrington, E. A., Bennett, M. R., Fanidi, A., Evan, G. (1994). c-Myc-induced apoptosis in fibroblasts is inhibited by specific cytokines. EMBO J. *13*, 3286–3295.

Heikkila, R., Schwab, G., Wickstrom, E., Loke, S. L., Pluznik, D. H., Watt, R., Neckers, L. M. (1987). A c-myc antisense oligodeoxynucleotide inhibits entry into S phase but not progress from G0 to G1. Nature *328*, 445–449.

Henriksson, M., Lüscher, B. (1996). Proteins of the Myc network: Essential regulators of cell growth and differentiation. Cancer Research *68*, 109–182.

Herber, B., Truss, M., Beato, M., Muller, R. (1994). Inducible regulatory elements in the human cyclin D1 promoter. Oncogene *9*, 1295–1304.

Hermeking, H., Eick, D. (1994). Myc-induced apoptosis is mediated by p53 protein. Science *265*, 2091–2093.

Hurlin, P. J., Queva, C., Eisenman, R. N. (1997). Mnt, a novel Max-interacting protein is coexpressed with Myc in proliferating cells and mediates repression at Myc binding sites. Genes Dev *11*, 44–58.

Hurlin, P. J., Queva, C., Koskinen, P. J., Steingrimsson, E., Ayer, D. E., Copeland, N. G., Jenkins, N. A., Eisenman, R. N. (1995). Mad3 and Mad4: novel Max-interacting transcriptional repressors that suppress c-myc dependent transformation and are expressed during neural and epidermal differentiation. EMBO J. *14*, 5646–5659.

Inghirami, G., Grignani, F., Sternas, L., Lombardi, L., Knowles, D. M., Dalla-Favera, R. (1990). Down-regulation of LFA-1 adhesion recptors by c-myc oncogene in human B lymphoblastoid cells. Science *250*, 682–686.

Jansen-Dürr, P., Meichle, A., Steiner, P., Pagano, M., Finke, K., Botz, J., Wessbecher, J., Draetta, G., Eilers, M. (1993). Differential modulation of cyclin gene expression by MYC. Proc Natl Acad Sci USA*90*, 3685–3689.

Jones, R. M., Branda, J., Johnston, K. A., Polymenis, M., Gadd, M., Rustgi, A., Callanan, L., Schmidt, E. V. (1996). An essential E box in the promoter of the gene encoding the mRNA cap-binding protein (eukaryotic initiation factor 4E) is a target for activation by c-myc. Mol Cell Biol *16*, 4754-64 Issn: 0270–7306.

Kato, G., Lee, W. M. F., Chen, L., Dang, C. V. (1992). Max: functional domains and interaction with c-myc. Genes Dev. *6*, 81–92.

Kato, G. J., Barrett, J., Villa-Garcia, M., Dang, C. V. (1990). An amino-terminal c-myc domain required for neoplastic transformation activates transcription. Mol. Cell. Biol. *10*, 5914–5920.

Keath, E. J., Caimi, P. G., Cole, M. D. (1984). Fibroblast lines expressing activated c-myc oncogenes are tumorigenic in nude mice and syngeneic animals. Cell *39*, 339–348.

Kelly, K., Cochran, B. H., Stiles, C. D., Leder, P. (1983). Cell-specific regulation of the c-myc gene by lymphocyte mitogens and platelet-derived growth factor. Cell 35, 603–610.

Kelly, K., Siebenlist, U. (1986). The regulation and expression of c-myc in normal and malignant cells. Ann.Rev.Immunol. *4*, 317–338.

Koff, A., Giordano, A., Desai, D., Yamashita, K., Harper, J. W., Elledge, S., Nishimoto, T., Morgan, D. O., Franza, B. R., Roberts, J. M. (1992). Formation and activation of a cyclin E-cdk2 complex during the G1 phase of the human cell cycle. Science *257*, 1689-94 Issn: 0036–8075.

Laherty, C. D., Yang, W.-M., Sun, J.-M., Davie, J. R., Seto, E., Eisenman, R. N. (1997). Histone Deacetylases associated with the mSin3 corepressor mediate mad transcriptional repression. Cell *89*, 349–356.

Langdon, W. Y., Harris, A. W., Cory, S., Adams, J. M. (1986). The c-myc oncogene perturbs B lymphocyte development in Em-myc transgenic mice. Cell *47*, 11–18.

Langer, S. J., Bortner, D. M., Roussel, M. F., Sherr, C. J., Ostrowski, M. C. (1992). Mitogenic signaling by colony-stimulating factor 1 and ras is suppressed by the ets-2 DNA-binding domain and restored by myc overexpression. Mol Cell Biol *12*, 5355–62.

Larsson, L.-G., Ivhed, I., Gidlund, M., Pettersson, U., Vennstroem, B., Nilsson, K. (1988). Phorbol ester-induced terminal differentiation is inhibited in human U-937 monoblastic cells expressing a v-myc oncogene. Proc Natl Acad Sci USA*85*, 2638–2642.

Larsson, L.-G., Pettersson, M., Öberg, F., Nilsson, K., Lüscher, B. (1994). Expression of mad, mxi1, max and c-myc during induced differentiation of hematopoietic cells: opposite regulation of mad and c-myc. Oncogene *9*, 1247–1252.

Li, L., Nerlov, C., Prendergast, G., MacGregor, D., Ziff, E. B. (1994). c-Myc represses transcription in vivo by a novel mechanism dependent on the initiator element and Myc box II. EMBO J. *13*, 4070–4079.

Lovec, H., Sewing, A., Lucibello, F. C., Muller, R., Moroy, T. (1994). Oncogenic activity of cyclin D1 revealed through cooperation with Ha-ras: link between cell cycle control and malignant transformation. Oncogene *9*, 323–326.

Marhin, W. W., Hei, Y.-J., Chen, S., Jiang, Z., Gallie, B., Phillips, R. A., Penn, L. Z. (1996). Loss of Rb and Myc activation co-operate to suppress cyclin D1 and contribute to transformation. Oncogene *12*, 43–55.

Miltenberger, R. J., Sukow, K. A., Farnham, P. J. (1995). An E-box-mediated increase in cad transcription at the G(1)/S-phase boundary is suppressed by inhibitory c-Myc mutants. Mol.Cell.Biol. *15*, 2527–2535.

Miner, J. H., Wold, B. J. (1991). c-myc inhibition of MyoD and myogenin-initiated myogenic differentiation. Mol. Cell. Biol. *11*, 2842–2851.

Müller, D., Bouchard, C., Rudolph, B., Steiner, P., R.Saffrich, Ansorge, W., Eilers, M. (1997). Phosphorylation of p27 by cyclin E/cdk2 facilitates ist Myc-induced release from release from cdk2 complexes Oncogene, (in press)

Munn, A. L., Stevenson, B. J., Geli, M. I., Riezman, H. (1995). end5, end6, and end7: mutations that cause actin delocalization and block the internalization step of endocytosis in Saccharomyces cerevisiae. Mol Biol Cell *6*, 1721–42.

Oswald, F., Lovec, H., Möröy, T., Lipp, M. (1994). E2F-dependent regulation of human MYC: transactivation by cyclins D1 and A overrides tumour suppressor protein functions. Oncogene *9*, 2029–2036.

Packham, G., Cleveland, J. L. (1994). Ornithine decarboxylase is a mediator of c-Myc-induced apoptosis. Mol. Cell Biol. *14*, 5741–5747.

Penn, L. J. Z., Brooks, M. W., Laufer, E. M., Land, H. (1990). Negative autoregulation of c-myc transcription. EMBO J. *9*, 113–121.

Perez-Roger, I., Solomon, D. L., Sewing, A., and Land, H. (1997). Myc activation of cyclin E/Cdk2 kinase involves induction of cyclin E gene transcription and inhibition of p27(Kip1) binding to newly formed complexes. Oncogene *14*, 2373–81.

Philipp, A., Schneider, A., Väsrik, I., Finke, K., Xiong, Y., Beach, D., Alitalo, K., Eilers, M. (1994). Repression of Cyclin D1: a Novel Function of MYC. Mol. Cell. Biol. *14*, 4032–4043.

Prendergast, G. C., Lawe, D., Ziff, E. B. (1991). Association of myn, the murine homolog of max, with c-myc stimulates methylation-sensitive DNA binding and ras cotransformation. Cell *65*, 395–407.

Pusch, O., Bernaschek, G., Eilers, M., Hengstschläger, M. (1997). Uncoupling of G1 cyclin-dependent kinase activity from cell cycle progression upon activation of Myc. Oncogene (in press).

Revardel, E., Bonneau, M., Durrens, P., Aigle, M. (1995). Characterization of a new gene family developing pleiotropic phenotypes upon mutation in Saccharomyces cerevisiae. Biochim Biophys Acta *1263*, 261–5.

Roussel, M. F., Ashmun, R. A., Sherr, C. J., Eisenman, R. N., Ayer, D. E. (1996). Inhibition of cell proliferation by the Mad1 transcriptional repressor. Mol Cell Biol *16*, 2796–801.

Roussel, M. F., Cleveland, J. L., Shurtleff, S. A., Sherr, C. J. (1991). Myc rescue of a mutant CSF-1 receptor impaired in mitogenic signalling. Nature *353*, 361–363.

Roussel, M. F., Davis, J. N., Cleveland, J. L., Ghysdael, J., Hiebert, S. W. (1994). Dual control of myc expression through a single DNA binding site targeted by ets family proteins and E2F-1. Oncogene *9*, 405–15.

Roussel, M. F., Theodoras, A. M., Pagano, M., Sherr, C. J. (1995). Rescue of defective mitogenic signaling by D-type cyclins. Proc. Natl. Acad. Sci. U.S.A. *92*, 6837–6841.

Roy, A. L., Carruthers, C., Gutjahr, T., Roeder, R. G. (1993). Direct role for Myc in transcription initiation mediated by interactions with TFII-I. Nature *365*, 359–361.

Rudolph, B., Saffrich, R., Zwicker, J., Henglein, B., Muller, R., Ansorge, W., Eilers, M. (1996). Activation of cyclin-dependent kinases by Myc mediates induction of cyclin A, but not apoptosis. Embo J *15*, 3065–76.

Sakamuro, D., Elliott, K. J., Wechsler-Reya, R., Prendergast, G. C. (1996). Bin1 is a novel Myc-interacting protein with features of a tumours suppressor. Nature genetics *14*, 69–76.
Schreiber Agus, N., Chin, L., Chen, K., Torres, R., Rao, G., Guida, P., Skoultchi, A. I., DePinho, R. A. (1995). An amino-terminal domain of Mxi1 mediates anti-Myc oncogenic activity and interacts with a homolog of the yeast transcriptional repressor SIN3. Cell *80*, 777–786.
Sherr, C. J. (1994). G1 phase progression: cycling on cue. Cell *79*, 551–555.
Shrivastava, A., Saleque, S., Kalpana, G. V., Artandi, S., Goff, S. P., Calame, K. (1993). Inhibition of transcriptional regulator Yin-Yang-1 by association with c-Myc. Science *262*, 1889–1892.
Sommer, A., Hilfenhaus, S., Menkel, A., Kremmer, E., Seiser, C., Loidl, P., Lüscher, B. (1997). Cell growth inhibition by the Mad/Max complex through recruitment of histone deacetylase activity. Curr. Biol. *7*.
Steiner, P., Philipp, A., Lukas, J., Godden-Kent, D., Pagano, M., Mittnacht, S., Bartek, J., Eilers, M. (1995). Identification of a Myc-dependent step during the formation of active G1 cyclin/cdk complexes. EMBO J. *14*, 4814–4826.
Stone, J., de Lange, T., Ramsay, G., Jakobovits, E., Bishop, J. M., Varmus, H., Lee, B. (1987). Definition of regions in human *c-myc* that are involved in transformation and nuclear localization. Mol. Cell. Biol. *7*, 1697–1709.
Vlach, J., Hennecke, S., Alevizopoulos, K., Conti, D., Amati, B. (1996). Growth arrest by the cyclin-dependent kinase inhibitor p27Kip1 is abrogated by c-Myc. EMBO J *15*, 6595–6604.
Wagner, A. J., Kokontis, J. M., Hay, N. (1994). Myc-mediated apoptosis requires wild-type p53 in a manner independent of cell cycle arrest and the ability of p53 to induce p21waf1/cip1. Genes Dev *8*, 2817–30.
Wagner, A. J., Small, M. B., Hay, N. (1993). Myc-mediated apoptosis is blocked by ectopic expression of Bcl-2. Mol Cell Biol *13*, 2432–40.
Zervos, A. S., Gyuris, J., Brent, R. (1993). Mxi1, a protein that specifically interacts with Max to bind Myc-Max recognition sites. Cell *72*, 223–232..

Growth Regulation by the E2F and DP Transcription Factor Families

L. Yamasaki[1]

1 Introduction

What we know about E2F transcriptional activity has been derived from several lines of investigation. Each line of study gives us greater insight into understanding the function of E2F transcriptional activity in growth control. In vivo and in vitro, E2F transcriptional activity appears to be complex, often capable of completely opposing actions. The basis for positive and negative action at many levels appears to be the ability of E2F transcriptional activity to transactivate and repress gene expression, which both require the same initial step, i.e., recognition of target DNA. When bound to the retinoblastoma tumor suppressor (pRB) or other pRB family members, E2F functions as a repressor, and when "free" of pRB family interaction, E2F functions as an activator. It is this positive and negative, up and down nature of E2F activity that fascinates so many investigators.

This review covers the important points concerning E2F: (i) the molecular basis for heterogeneous E2F activity (E2F family and DP family heterodimers), (ii) interaction of E2F with pRB family members, (iii) E2F target genes and cell-cycle dependent transcription, (iv) induction of proliferation vs. apoptosis, and (v) animal models for studying the function of E2F activity in vivo. The E2F literature is interlaced with studies of the pRB family members, cyclin-dependent kinases and their inhibitors, since E2F activity was the first identified target of growth suppression by the pRB family. The overwhelming number of mutations in human tumors that deregulate growth control by pRB and the impressive body of studies implicating E2F activity in growth control suggest that loss of pRB function during oncogenesis leads to deregulated E2F activity. However, we now understand that E2F and DP heterodimers are only one subset of the many transcription factors and cellular proteins reported to interact with pRB (Riley et al. 1994; Horowitz and Udvadia 1995). Although it is tempting to equate pRB family function with E2F action, this move is unwise. Instead, the hypothesis that pRB family function reflects regulation of E2F function must be tested rigorously. When it is possible, this review tries to distinguish between more global pRB family effects and observations derived directly from E2F activity.

[1] Columbia University, Department of Biological Sciences, 1212 Amsterdam Avenue, New York, New York 10027, USA, E-mail: ly63@columbia.edu

2 The Molecular Basis for Heterogeneous E2F Activity

2.1 A Brief Tumor Virus History

Our first look at E2F activity stems from studies of the DNA tumor viruses (Nevins 1992; Moran 1993; La Thangue 1994). Initially, E2F transcriptional activity was defined as the host cellular activity required by adenoviral E1A to transactivate the viral E2 promoter (Kovesdi et al. 1986). This cellular E2F activity binds to the sequence 5'TTTCGCGC3' found in two head-to-head repeats within the adenoviral E2 promoter (and E1A promoter) following infection. The expression of several host genes important for replication and bearing E2F sites in their promoters is responsive to E1A. Thus, E1A-induced E2F activity propels cells into S phase, and thereby facilitates viral replication. Interestingly, E1A is required to generate adenoviral E2 promoter activity only in differentiated cells, but not in undifferentiated F9 embryonal carcinoma cells, suggesting that differentiation somehow inhibited or masked this E2F activity, also referred to as DRTF for differentiation-regulated transcriptional factor (Imperiale et al. 1984; La Thangue and Rigby 1987). The nature of this masking or inhibition of E2F activity became clear as investigators discovered that pRB forms complexes with this E2F activity which represses E2F-dependent transactivation. In the presence of E1A, E2F activity is released from pRB and cells enter S phase. The SV40-T and HPV-E7 proteins similarly free E2F activity from pRB repression. However, E2F activity appears to be heterogeneous in numerous ways. E2F-DNA binding experiments demonstrated the presence of various E2F complexes: "free" E2F activity, E2F activity bound to pRB or the pRB-related protein p107, and E2F activity bound to cyclin-dependent kinases. Finally, in serum restimulation experiments using quiescent fibroblasts, the timing of these E2F complexes differed relative to entry into S phase. Furthermore, the purification of E2F activity from HeLa cells yielded multiple protein species which bound DNA most efficiently when individual purified species were recombined (Huber et al. 1993). This suggested that total E2F activity was a complex mixture of individual proteins that required heterodimerization for activity.

2.2 Cloning of the E2F and DP Families

The rapid cloning of E2F and DP family members gave us our first insight into the structure and biochemical function of a set of transcription factors, all of which interact with pRB or the pRB family members p107 and p130. Currently, there are five cloned mammalian E2F family members (E2F-1,-2,-3, -4, -5) (Slansky and Farnham 1996). *E2F* genes have been identified in human: *E2F-1* (Helin et al. 1992; Kaelin et al. 1992; Shan et al. 1992), *E2F-2* (Ivey-Hoyle et al.

1993; Lees et al. 1993), *E2F-3* (Lees et al. 1993), *E2F-4* (Beijersbergen et al. 1994; Ginsberg et al. 1994; Sardet et al. 1995), *E2F-5* (Buck et al. 1995; Hijmans et al. 1995; Sardet et al. 1995); in mouse: *mE2F-1* (Li et al. 1994; A. Talis and K. Helin, unpubl. observ.); in chicken (Pasteau et al. 1995); in *Xenopus* (Philpott and Friend 1994); and in *Drosophila*: *dE2F* (Dynlacht et al. 1994a; Ohtani and Nevins 1994). Two human DP members including DP-1 (Girling et al. 1993; Helin et al. 1993b) and *DP-2* (Wu et al. 1995; Zhang and Chellappan 1995; Rogers et al. 1996) have been identified. DP-2 appears to exist as several different spliced forms in different tissues (Ormondroyd et al. 1995; Zhang and Chellappan 1995; Rogers et al. 1996), and its murine homologue has been referred to as DP-3 based on the existence of three DP members in *Xenopus* (Ormondroyd et al. 1995). A DP homologue has also been identified in *Drosophila*, *dDP* (Dynlacht et al. 1994a; Hao et al. 1995). E2F/DP or pRB homologues have not been identified in yeast, although other cell cycle-regulated transcription factors do exist and ectopically expressed pRB can be phosphorylated by various Cln/Cdc28 complexes in yeast (Hatakeyama et al. 1994). There is limited homology between the E2F and the DP families within the DNA binding regions, suggesting these genes were derived from a common ancestor (Girling et al. 1993).

2.3 Heterodimerization

Although individual E2F or DP species can bind DNA weakly (Helin et al. 1992; Kaelin et al. 1992; Shan et al. 1992; Girling et al. 1993), robust DNA binding of E2F site-containing oligonucleotides and transactivation of E2F reporter genes require heterodimerization of an E2F member with a DP member (Bandara et al. 1993; Helin et al. 1993b; Ivey-Hoyle et al. 1993; Krek et al. 1993; Beijersbergen et al. 1994; Ginsberg et al. 1994; Hijmans et al. 1995; Sardet et al. 1995; Wu et al. 1995; Zhang and Chellappan 1995). Heterodimerization requires conserved domains that lie downstream of the DNA-binding domains in the E2F and DP proteins. During adenoviral infection, cooperative binding and subsequent transactivation of the E2 promoter occur between multiple E2F/DP heterodimers and the small (19 kDa) adenoviral E4-6/7 proteins (Bandara et al. 1994; Helin and Harlow 1994; Obert et al. 1994). Interaction with E4-6/7 requires a domain of E2F-1 (amino acids 284-358) that lies downstream of the DP dimerization domain and includes sequences conserved in E2F-2 and E2F-3 (Helin and Harlow 1994), which are referred to as the "marked box." The binding of E4-6/7 to E2F-1 precludes the subsequent binding of pRB, possibly through steric interference. Phosphorylations at Ser 332 and Ser 337 of E2F-1 are required for interaction with E4-6/7, although these modifications prevent its interaction with pRB (Fagan et al. 1994). The observation that viral proteins have evolved to aid the cooperative and stable interaction with and high level transactivation by E2F/DP complexes at a viral promoter raises the possibility that unidentified cellular proteins may exist that normally function in uninfected cells in a manner similar to E4-6/7 through the marked box. Interestingly, while pRB/E2F/DP complexes cannot distort or bend DNA structures containing E2F sites, E2F/DP heterodimers can do so with the DNA bending activity also requiring the marked box (Cress and Nevins 1996).

2.4
Association with and Phosphorylation by Cyclins/Cyclin-Dependent Kinases

Most of the E2F family members and both DP family members exist as phosphoproteins within the cell: E2F-1 (Dynlacht et al. 1994b; Fagan et al. 1994; Krek et al. 1994; Kitagawa et al. 1995; Peeper et al. 1995); E2F-2 and E2F-3 (Lees et al. 1993), E2F-4 (Beijersbergen et al. 1994; Ginsberg et al. 1994), DP-1 (Bandara et al. 1994; Dynlacht et al. 1994b), and DP-2 (Rogers et al. 1996). Phosphorylation of E2F-1 and DP-1 has been shown to be due to the direct action of a bound cyclin A/cdk2 complex (Dynlacht et al. 1994b; Krek et al. 1994; Xu et al. 1994). Residues 67-108 of E2F-1 are sufficient for binding cyclin A, and this region contains a basic domain, P(V/A)KR(R/K)L(D/E)L, conserved between E2F-1, E2F-2 and E2F-3 (Krek et al. 1994). E2F-1/DP-1 heterodimers which have been phosphorylated by cyclin A/cdk2 are no longer capable of DNA binding or transactivation. An E2F-1 mutant which no longer binds cyclin A/cdk2 in fact shows prolonged DNA binding activity in S phase (Krek et al. 1995), supporting the notion that E2F-1/DP-1 association with cyclin A/cdk2 is important in vivo. In fact, the small conserved basic region within the cyclin A binding domain is similar to the sequences RXLF(G/V) involved in cyclin A binding which are found in p107 and p130, as well as in the cyclin-dependent kinase inhibitors p21, p27 and p57 (Hannon et al. 1993; Li et al. 1993b; Luo et al. 1995; Zhu et al. 1995a; Adams et al. 1996; Chen et al. 1996) and in cdc25A (Saha et al. 1997). Phosphorylation of E2F-4 in contrast to that of E2F-1, results from the association with p107, to which cyclin A/k2 or cyclin E/k2 is bound (Zhu et al. 1995a). In contrast to the cdk2 association with E2F-1/DP-1 complexes, an overexpressed cdk3 dominant negative mutant can interact with E2F-1/DP-1 heterodimers, most likely through DP-1 interaction, and inhibit transactivation without affecting DNA-binding activity (Hofmann and Livingston 1996). Thus, the association of E2F/DP with and subsequent phosphorylation by different cyclin/cdk complexes may be critical to coordinating E2F/DP transcriptional activity throughout the cell cycle.

2.5
Localization of E2F/DP Complexes

Recently several groups have demonstrated that the cellular localizations of different E2F and DP members are distinct (Magae et al. 1996; de la Luna et al. 1996; Lindeman et al. 1997; Müller et al. 1997; Verona et al. 1997). Overexpression of E2F-1, E2F-2, or E2F-3 results directly in nuclear localization, while overexpression of E2F-4 or E2F-5 results in a predominantly cytoplasmic location. Similarly, ectopic expression of DP-1 gives a cytoplasmic localization, while DP-2 gives a nuclear localization. Differential localization of E2F and DP proteins may be due to: (1) the size of the protein; (2) the presence of a nuclear localization signal (NLS); (3) the presence of dimerization domains that allow one protein to "piggyback" into the nucleus through association with another protein; and finally (4) the expression of the "piggyback" partner. The predicted molecular mass for each of the E2F and DP proteins (~38 KDa - 52KDa) is very close to the nuclear exclusion limit (~45 KDa), although their anomalous migration on SDS-PAGE (up to ~60 KDa perhaps due to post-translational modifications) suggests an NLS may be required. E2F-1 has a functional NLS since deletion of amino acids 85-91 inhibits nuclear entry, and the first 126 amino acids of E2F-1 can direct the nuclear localization of β-galactosidase (Müller et al. 1997). Likewise E2F-2 has a functional NLS since deletion or mutation of amino acids 83-88 inhibits nuclear entry (Verona et al. 1997). Although amino acids 85-91 of E2F-1 and amino acids 83-88 of E2F-2 lie in the cyclin A binding domain, the cell cycle-dependent expression of cyclin A makes it unlikely that E2F-1 or E2F-2 requires cyclin A for nuclear entry. Fusion of the aminoterminus of E2F-1 or E2F-2 or the SV40T NLS to E2F-4 can force nuclear entry, and allow the E2F-4 fusion protein to stimulate E2F-dependent transactivation and S phase entry (Lindeman et al. 1997; Müller et al. 1997; Verona et al. 1997). In light of the small predicted masses of E2F-4 (45 KDa) and E2F-5 (38 KDa), it is more surprising that these proteins are excluded from the nucleus upon overexpression, and this suggests that they contain cytoplasmic retention signals.

Several domains have been identified that allow "piggybacking" of E2F and DP proteins into the nucleus, allowing one subunit to enter the nucleus via movement of the other subunit. For instance, expression of E2F-1, E2F-2 or E2F-3 can induce the nuclear accumulation of co-expressed DP-1 and requires the heterodimerization domains on the E2F subunit and DP-1 (Magae et al. 1996). Additionally, co-expression of E2F-4 with DP-2, but not DP-1, results in the nuclear accumulation of E2F-4 (Magae et al. 1996; Lindeman et al. 1997; Verona et al. 1997). Furthermore, association of E2F-4 protein with p107/p130 results in nuclear localization of E2F-4 and requires the pRB family binding epitope. The association of E2F-1, -2, and -3 with cyclin A also may affect the nuclear localization of E2F.

To determine whether the localization of the E2F and DP proteins changes with cell cycle position, it is critical to analyze endogenous levels of these proteins. Studies have shown that endogenous E2F-4 localization is cell-cycle regulated, being predominantly nuclear in G0/G1 and becoming increasingly cytoplasmic as cells enter S phase (Lindeman et al. 1997; Müller et al. 1997; Verona et al. 1997). Although the nuclear pool of E2F-4 in G0/G1 may function to repress E2F target genes, the function of the cytoplasmic pool of E2F-4 is not clear at this time. Clearly, differential cellular localization adds another level of complexity to the regulation of E2F/DP activity.

3 Interaction of E2F Transcription Factors with and Repression by pRB Family Members

3.1 Binding pRB Family Members

Mammalian E2F family members were identified and grouped initially by their interaction with either pRB (for E2F-1,-2,-3 see (Helin et al. 1992; Kaelin et al. 1992; Shan et al. 1992; Ivey-Hoyle et al. 1993; Lees et al. 1993) or p107/p130 (for E2F-4 and -5 see (Ginsberg et al. 1994; Sardet et al. 1995; Vairo et al. 1995). pRB/E2F complexes are detected in G1, becoming less apparent as cells reach the S phase, which coincides with the accumulation of hyperphosphorylated pRB (Buchkovich et al. 1989; Chen et al. 1989; DeCaprio et al. 1989; Chellappan et al. 1991; Mittnacht and Weinberg 1991). p130/E2F complexes are prominent in G0 (Cobrinik et al. 1993) and then associate with cyclin A/cdk2 and cyclin E/cdk2 later in G1 (Hannon et al. 1993; Li et al. 1993b; Claudio et al. 1996), but are not reformed as cycling cells re-enter G1 from M phase (Smith et al. 1996). p107/E2F complexes are detected at low levels in G0/G1, but become prominent at the G1/S transition and in S phase when they contain cyclin E/cdk2 and cyclin A/cdk2 complexes, respectively (Cao et al. 1992; Devoto et al. 1992; Faha et al. 1992; Pagano et al. 1992; Shirodkar et al. 1992; Schwarz et al. 1993). Recently using specific antibodies, it has been possible to identify the particular E2F complexed to a single pRB family member at different points in the cell cycle (Ikeda et al. 1996; Moberg et al. 1996b). These studies have indicated that E2F-4 does exist in complexes with pRB as well as with p107 and p130. This approach has placed E2F-4 in the p130 complexes in G0; E2F-4 and E2F-3 in the pRB complexes in G1 and at the G1/S transition; and E2F-4 in the p107 complexes at the G1/S transition and in S-phase. E2F-1 appeared to be mainly in the "free" E2F fraction at the G1/S transition and in S-phase. Besides the E2F-4 found complexed to the pRB family members, the bulk of "free" E2F appears to be E2F-4 (Ikeda et al. 1996; Moberg et al. 1996b). So far it has been difficult to find endogenous E2F-2 and E2F-5 in higher order complexes in these timing studies.

Each E2F interacts with a pRB family member directly through a small epitope (amino acids 409-426 in E2F-1) in the extreme carboxyterminus of E2F, which permits the specific interaction with the large bipartite conserved domain of pRB family members generally referred to as the "pocket." In fact, the expression cloning of E2F-1 relied on this interaction of E2F-1 with pRB (Helin et al. 1992; Kaelin et al. 1992; Shan et al. 1992). The carboxyterminus of each E2F member also contains a transactivation domain which allows E2F to activate the adenoviral E2 promoter or synthetic E2F promoters when co-expressed. Since the pRB family-binding domain is embedded within the larger carboxyterminal transactivation domain, the binding of a pRB family member to each E2F prevents or represses E2F-mediated transactivation when co-expressed. Thus, the capacity of pRB family members to repress transcription is due in part to masking the transactivation domain of E2F.

It is important to note that E2F family members may transactivate target genes directly through an interaction with the basal transcription machinery and/or with other transcription factors. Several such interesting interactions have been reported for E2F-1. An interaction of E2F-1 with TBP (Hagemeier et al. 1993; Emili and Ingles 1995) and one of E2F-1 with CBP/p300 via sequences in the transactivation domain of E2F-1 which share limited homology with E1A (Trouche and Kouzarides 1996) have been reported. Sp1 increases E2F-dependent transactivation through a physical interaction with E2F-1, which appears to require the aminoterminus of E2F-1 (Karlseder et al. 1996; Lin et al. 1996). v-Abl, BCR-Abl and c-Abl have been shown to stimulate E2F-1-mediated transcription, and v-Abl interacts directly with E2F-1 (Birchenall-Roberts et al. 1997). Therefore, the ability of pRB to repress E2F-1-mediated transcription may stem from its masking the interaction of E2F-1 with the basal transcription machinery and/or with other transcription factors.

Heterodimers composed of an E2F family member and a DP family member form stable trimeric complexes with pRB family members, as first shown for E2F-1 (Helin et al. 1993b; Krek et al. 1993; Wu et al. 1995). This principally is due to the direct association of the carboxyterminus of E2F with the pRB family member, but can also be influenced by weaker contacts between the carboxyterminus of DP with pRB family member (Bandara et al. 1994). Association of E2F-1 with pRB can be precluded by mutating Ser 332 and Ser 337 within E2F-1 (Fagan et al. 1994) and enhanced by phosphorylation of E2F-1 at Ser 375 by cyclin A/cdk2 (Peeper et al. 1995), suggesting that additional residues outside of the pRB binding epitope can influence the interaction with pRB. Interestingly, interaction of pRB with E2F-1 can stabilize E2F-1 and, similarly, that of p107 or p130 with E2F-4 can stabilize E2F-4 by preventing ubiquitin-mediated degradation (Hateboer et al. 1996; Hofmann et al. 1996; Campanero and Flemington 1997). Since carboxyterminal truncations of E2F-1 or E2F-4 also stabilize these proteins, a signal for degradation may exist near the pRB binding epitope in these E2F members.

The release of E2F/DP activity from complexes with pRB family members is thought to be regulated by phosphorylation of pRB family members. In the case of pRB complexes, the increase in E2F/DP activity correlates with a conversion of pRB from hypophosphorylated to hyperphosphorylated forms at the G1/S transition. In fact, a mutant of pRB which cannot be phosphorylated at eight putative cyclin-dependent kinase (CDK) sites functions as a super-repressor of E2F-dependent transactivation (Hamel et al. 1992a; Hamel et al. 1992b). The p107 and p130 proteins also undergo cell cycle-dependent phosphorylations (Baldi et al. 1995; Beijersbergen et al. 1995; Mayol et al. 1995; Studbal et al. 1996) with timing that suggests the release of E2F/DP activity may also be sensitive to phosphorylation of these pRB family members. Recently, one study has demonstrated that the hypo- or hyperphosphorylation state of pRB cannot always predict the ability of pRB to halt cell cycle progression (Connell-Crowley et al. 1997). However, at least in vitro, phosphorylation of pRB by either cyclin D1/cdk4 or cyclin E/cdk2 complexes prevent subsequent complex formation with E2F-1/DP-1 heterodimers (Suzuki-Takahasi et al. 1995; Connell-Crowley et al. 1997).

3.2 Repression by pRB Family Members

The association of pRB family members with E2F/DP heterodimers exerts repression over genes bearing E2F sites in their promoters. Initially, silencing by pRB was proposed by the observation that E2F sites could "switch from a positive element to a negative element" if pRB was present, suggesting that pRB/E2F/DP complexes could bind promotors and directly repress transcription (Weintraub et al. 1992). Mutation of these E2F sites eliminated activation and/or this repression by pRB, since it prevented the binding of a pRB/E2F/DP complex. This pRB-mediated repression can be transferred to heterologous promoters containing Gal4 sites when Gal4/E2F-1 carboxyterminal fusion proteins are co-expressed with pRB (Flemington et al. 1993; Helin et al. 1993a; Weintraub et al. 1995). Mutations in either the carboxyterminal domain of E2F or the pocket region of pRB family members abrogate this repression. This active repression by pRB occurs for several reasons, some of which are: (1) masking of the transactivation domain of E2F by pRB; (2) inhibition of transactivation by other transcription factors acting within bound promoters; and (3) inhibition of the basal transcription machinery by pRB. There are several experiments that suggest (2) and (3) may occur simultaneously with (1). Expression of pRB fusion proteins with the Gal4- or tetracycline repressor-DNA binding domain obviates the need for E2F sites and pRB/E2F complexes by simultaneously tethering pRB to the promoter bearing the appropriate DNA binding site and imposing repression (Adnane et al. 1995; Bremner et al. 1995; Sellers et al. 1995; Weintraub et al. 1995). Additionally, transactivation stimulated by transcription factors bound to promoters near E2F sites could be repressed by the binding of pRB/E2F/DP complexes in the vicinity (Weintraub et al. 1995). Furthermore, Gal4/pRB or LexA/pRB fusion proteins could repress transactivation of numerous transcription factors (e.g., Sp-1, AP-1, p53, c-myc, Elf-1 and PU.1) bound nearby (Adnane et al. 1995; Weintraub et al. 1995). Concerning possibility (3), it is provocative that sequence similarity between two regions of pRB and two basal transcription factors, TFIID and TFIIB, exists (Hagemeier et al. 1993). It remains to be seen whether pRB actively represses transcription by inhibiting the function of the basal transcription machinery through mimicry.

4
E2F/DP Target Genes and Cell-Cycle Dependent Transcription

4.1
Definition of E2F/DP Targets

Extensive effort has been expended to define cellular E2F/DP target genes beyond the originally identified adenoviral E2 and E1A targets (Slansky and Farnham 1996). Definition of E2F/DP target genes involves the delineation of the target promoter conferring proper growth regulation to a reporter gene and the subsequent identification of functional E2F binding sites within this promoter. Functional E2F sites are identified by their deletion or mutation which affects the expression of the target gene. Mutation of a functional E2F site may lead to a deregulated level or timing of expression across the cell cycle. Many of these functional E2F sites that have been identified closely resemble the 5'TTTCGCGC3' found in the adenovirus E2 promoter. However, gel mobility shift experiments using probes spanning the putative E2F site (and the wild-type and mutant adenoviral E2 oligonucleotides as competitors) are needed to verify the specificity of binding, especially for putative E2F sites which differ substantially from the adenoviral E2 sequence (Zwicker and Muller 1997). Often functional E2F sites lie in tandem repeats (direct or inverted and often palindromic) at or near the transcriptional start site of TATA-less promoters. Although one E2F site has been shown to function 2 kb away from the site of initiation (Weintraub et al. 1995), other E2F sites (e.g. the DHFR promoter) show position-dependent activation (Fry et al. 1997). Thus, the promoter context in which a functional E2F site resides influences the activity of that E2F site. In isolated cases, investigators have been able to identify biochemical differences between the closely positioned E2F sites within a given promoter (Zhu et al. 1995b; Wells et al. 1996). Recently, differences in sequence-specific binding have been defined between different E2F family members using random oligonucleotide-based polymerase chain reaction (CASTing) approaches (Tao et al. 1997).

E2F/DP target genes can be classified according to their apparent function as follows:

(1) **S phase genes** important for replication, e.g., *DHFR* (Blake and Azizkhan 1989; Means et al. 1992), *RNR* (Thelander and Thelander 1989; Bjorklund et al. 1993; DeGregori et al. 1995a), *POLα* (Pearson et al. 1991), *TS*, *TK* (Li et al. 1993a; Ogris et al. 1993; Dou et al. 1994), *CAD* (Miltenberger et al. 1995), *ORC1* (Ohtani et al. 1996), *H2A* (Yagi et al. 1995; Oswald et al. 1996); *PCNA* (Yamaguchi et al. 1992); (2) **proto-oncogenes**, e.g., *c-myc* (Hiebert et al. 1989; Thalmeier et al. 1989; Hiebert et al. 1991), *N-myc* (Mudryj et al. 1990; Hiebert et al. 1991), *B-myb* (Lam and Watson 1993)]; (3) genes encoding **cyclins** and **CDKs**, e.g., *cdc2* (Dalton 1992; Furukawa et al. 1994), *cyclin A* (Henglein et al. 1994; Schulze et al. 1995), *cyclin E* (Ohtani et al. 1995; Botz et al. 1996; Geng et al. 1996), *cyclin D*

(Müller et al. 1994)]; (4) **E2F** genes and **pRB family** genes, e.g., *RB* (Hamel et al. 1992b), *p107* (Zhu et al. 1995b), *E2F-1* (Hsiao et al. 1994; Johnson et al. 1994b; Neuman et al. 1994), *E2F-2* (Sears et al. 1997)]. Certainly the spectrum of functions attributed to these target genes suggests that E2F/DP deregulation may be quite pleiotropic.

The identification of the *E2F* genes themselves as E2F/DP targets (4) underscores the potential complexity involved in normal cell cycle progression via E2F/DP-dependent transcriptional regulation. E2F-1 mRNA begins to accumulate at the G1/S transition and continues throughout S phase (Slansky et al. 1993) due to the presence of E2F sites in the *E2F-1* promoter itself which allows E2F/DP activity to regulate the abundance of E2F-1. The identity of the E2F which is normally responsible for regulating the expression of E2F-1 is not known at this time, but overexpression of E2F-1, -2 and -3 can stimulate the *E2F-1* promoter (Hsiao et al. 1994; Johnson et al. 1994b; Neuman et al. 1994). The *E2F-2* promoter is under similar E2F-dependent regulation (Sears et al. 1997). Serum restimulation experiments have shown that the levels of other E2F mRNAs fluctuate uniquely throughout the cell cycle. In particular, E2F-4 mRNA expression is high in G0, while E2F-5 mRNA expression is induced in early G1 (Ginsberg et al. 1994; Sardet et al. 1995). Distinct protein expression patterns are also observed for the various E2F members (Moberg et al. 1996). Taken together, these observations suggest that the timed expression of one E2F may control the subsequent expression of other *E2F* genes.

4.2 Cell Cycle-Dependent E2F/DP-Mediated Transcription

E2F/DP target genes can also be classified according to the nature (activation vs. repression) of the E2F regulation at their promotors. Mutation of the functional E2F site may lead to either increased or decreased promoter activity relative to that obtained from the wild-type promoter using quiescent fibroblasts stimulated to re-enter the cell cycle. Loss of activation due to the mutation of an E2F site has been observed for the *N-myc* promoter (Hiebert et al. 1991) and *DHFR* promoter (Means et al. 1992), and leads to a low level of promoter activity as cells re-enter the cell cycle. Loss of repression due to the mutation of an E2F site has been observed for the *RB* promoter (Hamel et al. 1992b), *B-myb* promoter (Lam and Watson 1993), *E2F-1* promoter (Hsiao et al. 1994; Johnson et al. 1994b; Neuman et al. 1994), and *E2F-2* promoter (Sears et al. 1997), and leads to a constitutively high level of promoter activity as cells re-enter the cell cycle. The ability of pRB family members to repress these promoters depends on the presence of the identified E2F site. Therefore, repression of target promoters containing E2F sites suggests that the function of E2F/DP transcription factors may be to select target genes for cell cycle-dependent silencing. Of course, the proper growth regulation of a given promoter is due to the sum of activation and repression across the cell cycle, and both activation and repression may only be due in part to the direct action of E2F/DP. For instance, the frequent presence of Sp-1 sites within the promoters of E2F/DP target genes and the decreased activity of these promoters upon mutation of the Sp-1 sites suggest that E2F/DP transcriptional activity confers only a portion of the proper growth regulation to a target promoter.

Other elements which may contribute to the proper growth regulation of these promoters have been defined (Zwicker and Muller 1997). Sequence alignments of E2F sites within target promoters which are activated by E2F (e.g., adenoviral *E2*, *c-myc*, *cyclin E* and *TK*) vs. variant E2F sites in target promoters which are repressed (e.g., *cyclin A*, *cdc2*, and *cdc25C*) have identified a *c*ell-cycle *d*ependent *e*lement (CDE), CGCGG, and a downstream *c*ell-cycle *h*omology region (CHR) element, TTGAA, which is required for repression. Similarly, alignment of the consensus E2F sites in promoters which are repressed (e.g., *E2F-1* and *B-myb*) have identified a downstream CHR-like element, AGGA(T/A). Mutation of these new elements destroys proper cell cycle-dependent expression, suggesting that other important factors may be bound to these elements.

Defining which E2F/DP heterodimer and pRB family member bind to a particular E2F target is crucial for understanding E2F regulation of the cell cycle. Recently, two approaches have been taken to identify E2F/DP target gene expression from the endogenous target locus using quiescent fibroblasts stimulated to re-enter the cell cycle. The first approach examined target gene activation from overexpression of various E2F and DP family members via adenoviral expression vectors in quiescent REF52 cells, and subsequent monitoring of induction of various known E2F target genes (DeGregori et al. 1995a; DeGregori et al. 1997). E2F/DP-dependent target gene induction was compared to that obtained by serum re-stimulation alone. The ability of various E2F members to induce S phase did not necessarily correlate with their ability to induce specific target genes. Using this system, some known E2F targets (*POLα* and *cyclin E*) were induced by E2F-1, -2, -3 and -4, while other E2F targets were induced by smaller subsets of E2Fs (e.g., *RNR2*, *cyclin A* and *cdc2* by E2F-1 and -2). This approach suggests that E2F members are not equivalent, but every E2F target gene examined responded to more than one E2F member.

The second approach recently taken to identify E2F/DP target gene expression from the endogenous locus used mouse embryonic fibroblasts (MEFs) derived from pRB family knockout mice to examine target gene repression (Herrera et al. 1996; Hurford et al. 1997). Wild-type MEFs, *RB*(-/-), *p107*(-/-), *p130*(-/-) MEFs and *p107*(-/-); *p130*(-/-) double mutant MEFs were examined after serum-starvation and re-stimulation. Surprisingly, these various mutations only modestly affect the kinetics of cell cycle entry (Lukas et al. 1995; Herrera et al. 1996; Hurford et al. 1997). By this method, pRB loss induced the derepression of *cyclin E* and *p107*. Loss of only p107 or only p130 did not derepress known E2F targets. However, loss of both p107 and p130 derepresses *B-myb*, *cdc2*, *E2F-1*, *TS*, *RNR* and *cyclin A* genes and induces *DHFR* expression, suggesting overlapping function for p107 and p130 repression of E2F/DP target genes (Hurford et al. 1997). Furthermore, this study suggests that different E2F/DP target genes are regulated by either pRB complexes or p107 and p130 complexes.

This sophisticated analysis to examine known E2F/DP targets and the possibility of using knockout MEFs to screen for novel E2F/DP targets suggests that the identification of key E2F target genes is now the emerging question. Key E2F target genes would be defined as those genes whose expression is crucial for the subsequent action (e.g., arrest, proliferation, apoptosis, etc). Still unaddressed is the possibility that selective activation or repression of E2F/DP target genes in vivo may require a combination of specific E2F, DP and pRB family member expression in addition to specific E2F target gene responsiveness.

5 Induction of Proliferation vs. Apoptosis

5.1 Entry into S-phase

Growth suppression by pRB and a portion of growth suppression by p107/p130 requires the conserved pocket domain which binds E2F/DP heterodimers and many other factors (see Chapter by M.Ewen on pRB). Given the role of E2F target genes in S-phase, it is thought that pRB family members suppress growth in part by repressing E2F-dependent transactivation in G0 and G1. The release and accumulation of E2F/DP transcriptional activity at the G1/S transition is thought to facilitate entry of cells into S-phase and thus, stimulate proliferation. To test the role of E2F/DP transcription factors in S-phase entry, several groups induced the expression of E2F-1 in quiescent fibroblasts and monitored cell cycle progression. Overexpression of E2F-1 alone (Johnson et al. 1993; Qin et al. 1994; Wu and Levine 1994) or in combination with DP members (Shan and Lee 1994) can induce S phase entry.

Consistent with the capacity of E2F overexpression to stimulate entry into S phase is the observation that dominant negative mutants of E2F-1 or DP-1 prevent S phase entry (Dobrowolski et al. 1994; Logan et al. 1995; Wu et al. 1996; Fan and Bertino 1997). Additionally, proliferation can also be inhibited by RNA ligands that selectively block E2F function (Ishizaki et al. 1996). Overexpression of other E2F members also can induce S phase entry (Lukas et al. 1996; DeGregori et al. 1997). Ectopic expression of E2F-2 or E2F-3 alone induces S phase entry, but E2F-4 requires co-expression of DP-1 to induce S phase. The ability of E2F-5 to induce S phase appears to vary (DeGregori et al. 1997), and in at least one setting appears to require the co-expression of DP-1 (Lukas et al. 1996). G1 blocks imposed by the overexpression of numerous CDK inhibitors (e.g. p16, p21, p27) can be overcome by E2F-1 (DeGregori et al. 1995b). Furthermore, the

G1 block imposed by overexpression of p16 can be overcome by E2F-1, E2F-2 and E2F-3, but not E2F-4 or E2F-5 alone (Lukas et al. 1996; Mann and Jones 1996).

5.2 Action as Oncogenes

In agreement with the ability of overexpressed E2F members to drive entry into S phase, several groups have reported that *E2Fs* and *DP-1* act as bona fide oncogenes. In combination with activated *ras*, co-expression of E2F-1 and DP-1 transforms rat embryo fibroblasts as judged by colony formation and growth in soft agar assays, and these fibroblasts can form tumors when injected into nude mice (Johnson et al. 1994a). Although several investigators report that stable cell lines constitutively overexpressing E2F-1 have been hard to obtain, two groups were able to establish that overexpression of E2F-1 alone can transform fibroblasts (for REFs (Singh et al. 1994) and for NIH3T3s (Xu et al. 1995)). One of these groups reported that an E2F-1 mutant unable to bind pRB is more effective at transformation (Xu et al. 1995). Similar to E2F-1, overexpression of E2F-2 alone or E2F-3 alone results in transformation of NIH3T3s. E2F-4 and DP-1 overexpression can also cooperate with activated *ras* to transform rat embryo fibroblasts as judged by colony formation and growth in soft agar assays (Beijersbergen et al. 1994). Overexpression of only an *E2F-4* mutant which no longer interacts with pRB family members (but not intact E2F-4) can transform NIH3T3 fibroblasts as judged by growth in soft agar, which readily form tumors in nude mice (Ginsberg et al. 1994). Overexpressed DP-1 also can cooperate with activated *ras* to transform rat embryo fibroblasts using the colony formation assay (Jooss et al. 1995).

A separate cooperation assay using human foreskin keratinocytes has demonstrated that *E2F-1* has only a weakly oncogenic capacity (Melillo et al. 1994). This system uses HPV-E6 (by mediating p53 degradation) and E7 (by pRB family inactivation and p300 binding) proteins for immortalization. In this system, E2F-1 is not able to complement HPV-E6 to immortalize keratinocytes for two reasons. First, high level expression of E2F-1 prevents E6 and E7 from achieving immortalization, presumably by inducing apoptosis. Second, low level expression of E2F-1 cannot replace all of the immortalization function of E7, but can complement mutants of E7 (which no longer bind pRB, but bind p300) to immortalize keratinocytes. Taken together, these studies suggest that *E2F-1* functions as an oncogene by collaborating with other mutations such as activated *ras* and that this transformation may be subject to cell-type differences.

5.3
E2F Mutations in Human Tumors

Given the capacity of many E2F family members to act as oncogenes when overexpressed, one obvious prediction is that E2F overexpression or mutations may be found in human tumors. Several groups have tried to test this prediction by determining if the human chromosomal locations for the various *E2F* genes correlate with genomic regions frequently mutated in human tumors. The human chromosomal locations are 20q11 for the *E2F-1* gene (Saito et al. 1995), 1p36 for the *E2F-2* gene (Lees et al. 1993), 6q22 for the *E2F-3* gene (Lees et al. 1993) and 16q22 for the *E2F-4* gene (Ginsberg et al. 1994). Upon fine mapping, these E2F loci do not correlate with regions that are commonly associated with deletions or mutations in human cancer. Although no *E2F-1* gene rearrangements were detected in primary acute leukemia or myelodysplasia cases by Southern blotting, the *E2F-1* locus is amplified (nine copies) and its protein is overexpressed in the human erythroleukemia cell line HEL (Saito et al. 1995). The *p107* locus which lies telomeric to the *E2F-1* locus in band 20q11 is not amplified in the HEL cell line. Moreover, several leukemia and lymphoma cell lines show deregulated E2F-1 mRNA and/or protein expression. Recently, E2F-1 protein overexpression and E2F-1 PCR-SSCP shifts have been detected in transitional cell carcinomas of the bladder (F. Rabbani, V. Richon, M. Lu, M. Drobnjak, Z.-F. Zhang, C. Cordon Cardo, unpubl. observ.). Alterations of the *E2F-4* gene, specifically in the polyserine repeat, are frequently detected in gastrointestinal tumors (3 % gastric tumors, 33 % ulcerative colitis-associated neoplasms, 42 % colorectal carcinomas; replication error-positive), but not in endometrial, prostate or a subset of colorectal tumors (replication error-negative) (Souza et al. 1997). However, heterogeneity in the number of serines (7, 9, 11, 14 or 16) within this polyserine repeat from apparently normal tissue was identified (Ginsberg et al. 1994) when *E2F-4* was first cloned. No detectable difference in DNA binding or transactivation by E2F-4 is known to correlate with changes in the polyserine repeat.

5.4
Triggering p53-Dependent Apoptosis

Although ectopic expression of many E2F family members induces S phase entry, this cell cycle progression does not necessarily lead to proliferation. In fact, in many settings the forced overexpression of E2F-1 leads to apoptosis. For example, fibroblasts transiently expressing E2F-1 (with or without DP-1) proliferate unless grown in low serum concentrations (Qin et al. 1994; Shan and Lee 1994; Kowalik et al. 1995) or subjected to γ-irradiation (Kowalik et al. 1995) or arrested by wild type p53 (Wu and Levine 1994). Myeloid progenitors expressing E2F-1 alone proliferate unless they undergo IL-3 withdrawal (Hiebert et al. 1995). These studies led to the model that conflicting signals sent by p53 (arrest) and E2F-1 (proliferation) resulted in apoptosis due to the incompatibility of both of these pathways. The ability of E2F-1 to induce apoptosis appears to be an exclusive property of this E2F family member, since adenoviral-based overexpression of any other E2F member does not lead to apoptosis (DeGregori et al. 1997). However, it has not been possible to establish stable cell lines constitutively overexpressing other E2F family members.

Apoptosis induced by overexpression of E2F-1 is dependent on p53 status. Using either temperature-sensitive *p53* mutant cells (Wu and Levine 1994) or fibroblasts from *p53* (-/-) mice (Kowalik, 1995) or fibroblasts expressing dominant negative *p53* mutants (Qin et al. 1994; Kowalik et al. 1995), E2F-1 overexpression induces apoptosis when p53 is wild-type. This p53-dependent apoptosis requires the DNA binding and transactivation domains of E2F-1. Co-expression of pRB can inhibit E2F-1 from promoting p53-dependent apoptosis, and *E2F-1* mutations which prevent pRB association abrogate this effect (Qin et al. 1994; Shan et al. 1996). p53 interacts with mdm2 and results in less p53-mediated apoptosis, less p53-dependent transactivation (Picksley and Lane 1993; Chen et al. 1995; Thut et al. 1997), and enhanced p53 degradation (Haupt et al. 1997; Kubbutat et al. 1997). Thus, co-expression of mdm2 also can inhibit apoptosis driven by overexpression of E2F-1. The requirement for the transactivation domain of E2F-1 suggests that prolonged activation by E2F-1 of an apoptotic target gene may be responsible for subsequent apoptosis.

In light of the p53 dependence of apoptosis induced by E2F-1, there are several relevant reports concerning an interaction of E2F-1 with mdm2 (Martin et al. 1995), an interaction of E2F-1 with p53 (O'Connor et al. 1995) and an interaction of DP-1 with p53 (O'Connor et al. 1995; Sorensen et al. 1996). The in vitro association of E2F-1 with mdm2 relies on a small p53-like domain of E2F-1 lying upstream of the pRB binding epitope and does not prevent dimerization of E2F-1 with DP-1. Although overexpression of mdm2 results in greater E2F/DP-dependent transcription, its functional significance is unclear and at least requires the demonstration of in vivo association of E2F-1 with mdm2. Interaction of E2F-1 and p53 has been seen in vitro and in vivo in C6 cells, although the domains responsible for E2F-1 binding to p53 were not mapped (O'Connor et al. 1995). The binding of DP-1 to p53 requires a carboxyterminal domain of DP-1 and an aminoterminal domain of p53(Sorensen et al. 1996). Whether the interaction of p53 with DP-1 interferes with (Sorensen et al. 1996) or does not affect (O'Connor et al. 1995) the dimerization of DP-1 to E2F-1 is unclear. Interaction with p53 can decrease E2F/DP-dependent transcription. E2F-1/DP-1 can also down-regulate p53-dependent transcription independent of mdm2 (O'Connor et al. 1995). While these data suggest that E2F/DP activity may be regulated in opposite ways by interaction with p53 or mdm2, it is difficult to imagine that an increase in E2F-dependent transactivation by mdm2 is consistent with a requirement for E2F-1 dependent transactivation in p53-dependent apoptosis.

When does E2F-1 drive cell cycle progression and when does it induce apoptosis? Following serum restimulation, NIH3T3 fibroblasts re-enter the cell cycle and E2F-1 protein and transcriptional activity accumulates across late G1 and into S phase, when it stimulates the expression of S phase genes. In late S phase, this E2F-1/DP-1-dependent transactivation is down-regulated, allowing the cell to complete S and G2 phases and then divide (Krek et al. 1995). This decrease in E2F-1/DP-1 dependent transcriptional activity occurs through the association of E2F-1/DP-1 complexes with and phosphorylation by cyclin A/cdk2 complexes (see Section 2 above). Subsequent to the phosphorylation of DP-1 by bound cyclin A/cdk2, E2F-1/DP-1 complexes lose the ability to bind DNA, and thus E2F-1/DP-1-dependent transactivation ceases. Consistent with this model are two observations made using E2F-1 mutants, which no longer associate with the cyclin A/cdk2 kinase. First, E2F-1 mutants which are defective for cyclinA/cdk2 binding, retain high levels of DNA binding and transactivation activity late in S phase, which increases their ability to induce apoptosis (Krek et al. 1995). Second, E2F-1 mutants defective for cyclinA/cdk2 binding and transactivation cause S phase arrest, but not apoptosis (Qin et al. 1995).

5.5 p53-Independent Apoptosis

In contrast to studies that have reported the p53 dependence of apoptosis by E2F-1, several groups have recently reported that overexpression of E2F-1 can also induce apoptosis, which is p53-independent (Hsieh et al. 1997; Nip et al. 1997; Phillips et al. 1997). In this case the requirements for p53-independent apoptosis differ substantially from those for p53-dependent apoptosis. Such apoptosis still requires DNA binding, but does not require the transactivation domain of E2F-1. Coexpression of pRB can inhibit apoptosis driven by E2F-1 overexpression even in the absence of p53, while coexpression of mdm2 fails to inhibit p53-independent apoptosis. The ability of E2F-1 to induce p53-independent apoptosis does not require entry into S phase (Phillips et al. 1997). In contrast to p53-dependent apoptosis through E2F-1, the requirement for the DNA binding domain, but not for the transactivation domain of E2F-1 to induce p53-independent apoptosis suggests that E2F-1 also may repress an apoptotic target gene.

Taken together, these studies suggest that overexpression of E2F-1 can indeed induce apoptosis, via a p53-dependent route as well as a p53-independent route. It is also possible that abnormally high levels of E2F-1 send an apoptotic signal to the cell that can be amplified by p53 function. However, given the distinct characteristics of p53-dependent and p53-independent apoptosis through E2F-1, this seems unlikely. It is too soon to tell whether E2F/DP-dependent activation and/or repression of an apoptotic gene is the mechanism for E2F-1-dependent apoptosis.

6
Animal Models for Studying E2F/DP Activity

Animal models in *Drosophila* and mouse have been established to study the roles of the E2F and DP families of transcription factors in vivo. There are advantages to using either organism. In mammals, there are five *E2F* and two *DP* genes to which function must be ascribed. In *Drosophila*, the situation is much less complex. In mammals, the *RB* gene was identified prior to the cloning of the *E2F* and *DP* genes, while in *Drosophila*, the pRB homologue, *RBF*, was identified subsequent to the cloning of the *dE2F* and *dDP* genes. The order in which questions have been asked in either system has differed, but it is clear that in either mouse or *Drosophila*, the action of E2F and DP is complex. Since both mammalian *RB* and the *Drosophila RBF* genes exist, we eventually will be able to compare the phenotypes derived from mutation of *RB* or *RBF* with phenotypes from mutation of *E2F* or *DP*. Furthermore, by combining mutations we can look for genetic interactions between *RB* and *E2F*, and test whether loss of *RB* deregulates *E2F*. This is an increasingly important question given the growing list of proteins that have been reported to interact with pRB and the low number of *E2F* mutations detected in human tumors. *Drosophila* genetics has the further advantage of allowing investigators to screen for unknown genetic modifiers of E2F/DP activity.

6.1
The *Drosophila* Model

The *Drosophila* genes encoding the dE2F and dDP homologues were identified by low stringency hybridization using probes derived from the DNA binding domains of mammalian counterparts (Dynlacht et al. 1994a; Ohtani and Nevins 1994; Hao et al. 1995). The dE2F and dDP proteins display substantial homology in DNA binding and heterodimerization domains, and thus can heterodimerize and transactivate *E2F* reporter genes when overexpressed. *Drosophila E2F* is an essential gene and null *dE2F* mutants display impaired transcription of *RNR2* and *PCNA* genes (Duronio et al. 1995). The *dPCNA* gene has been shown to contain functional E2F sites in its promoter (Yamaguchi et al. 1995). *Drosophila E2F* mutants have also enhanced position-effect variegation, a special form of genomic silencing (Seum et al. 1996). Additionally, cyclin E is an essential downstream target of dE2F for S phase entry, except in certain cell types where cyclin E is constitutively expressed (Duronio and O'Farrell 1995). Ectopic expression of dE2F and dDP can induce S phase and apoptosis in imaginal disc cells (Asano et al. 1996; Du et al. 1996b) similar to that observed for overexpression of E2F-1 in mammalian cells. High expression of endogenous dE2F is seen in the postmitotic cells of the larval eye disc, and loss of dE2F at these later stages interferes with proliferation and survival of various cell types within this compartment (Brook et al. 1996). The presence of a pRB family-binding epitope in dE2F has been exploited to identify *Drosophila* RBF, which can repress dE2F/dDP-dependent transcription and suppress the eye phenotype due to ectopic dE2F/dDP expression (Du et al. 1996a). Recently, *dDP* mutants have been isolated that have abnormal transcription of *PCNA* and *RNR2* genes (Royzman et al. 1997). Although essential, *dE2F* and *dDP* genes do not appear to be essential for embryonic S phase entry (Royzman et al. 1997), which is in partial contrast to an earlier report (Duronio et al. 1995). Furthermore *dE2F* mutants and *dDP* mutants display abnormal proliferative tissue in pupal stages, suggesting abnormal growth control (Royzman et al. 1997). The survival of *dE2F* mutants and dDP mutants to pupae may be due to the presence of an additional dE2F, which recently has been identified (D. Huen, W. Du, and N. Dyson, unpubl. observ.).

6.2 The Murine Model

The complexity of the pRB family and E2F/DP families in mammals suggests that the functions of different E2F members may be unique. An examination of the pRB family member deficient-mice tells us that loss of each family member is not equivalent, as pRB-deficient mice die in utero (Clarke et al. 1992; Jacks et al. 1992; Lee et al. 1992) and p107-deficient mice (Lee et al. 1996) or p130-deficient mice (Cobrinik et al. 1996) are viable and normal. The expression patterns of E2F and DP homologues in mouse support the view that E2F members may function differently from one another. Using RNAse protection and in situ hybridization, abundant E2F-1 and DP-1 mRNA is detected from midgestation (E14.5) to post-natal day 3 in specific cell types and tissues (Tevosian et al. 1996; Gropalkrishnan et al. 1996). Furthermore, the expression patterns for E2F-2, E2F-4 and E2F-5 mRNAs have also been determined during murine development (Dagnino et al. 1997). In both the developing epidermis and gastrointestinal tract, differences in E2F-4 and E2F-2 levels versus E2F-5 levels suggest that E2F-4 and E2F-2 are more abundant in proliferating and undifferentiated cell types, while E2F-5 is more prominent in differentiated cell types.

Two groups have reported the result of E2F-1 overexpression in transgenic mice. The megakaryocyte-specific platelet factor 4 promoter has been used to target ectopic E2F-1 expression specifically to post-mitotic megakaryocytes in transgenic mice (Guy et al. 1996). Ectopic E2F-1 causes megakaryocyte proliferation and blocks terminal differentiation in bone marrow, spleen and lymph node, which leads to thrombocytopenia. Apoptosis was also more apparent. The ubiquitously expressed HMG housekeeping promoter has been used to express ectopic E2F-1 or DP-1 in a wide variety of tissues in transgenic mice (C. Holmberg et al. 1998). Although overexpression of DP-1 was tolerated in a number of tissues without consequence, overexpression of E2F-1 was only obtained in the testes, where it resulted in testicular atrophy. This atrophy was due to apoptosis and was more apparent when E2F-1 and DP-1 were both overexpressed. However, when transgenics expressing both E2F-1 and DP-1 were crossed with p53-deficient animals, the testicular atrophy still occurred, demonstrating that this apoptosis was not p53-dependent. These studies suggest that several observations made using tissue culture models may be useful for interpreting phenotypes in vivo.

To understand the normal function of *E2F* and *DP* genes in vivo, many investigators have begun to inactivate these loci using homologous recombination in mouse embryonic stem cells. E2F-1-deficient mice have been generated by two groups, which observed that the loss of E2F-1 resulted in viable mice which developed tissue-specific abnormalities (Field et al. 1996; Yamasaki et al. 1996). For instance, the thymus of *E2F-1*-deficient mice is more prominent due to hyperproliferation and the inability of thymocytes to undergo p53-dependent apoptosis (Field et al. 1996). E2F-1-deficient mice develop testicular atrophy and exocrine gland (e.g., salivary glands and pancreas) dysplasia and are predisposed to neoplasia (e.g., lung adenocarcinoma, reproductive tract sarcoma and lymphoma) (Yamasaki et al. 1996). This tumorigenesis may be due to loss of pRB-mediated repression or loss of p53-mediated apoptosis or other mechanisms, but demonstrates that E2F-1 functions in some tissues as a tumor suppressor. Furthermore, inactivation of E2F-1 extends lifespan and reduces pituitary and thyroid tumorigenesis (L. Yamasaki et al. 1998) that occurs in the *RB*(+/-) mice (Jacks et al. 1992; Hu et al. 1994; Maandag et al. 1994; Harrison et al. 1995), suggesting that *E2F-1* is downstream of pRB control in these tissues, acting as an oncogene. Thus, E2F-1 can control growth positively and negatively in a tissue-dependent manner. Additionally, *DP-1*-deficient mice die as embryos midway through gestation, suggesting that the requirement for individual members of the E2F or DP families can vary tremendously (L. Yamasaki et al., unpubl. observ.). Inactivation of the remaining *E2F* and *DP* genes is currently underway by numerous groups. *E2F*- or *DP*-deficient mice test the requirement for a single E2F or DP member throughout development and adult life, and therefore will be invaluable to understand the functions of the E2F and DP transcription factor families.

7 Viewpoint

"Redundancy" is an often used to describe members of gene families for which few if any distinctions can be made. Although groups of E2F members and both DP members sometimes appear capable of executing similar tasks in tissue culture models, they may not function similarly in animal models. This may be due to the temporal- or tissue-specific expression patterns of the E2F and DP genes, or their regulators. These factors, in combination with key E2F/DP target genes, could produce a spectrum of cell biology (e.g., proliferation, apoptosis, and differentiation) and a range of histopathology resulting from the action of the E2F/

DP families. Amazingly, this range of function exists in even a single member, E2F-1. Thus, several important questions lie ahead. How can these broadly expressed transcription factors give temporal- and tissue-specific effects? Do they accomplish this as activators or repressors of key target genes? What are these key targets? These questions will be answered only by combining several lines of in vitro and in vivo study. As always and as with E2F/DP and pRB, it will be a question of balance and how to maintain that balance.

Note Added in Proof:
Recently another E2F has been described (referred to as EMA by Morkel M, Wenkel J, Bannister AJ, Kouzarides T, Hagemeier C (1997) Nature 390: 567–568 or as E2F-6 by Cartwright P, Muller H, Wagner C, Holm K, Helin K, manuscript submitted) that heterodimerizes with DP-1, binds E2F sites, fails to bind pRB family members and yet inhibits E2F-dependent transcription.

Acknowledgments Yamasaki wishes to thank K. Helin, J. Horowitz, J.A. Lees, N. Dyson and V. Richon as well as their collaborators for sharing information prior to publication. L. Y. apologizes in advance for any oversight of published data not in this chapter. She also thanks Nick Dyson, Ed Harlow and the members of the Laboratory of Molecular Oncology at the Massachusetts General Hospital Cancer Center for many helpful and lively discussions. She is most grateful for the careful reading of this chapter and helpful comments, as well as the invaluable and constant support from Michele Pagano. This work has been supported in part by a Special Fellowship to L. Y. from the Leukemia Society of America.

References

Adams PD, Sellers WR, Sharma SK, Wu AD, Nalin CM, Kaelin WG (1996) Identification of a cyclin-cdk2 recognition motif present in substrates and p21-like cyclin-dependent kinase inhibitors. Mol. Cell. Biol. 16: 6623–6633

Adnane J, Shao Z, Robbins PD (1995) The retinoblastoma susceptibility gene product represses transcription when directly bound to the promoter. J. Biol. Chem. 270: 8837–8843

Asano M, Nevins JR, Wharton RP (1996) Ectopic E2F expression induces S phase and apoptosis in *Drosophila* imaginal discs. Genes Dev. 10: 1422–1432

Baldi A, De Luca A, Claudio PP, Baldi F, Giordano GG, Tommasino M, Paggi MG, Giordano A (1995) The Rb2/p130 gene product is a nuclear protein whose phosphorylation is cell cycle regulated. J. Cell. Biochem. 59: 402–408

Bandara LR, Buck VM, Zamanian M, Johnston LH, La Thangue NB (1993) Functional synergy between DP-1 and E2F-1 in the cell cycle-regulating transcription factor DRTF1/E2F. EMBO J. 12: 4317–4324

Bandara LR, Lam EW-R, Sorensen TS, Zamanian M, Girling R, La Thangue NB (1994) DP-1: a cell cycle-regulated and phosphorylated component of transcription factor DRTF1/E2F which is functionally important for recognition by pRB and the adenovirus E4 orf 6/7 protein. EMBO J. 13: 3104–3114

Beijersbergen RL, Carlee L, Kerkhoven RM, Bernards R (1995) Regulation of the retinoblastoma protein-related p107 by G1 cyclin complexes. Genes Dev. 9: 1340–1353

Beijersbergen RL, Kerkhoven RM, Zhu L, Carlee L, Voorhoeve PM, Bernards R (1994) E2F-4, a new member of the E2F gene family, has oncogenic activity and associates with p107 in vivo. Genes Dev. 8: 2680–2690

Birchenall-Roberts MC, Yoo YD, Bertolette IDC, Lee K-H, Turley JM, Bang O-S, Ruscetti FW, Kim S-J (1997) The p120-v-Abl protein interacts with E2F-1 and regulates E2F-1 transcriptional activity. J. Biol. Chem. 272: 8905–8911

Bjorklund S, Hjortsbert K, Johansson E, Thelander L (1993) Structure and promoter characterization of the gene encoding the large subunit (R1 protein) of mouse ribonucleotide reductase. Proc. Natl. Acad. Sci. USA 90: 11322–11326

Blake MC, Azizkhan JC (1989) Transcription factor E2F is required for efficient expression of the hamster dihydrofolate reductase gene in vitro and in vivo. Mol. Cell. Biol. 9: 4994–5002

Botz J, Zerfass-Thome K, Spitkovsky D, Delius H, Vogt B, Eilers M, Hatzigeorgiou A, Jansen-Durr P (1996) Cell cycle regulation of the murine cyclin E gene depends on an E2F binding site in the promoter. Mol. Cell. Biol. 16: 3401–3409

Bremner R, Cohen BL, Sopta M, Hamel PA, Ingles CJ, Gallie BL, Phillips RA (1995) Direct transcriptional repression by pRB and its reversal by specific cyclins. Mol. Cell. Biol. 15: 3526–3265

Brook A, Xie J-E, Du W, Dyson N (1996) Requirements for dE2F function in proliferating cells and in post-mitotic differentiating cells. EMBO J. 15: 3676–3683

Buchkovich K, Duffy LA, Harlow E (1989) The retinoblastoma protein is phosphorylated during specific phases of the cell cycle. Cell 58: 1097–1105

Buck V, Allen KE, Sorensen T, Bybee A, Hijmans EM, Voorhoeve PM, Bernards R, La Thangue NB, Sorensen T (1995) Molecular and functional characterisation of E2F-5, a new member of the E2F family. Oncogene 11: 31–38

Campanero MR, Flemington EK (1997) Regulation of E2F through ubiquitin-proteasome-dependent degradation: Stabilization by the pRB tumor suppressor protein. Proc. Natl. Acad. Sci. USA 94: 2221–2226

Cao L, Faha B, Dembski M, Tsai L-H, Harlow E, Dyson N (1992) Independent binding of the retinoblastoma protein and p107 to the transcription factor E2F. Nature 355: 176–179

Chellappan S, Hiebert S, Mudryj M, Horowitz J, Nevins J (1991) The E2F transcription factor is a cellular target for the RB protein. Cell 65: 1053–1061

Chen J, Saha P, Kornbluth S, Dynlacht B D, Dutta A (1996) Cyclin-binding motifs are essential for the function of p21CIP1. Mol. Cell. Biol. 16: 4673–4682

Chen JD, Lin JY, Levine AJ (1995) Regulation of transcription functions of the p53 tumor suppressor by the mdm2 oncogene. Mol. Med. 1: 141–142

Chen P-L, Scully P, Shew J-Y, Wang J, Lee W-H (1989) Phosphorylation of the retinoblastoma gene product is modulated during the cell cycle and cellular differentiation. Cell 58: 1193–1198

Clarke A, Maandag E, van Roon M, van der Lugt N, van der Valk M, Hooper M, Berns A, te Riele H (1992) Requirement for a functional *Rb-1* gene in murine development. Nature 359: 328–330

Claudio PP, De Luca A, Howard CM, Baldi A, Firpo EJ, Koff A, Paggi MG, Giordano A (1996) Functional analysis of pRb2/p130 interaction with cyclins. Cancer Res. 56: 2003–2008

Cobrinik D, Whyte P, Peeper D S, Jacks T, Weinberg R A (1993) Cell cycle-specific association of E2F with the p130 E1A-binding domain. Genes Dev. 7: 2392–2404

Cobrinik D, Lee M-H, Hannon G, Mulligan G, Bronson RT, Dyson N, Harlow E, Beach D, Weinberg RA, Jacks T (1996) Shared role of the pRB-related p130 and p107 proteins in limb development. Genes Dev. 10: 1633–1644

Connell-Crowley L, Harper JW, Goodrich DW (1997) Cyclin D1/Cdk4 regulates retinoblastoma protein-mediated cell cycle arrest by site-specific phosphorylation. Mol. Biol. Cell. 8: 287–301

Cress WD, Nevins JR (1996) A role for a bent DNA structure in E2F-mediated transcriptional activation. Mol. Cell. Biol. 16: 2119–2127

Dagnino L, Fry CJ, Bartley SM, Farnham P, Gallie BL, Phillips RA (1997) Expression patterns of the E2F family of transcription factors during murine epithelial development. Cell Growth Diff. 8: 553–563

Dalton S (1992) Cell cycle regulation of the human cdc2 gene. EMBO J. 11: 1797–1804

DeCaprio JA, Ludlow JW, Lynch D, Furukawa Y, Griffin J, Piwnica-Worms H, Huang CM, Livingston DM (1989) The product of the retinoblastoma susceptibility gene has properties of a cell cycle regulatory element. Cell 58: 1085–1095

DeGregori J, Kowalik T, Nevins J (1995a) Cellular targets for activation by the E2F1 transcription factor include DNA synthesis- and G1/S-regulatory genes. Mol. Cell. Biol. 15: 4215–4224

DeGregori J, Leone G, Ohtani K, Miron A, Nevins JR (1995b) E2F-1 accumulation bypasses a G1 arrest resulting from the inhibition of G1 cyclin-dependent kinase activity. Genes Dev. 9: 2873–2887

DeGregori J, Leone G, Miron A, Jakoi L, Nevins JR (1997) Distinct roles for E2F proteins in cell growth control and apoptosis. Proc. Natl. Acad. Sci. USA 94: 7245–7250

De la Luna S, Burden MJ, Lee C-W, La Thangue NB (1996) Nuclear accumulation of the E2F heterodimer regulated by subunit composition and alternative splicing of a nuclear localization signal. J Cell Sci. 109: 2443–2452.

Devoto SH, Mudryj M, Pines J, Hunter T, Nevins JR (1992) A cyclin A-protein kinase complex possesses sequence-specific DNA binding activity: p33cdk2 is a component of the E2F- cyclin A complex. Cell 68: 167–176

Dobrowolski SF, Stacey DW, Harter ML, Stine JT, Hiebert SW (1994) An E2F dominant negative mutant blocks E1A induced cell cycle progression. Oncogene 9: 2605–2612

Dou QP, Zhao S, Levin AH, Wang J, Helin K, Pardee AB (1994) G1/S-regulated E2F-containing protein complexes bind to the mouse thymidine kinase gene promoter. J. Biol. Chem. 269: 1306–1313

Du W, Vidal M, Xie J-E, Dyson N (1996a) RBF, a novel RB-related gene that regulates E2F activity and interacts with *cyclin E* in *Drosophila*. Genes Dev. 10: 1206–1218

Du W, Xie J-E, Dyson N (1996b) Ectopic expression of dE2F and dDP induces cell proliferation and death in the *Drosophila* eye. EMBO J. 15: 3684–3692

Duronio RJ, O'Farrell PH (1995) Developmental control of the G1 to S transition in Drosophila; cyclin E is a limiting downstream target of E2F. Genes Dev. 9: 1456–1468

Duronio RJ, O'Farrell PH, Xie J-E, Brook A, Dyson N (1995) The transcription factor E2F is required for S phase during *Drosophila* embryogenesis. Genes Dev. 9: 1445–1455

Dynlacht BD, Brook A, Dembski MS, Yenush L, Dyson N (1994a) DNA-binding and trans-activation properties of *Drosophila* E2F and DP proteins. Proc. Natl. Acad. Sci. USA 91: 6359–6363

Dynlacht B D, Flores O, Lees J A, Harlow E (1994b) Differential regulation of E2F *trans*-activation by cyclin-cdk2 complexes. Genes Dev. 8: 1772–1786

Emili A, Ingles CJ (1995) Promoter-dependent photocross-linking of the acidic transcriptional activator E2F-1 to the TATA-binding protein. J. Biol. Chem. 270: 13674–13680

Fagan R, Flint KJ, Jones N (1994) Phosphorylation of E2F-1 modulates its interaction with the retinoblastoma gene product and the adenoviral E4 19 kDa protein. Cell 78: 799–811

Faha B, Ewen ME, Tsai L-H, Livingston DM, Harlow E (1992) Interaction between human cyclin A and adenovirus E1A-associated p107 protein. Science 255: 87–90

Fan J, Bertino J R (1997) Functional roles of E2F in cell cycle regulation. Oncogene 14: 1191–1200

Field SJ, Tsai F-Y, Kuo F, Zubiaga AM, Kaelin WG, Livingston DM, Orkin SH, Greenberg ME (1996) E2F-1 functions in mice to promote apoptosis and suppress proliferation. Cell 85: 549–561

Flemington EK, Speck SH, Kaelin WG (1993) E2F-1 mediated transactivation is inhibited by complex formation with the retinoblastoma susceptibility gene product. Proc. Natl. Acad. Sci. USA 90: 6914–6918

Fry CJ, Slansky JE, Farnham PJ (1997) Position-dependent transcriptional regulation of the murine dihydrofolate reductase promoter by the E2F transactivation domain. Mol. Cell. Biol. 17: 1966–1976

Furukawa Y, Terui Y, Sakoe K, Ohta M, Saito M (1994) The role of cellular transcription factor E2F in the regulation of cdc2 mRNA expression and cell cycle control of human hematopoietic cells. J Biol. Chem. 269: 26249–26258

Geng Y, Eaton EN, Picon M, Roberts JM, Lundberg AS, Gifford A, Sardet C, Weinberg RA (1996) Regulation of cyclin E transcription by E2Fs and retinoblastoma protein. Oncogene 12: 1173–1180

Ginsberg D, Vairo G, Chittenden T, Xiao ZX, Xu G, Wydner KL, DeCaprio JA, Lawrence JB, Livingston DM (1994) E2F-4, a new member of the E2F transcription factor family, interacts with p107. Genes Dev, 8: 2665–2679

Girling R, Partridge JF, Bandara LR, Burden N, Totty NF, Hsuan JJ, La Thangue NB (1993) A new component of the transcription factor DRTF1/E2F. Nature 362: 83–87
Gopalkrishnan RV, Dolle P, Mattei M-G, La Thangue NB, Kedinger C (1996) Genomic structure and developmental expression of the mouse cell cycle regulatory transcription factor DP1. Oncogene 13: 2671–2680
Guy CT, Zhou W, Kaufman S, Robinson MO (1996) E2F-1 blocks terminal differentiation and causes proliferation in transgenic megakaryocytes. Mol. Cell. Biol. 16: 685–693
Hagemeier C, Bannister A J, Cook A, Kouzarides T (1993) The activation domain of transcription factor PU.1 binds the retinoblastoma (RB) protein and the transcription factor TFIID in vitro: RB shows sequence similarity to TFIID and TFIIB. Proc. Natl. Acad. Sci. USA 90: 1580 - 1584
Hamel PA, Gill RM, Phillips RA, Gallie BL (1992a) Regions controlling hyperphosphorylation and conformation of the retinoblastoma gene product are independent of domains required for transcriptional repression. Oncogene 7: 693–701
Hamel PA, Gill RM, Phillips RA, Gallie BL (1992b) Transcriptional repression of the E2-containing promoters EIIaE, c-myc, and RB1 by the product of the RB1 gene. Mol. Cell. Biol. 12: 3431–3438
Hannon GJ, Demetrick D, Beach D (1993) Isolation of the Rb-related p130 through its interaction with CDK2 and cyclins. Genes Dev. 7: 2378–2391
Hao XF, Alphey L, Bandara LR, Lam EW-F, Glover D, La Thangue NB (1995) Functional conservation of the cell cycle-regulating transcription factor DRTF1/E2F and its pathway of control in *Drosophila melanogaster*. J. Cell Sci. 108: 2945–2954
Harrison DJ, Hooper ML, Armstrong JF, Clarke AR (1995) Effects of heterozygosity for the Rb-1t19neo allele in the mouse. Oncogene 10: 1615–1620
Hatakeyama M, Brill JA, Fink GR, Weinberg RA (1994) Collaboration of G1 cyclins in the functional inactivation of the retinoblastoma protein. Genes Dev. 8: 1759–1771
Hateboer G, Kerkhoven RM, Shvarts A, Bernards R, Beijersbergen RL (1996) Degradation of E2F by the ubiquitin-proteasome pathway: regulation by retinoblastoma family proteins and adenovirus transforming proteins. Genes Dev. 10: 2960–2970
Haupt Y, Maya R, Kazaz A, Oren M (1997) Mdm2 promotes the rapid degradation of p53. Nature 387: 296–299
Helin K, Harlow E (1994) Heterodimerization of the transcription factors E2F1 and DP1 is required for binding to the adenovirus E4 (ORF6/7) protein. J. Virol. 68: 5027–5035
Helin K, Lees JA, Vidal M, Dyson N, Harlow E, Fattaey A (1992) A cDNA encoding a pRB-binding protein with properties of the transcription factor E2F. Cell 70: 337–350
Helin K, Harlow E, Fattaey AR (1993a) Inhibition of E2F-1 transactivation by direct binding of the retinoblastoma protein. Mol. Cell. Biol. 13: 6501–6508
Helin K, Wu C-L, Fattaey A, Lees J, Dynlacht B, Ngwu C, Harlow E (1993b) Heterodimerization of the transcription factors E2F-1 and DP-1 leads to cooperative transactivation. Genes Dev. 7: 1850–1861
Henglein B, Chenivesse X, Wang J, Eick D, Brechot C (1994) Structure and cell cycle-regulated transcription of the human cyclin A gene. Proc. Nat. Acad. Sci. USA 91: 5490–5494
Herrera RE, Sah VP, Williams BO, Makela TP, Weinberg RA, Jacks T (1996) Altered cell cycle kinetics, gene expression, and G1 restriction point regulation in Rb-deficient fibroblasts. Mol. Cell. Biol. 16: 2402–2407
Hiebert S W, Lipp M, Nevins J R (1989) E1A-dependent trans-activation of the human MYC promoter is mediated by the E2F factor. Proc. Natl. Acad. Sci. USA 86: 3594–3598
Hiebert SW, Blake M, Azizkhan J, Nevins JR (1991) Role of E2F transcription factor in E1A-mediated trans-activation of cellular genes. J. Virol. 65: 3547–3552
Hiebert SW, Packham G, Strom DK, Haffner R, Oren M, Zambetti G, Cleveland JL (1995) E2F-1:DP-1 induces p53 and overrides survival factors to trigger apoptosis. Mol. Cell. Biol. 15: 6864–6874
Hijmans EM, Voorhoeve PM, Beijersbergen RL, van 't Veer L, Bernards R (1995) E2F-5, a new E2F family member that interacts with p130 in vivo. Mol. Cell. Biol. 15: 3082–3089
Hofmann F, Livingston DM (1996) Differential effects of cdk2 and cdk3 on the control of pRB and E2F function during G1 exit. Genes Dev. 10: 851–861

Hofmann F, Martelli F, Livingston DM, Wang Z (1996) The retinoblastoma gene product protects E2F-1 from degradation by the ubiquitin-proteasome pathway. Genes Dev. 10: 2949–2959

Holmberg C, Helin K, Schested M, Karlstrom O (1998) E2F-1 induced p53-independent apoptosis in transgenic mice. Oncogene (in press)

Horowitz JM, Udvadia AJ (1995) Transcriptional regulation by the retinoblastoma (Rb) protein. Mol. Cell. Diff. 3: 275–314

Hsiao K-M, McMahon SL, Farnham PJ (1994) Multiple DNA elements are required for the growth regulation of the mouse E2F1 promoter. Genes Dev. 8: 1526–1537

Hsieh J-K, Fredersdorf S, Kouzarides T, Martin K, Lu X (1997) E2F1-induced apoptosis requires DNA binding but not transactivation and is inhibited by the retinoblastoma protein through direct interaction. Genes Dev. 11: 1840–1852

Hu N, Gutsmann A, Herbert DC, Bradley A, Lee W-H, Lee EY (1994) Heterozygous Rb-1 delta 20/+mice are predisposed to tumors of the pituitary gland with a nearly complete penetrance. Oncogene 9: 1021–1027

Huber HE, Edwards G, Goodhart PJ, Patrick DR, Huang PS, Ivey-Hoyle M, Barnett SF, Oliff A, Heimbrook DC (1993) Transcription factor E2F binds DNA as a heterodimer. Proc. Natl. Acad. Sci. USA 90: 3525–3529

Hurford R, Cobrinik D, Lee M-H, Dyson N (1997) pRB and p107/p130 are required for the regulated expression of different sets of E2F responsive genes. Genes Dev. 11: 1447–1463

Ikeda M-A, Jakoi L, Nevins J (1996) A unique role for the Rb protein in controlling E2F accumulation during cell growth and differentiation. Proc. Natl. Acad. Sci. USA 93: 3215–3220

Imperiale MJ, Kao HT, Feldman LT, Nevins JR, Strickland S (1984) Common control of the heat shock gene and early adenovirus genes: evidence for a cellular E1A-like activity. Mol. Cell. Biol. 4: 867–874

Ishizaki J, Nevins JR, Sullenger BA (1996) Inhibition of cell proliferation by an RNA ligand that selectively blocks E2F function. Nature Med. 2: 1386–1389

Ivey-Hoyle M, Conroy R, Huber H, Goodhart P, Oliff A, Heinbrook DC (1993) Cloning and characterization of E2F-2, a novel protein with the biochemical properties of transcription factor E2F. Mol. Cell Biol. 13: 7802–7812

Jacks T, Fazeli A, Schmitt EM, Bronson RT, Goodell MA, Weinberg RA (1992) Effects of an Rb mutation in the mouse [see comments]. Nature 359: 295–300

Johnson DG, Schwarz JK, Cress WD, Nevins JR (1993) Expression of transcription factor E2F1 induces quiescent cells to enter S phase. Nature 365: 349–352

Johnson DG, Cress WD, Jakoi L, Nevins JR (1994a) Oncogenic capacity of the E2F1 gene. Proc Natl Acad Sci USA 91: 12823–12827

Johnson DG, Ohtani K, Nevins JR (1994b) Autoregulatory control of E2F-1 expression in response to positive and negative regulators of cell cycle expression. Genes Dev. 8: 1514–1525

Jooss K, Lam EW-F, Bybee A, Girlin R, Muller R, La Thangue NB (1995) Proto-oncogenic properties of the DP family of proteins. Oncogene 10: 1529–1536

Kaelin WG, Krek W, Sellers WR, DeCaprio JA, Ajchenbaum F, Fuchs CS, Chittenden T, Li Y, Farnham PJ, Blanar MA, Livingston DM, Flemington EK (1992) Expression cloning of a cDNA encoding a retinoblastoma-binding protein with E2F-like properties. Cell 70: 351–364

Karlseder J, Rotheneder H, Wintersberger E (1996) Interaction of Sp1 with the growth- and cell cycle-regulated transcription factor E2F. Mol. Cell. Biol. 16: 1659–1667

Kitagawa M, Higashi H, Suzuki-Takahashi I, Segawa K, Hanks SK, Taya Y, Nishimura S, Okuyama A (1995) Phosphorylation of E2F-1 by cyclin A-cdk2. Oncogene 10: 229–236

Kovesdi I, Reichel R, Nevins JR (1986) Identification of a cellular transcription factor involved in E1A trans-activation. Cell 45: 219–228

Kowalik TF, DeGregori J, Schwarz JK, Nevins JR (1995) E2F1 overexpression in quiescent fibroblasts leads to induction of cellular DNA synthesis and apoptosis. J. Virol. 69: 2491–2500

Krek W, Livingston DM, Shirodkar S (1993) Binding to DNA and the retinoblastoma gene product promoted by complex formation of different E2F family members. Science 262: 1557–1560

Krek W, Ewen M, Shirodkar S, Arany Z, Kaelin WG, Livingston DM (1994) Negative regulation of the growth-promoting transcription factor E2F-1 by a stably bound cyclin A-dependent protein kinase. Cell 78: 161–172

Krek W, Xu G, Livingston DM (1995) Cyclin A-kinase regulation of E2F-1 DNA binding function underlies suppression of an S phase checkpoint. Cell 83: 1149–1158

Kubbutat MH, Jones SP, Vousden KH (1997) Regulation of p53 stability by Mdm2. Nature 387: 299–303

La Thangue NB (1994) DTRF1/E2F: an extending family of heteridimeric factors implicated in cell cycle control. Trends Biochem. Sci. 19: 108–114

La Thangue NB, Rigby PWJ (1987) An adenovirus E1A-like transcription factor is regulated during the differentiation of murine embryonal carcinoma stem cells. Cell 49: 507–513

Lam EW-F, Watson RJ (1993) An E2F-binding site mediates cell-cycle regulated repression of mouse B-myb transcription. EMBO J. 12: 2705–2713

Lee EY-HP, Chang C-Y, Hu N, Wang Y-CJ, Lai C-C, Herrup K, Lee W-H, Bradley A (1992) Mice deficient for Rb are nonviable and show defects in neurogenesis and haematopoiesis. Nature 359: 288–294

Lee M-H, Williams BO, Mulligan G, Mukai S, Bronson RT, Dyson N, Harlow E, Jacks T (1996) Targeted disruption of p107: functional overlap between p107 and Rb. Genes Dev. 10: 1621–1632

Lees JA, Saito M, Vidal M, Valentine M, Look T, Harlow E, Dyson N, Helin K (1993) The retinoblastoma protein binds to a family of E2F transcription factors. Mol. Cell. Biol. 13: 7813–7825

Li L-J, Naeve GS, Lee AS (1993) Temporal regulation of cyclin A-p107 and p33cdk2 complexes binding to a human thymidine kinase promoter element important for G1-S phase transcriptional regulation. Proc. Natl. Acad. Sci. USA 91: 3554–3558

Li Y, Graham C, Lacy S, Duncan AM, Whyte P (1993b) The adenovirus E1A-associated 130-kD protein is encoded by a member of the retinoblastoma gene family and physically interacts with cyclins A and E. Genes Dev. 7: 2366–2377

Li Y, Slansky JE, Myers DJ, Drinkwater NR, Kaelin WG, Farnham PJ (1994) Cloning, chromosomal location, and characterization of mouse E2F1. Mol Cell Biol. 14: 1861–1869

Lin S-Y, Black AR, Kostic D, Pajovic S, Hoover CN, Azizkhan JC (1996) Cell cycle-regulated association of E2F1 and Sp1 is related to their functional interaction. Mol. Cell. Biol. 16: 1668–1675

Lindeman GJ, Gaubatz S, Livingston DM, Ginsberg D (1997) The subcellular localization of E2F-4 is cell-cycle dependent. Proc. Natl. Acad. Sci. USA 94: 5095–5100

Logan TJ, Evans DL, Mercer WE, Bjornsti MA, Hall DJ (1995) Expression of a deletion mutant of the E2F transcription factor in fibroblasts lengthens S phase and increase sensitivity to S phase-specific toxin. Cancer Res. 55: 2883–2891

Lukas J, Bartkova J, Rohde M, Strauss M, Bartek J (1995) Cyclin D1 is dispensable for G1 control in retinoblastoma gene-deficient cells independently of cdk4 activity. Mol. Cell. Biol. 15: 2600–2611

Lukas J, Petersen BO, Holm K, Bartek J, Helin K (1996) Deregulated expression of E2F family members induces S-phase entry and overcomes p16INK4A-mediated growth suppression. Mol. Cell. Biol. 16: 1047–1057

Luo Y, Hurwitz J, Massague J (1995) Cell-cycle inhibition by independent CDK and PCNA binding domains in p21Cip1. Nature 375: 159–161

Maandag ECR, van der Valk M, Vlaar M, Feltkamp C, O'Brien J, Van Roon M, Van der Lugt N, Berns A, teRiele H (1994) Developmental rescue of an embryonic-lethal mutation in the retinoblastoma gene in chimeric mice. EMBO J. 13: 4260–4268

Magae J, Wu C-L, Illenye S, Harlow E, Heintz NH (1996) Nuclear localization of DP and E2F transcription factors by heterodimeric partners and retinoblastoma protein family members. J. Cell Sci. 109: 1717–1726

Mann DJ, Jones NC (1996) E2F-1 but not E2F-4 can overcome p16-induced G1 cell cycle arrest. Curr. Biol. 6: 474–483

Martin K, Trouche D, Hagemeier C, Sorensen TS, La Thangue NB, Kouzarides T (1995) Stimulation of E2F1/DP1 transcriptional activity by MDM2 oncoprotein. Nature 375: 691–694

Mayol X, Garriga J, Grana X (1995) Cell-cycle dependent phosphorylation of p130. Oncogene 11: 801–808

Means AL, Slansky JE, McMahon SL, Knuth MW, Farnham PJ (1992) The HIP binding site is required for growth regulation of the dihydrofolate reductase promoter. Mol. Cell. Biol. 12: 1054–1063

Melillo RM, Helin K, Lowy DR, Schiller JT (1994) Positive and negative regulation of cell proliferation by E2F-1: influence of protein level and human papillomavirus oncoproteins. Mol. Cell. Biol. 14: 8241–8249

Miltenberger RJ, Sukow KA, Farnham PJ (1995) An E-box-mediated increase in cad transcription at the G1/S-phase boundary is suppressed by inhibition by inhibitory c-myc mutants. Mol. Cell. Biol. 15: 2527–2535

Mittnacht S, Weinberg RA (1991) G1/S phosphorylation of the retinoblastoma protein is associated with altered affinity for the nuclear compartment. Cell 65: 381–393

Moberg K, Starz MA, Lees JA (1996) E2F-4 switches from p130 to p107 and pRB in response to cell cycle reentry. Mol. Cell. Biol. 16: 1436–1449

Moran E (1993) DNA tumor virus transforming proteins and the cell cycle. Curr. Opin. Genet. Dev. 3: 63–70

Mudryj M, Hiebert SW, Nevins JR (1990) A role for the adenovirus inducible E2F transcription factor in a proliferation dependent signal transduction pathway. EMBO J. 9: 2179–2184.

Müller H, Lukas J, Schneider A, Warthoe P, Bartek J, Eilers M, Strauss M (1994) Cyclin D1 expression is regulated by the retinoblastoma protein. Proc. Natl. Acad. Sci. USA 91: 2945–2949

Müller H, Moroni MC, Vigo E, Petersen BO, Bartek J, Helin K (1997) Induction of S-phase entry by E2F transcription factors depends on their nuclear localization. Mol. Cell. Biol. 17: 5508–5520

Neuman E, Flemington EK, Sellers WR, Kaelin WG (1994) Transcription of the E2F1 gene is rendered cell cycle-dependent by E2F DNA binding sites within its promoter. Mol. Cell. Biol. 14: 6607–6615

Nevins JR (1992) E2F: A link between the Rb tumor suppressor protein and viral oncoproteins. Science 258: 424–429

Nip J, Strom DK, Fee BE, Zambetti G, Cleveland JL, Hiebert SW (1997) E2F-1 cooperates with topoisomerase II inhibition and DNA damage to selectively augment p53-independent apoptosis. Mol. Cell. Biol. 17: 1049–1056

O'Connor DJ, Lam EW-F, Griffin S, Zhong S, Leighton LC, Burbidge SA, Lu X (1995) Physical and functional interactions between p53 and cell cycle co-operating transcription factors, E2F1 and DP1. EMBO J. 14: 6184–6192

Obert S, O'Connor RJ, Schmid S, Hearing P (1994) The adenovirus E4-6/7 protein transactivates the E2 promoter by inducing dimerization of a heterodimeric E2F complex. Mol. Cell. Biol. 14: 1333–1346

Ogris E, Rotheneder H, Mudrak I, Pichler A, Wintersberger E (1993) A binding site for transcription factor E2F is a target for transactivation of murine thymidine kinase by polyomavirus large T antigen and plays an important role in growth regulation of the gene. J Virol. 67: 1765–1771

Ohtani K, Nevins JR (1994) Functional properties of a Drosophila homolog of the E2F1 gene. Mol. Cell. Biol. 14: 1603–1612

Ohtani K, DeGregori J, Nevins JR (1995) Regulation of the cylcin E gene by transcription factor E2F1. Proc. Natl. Acad. Sci. USA 92: 12146–12150

Ohtani K, DeGregori J, Leone G, Herendeen DR, Kelly TJ, Nevins JR (1996) Expression of the HsOrc1 gene, a human ORC1 homolog is regulated by cell proliferation via the E2F transcription factor. Mol. Cell. Biol. 16: 6977–6984

Ormondroyd E, de la Luna S, La Thangue NB (1995) A new member of the DP family, DP-3, with distinct protein products suggests a regulatory role for alternative splicing in the cell cycle transcription factor DRTF1/E2F. Oncogene 11, 1437–1446

Oswald F, Dobner T, Lipp M (1996) The E2F transcription factor activates a replication-dependent human H2A gene in early S phase of the cell cycle. Mol. Cell. Biol. 16: 1889–1895

Pagano M, Draetta G, Jansen-Durr P (1992) Association of cdk2 kinase with the transcription factor E2F during S phase. Science 255: 1144–1147

Pasteau S, Loiseau L, Arnaud L, Trembleau A, Brun G (1995) Isolation and characterization of a chicken homolog of the E2F-1 transcription factor. Oncogene 11: 1475–1486

Pearson BE, Nasheuer HP, Wang TS (1991) Human DNA polymerase alpha gene: Sequences controlling expression in cycling and serum-stumulated cells. Mol. Cell. Biol. 11: 2081–2095

Peeper DS, Keblusek P, Helin K, Toebes M, van der Eb AJ, Zantema A (1995) Phosphorylation of a specific cdk site in E2F-1 affects its electrophoretic mobility and promotes pRB-binding in vitro. Oncogene 10: 39–48

Phillips AC, Bates S, Ryan KM, Helin K, Vousden KH (1997) Induction of DNA synthesis and apoptosis are separable functions of E2F-1. Genes Dev. 11: 1853–1863

Philpott A, Friend SH (1994) E2F and its developmental regulation in Xenopus laevis. Mol. Cell. Biol. 14: 5000–5009

Picksley SM, Lane DP (1993) The p53-mdm2 autoregulatory feedback loop-aparadigm for the regulation of growth control by p53? Bioessays 15: 689–690

Qin X-Q, Livingston DM, Kaelin WG, Adams P (1994) Deregulated transcription factor E2F-1 expression leads to S-phase entry and p53-mediated apoptosis. Proc. Natl. Acad. Sci. USA. 91: 10918–10922

Qin X-Q, Livingston DM, Ewen M, Sellers WR, Arany Z, Kaelin WG (1995) The transcription factor E2F-1 is a downstream target of RB action. Mol. Cell. Biol. 15: 742–755

Riley DJ, Lee EY-HP, Lee W-H (1994) The retinoblastoma protein: more than a tumor suppressor. Annual Rev. Cell Biol. 10: 1–29

Rogers KT, Higgins PDR, Milla MM, Phillips RS, Horowitz JM (1996) DP-2, a heterodimeric partner of E2F: Identification and characterization of DP-2 proteins expressed in vivo. Proc. Natl. Acad. Sci. USA 93: 7594–7599

Royzman I, Whittaker A J, Orr-Weaver T L (1997) Mutations in Drosophila DP and E2F distinguish G1-S progression from an associated transcriptional program. Genes Dev. 11: 1999–2011

Saha P, Eichbaum Q, Silberman ED, Mayer BJ, Dutta A (1997) p21Cip1 and CDC25A: Competition between an inhibitor and an activator of cyclin-dependent kinases. Mol. Cell. Biol. 17: 4338–4345

Saito M, Helin K, Valentine MB, Griffith BB, Willman CL, Harlow E, Look AT (1995) Amplification of the E2F1 transcription factor gene in the HEL erythroleukemia cell line. Genomics 25: 130–138

Sardet C, Vidal M, Cobrinik D, Geng Y, Onufryk C, Chen A, Weinberg RA (1995) E2F-4 and E2F-5, two novel members of the E2F family, are expressed in the early phases of the cell cycle. Proc. Natl. Acad. Sci. USA 92: 2403–2407

Schulze A, Zerfass K, Spitkovsky D, Middendorp S, Berges J, Helin K, Jansen-Durr P, Henglein B (1995) Cell cycle regulation of the cyclin A gene promoter is mediated by a variant E2F site. Proc. Natl. Acad. Sci. USA 92: 11264–11268

Schwarz JK, Devoto SH, Smith EJ, Chellappan SP, Jakoi L, Nevins JR (1993) Interactions of the p107 and Rb proteins with E2F during the cell proliferation response. EMBO J. 12: 1013–1020

Sears R, Ohtani K, Nevins JR (1997) Identification of positively and negatively acting elements regulating expression of the E2F2 gene in response to cell growth signals. Mol. Cell. Biol. 17: 5227–5235

Sellers WR, Rodgers JW, Kaelin WG (1995) A potent transrepression domain in the retinoblastoma protein induces a cell cycle arrest when bound to E2F sites. Proc. Natl. Acad. Sci. USA 92: 11544–11548

Seum C, Spierer A, Pauli D, Szidonya J, Reuter G, Spierer P (1996) Position-effect variegation in Drosophila depends on the dose of the gene encoding the E2F transcriptional activator and cell cycle regulator. Development 122: 1949–1956

Shan B, Zhu X, Chen P-L, Durfee T, Yang Y, Sharp D, Lee W-H (1992) Molecular cloning of cellular genes encoding retinoblastoma-associated proteins: identification of a gene with properties of the transcription factor E2F. Mol. Cell. Biol. 12: 5620–5631

Shan B, Lee W-H (1994) Deregulated expression of E2F-1 induces S-phase entry and leads to apoptosis. Mol. Cell. Biol. 14: 8166–8173

Shan B, Farmer AA, Lee W-H (1996) The molecular basis of E2F-1/DP-1-induced S-phase entry and apoptosis. Cell Growth Diff. 7: 689–697

Shirodkar S, Ewen M, DeCaprio JA, Morgan D, Livingston DM, Chittenden T (1992) The transcription factor E2F interacts with the retinoblastoma product and a p107-cyclin A complex in a cell cycle-regulated manner. Cell 68: 157–166

Singh P, Wong S H, Hong W (1994) Overexpression of E2F-1 in rat embryo fibroblasts leads to neoplastic transformation. EMBO J. 13: 3329–3338

Slansky J, Li Y, Kaelin WG, Farnham PJ (1993) A protein synthesis-dependent increase in E2F1 mRNA correlates with growth regulation of the dihydrofolate reductase promoter. Mol. Cell. Biol. 13: 1610–1618

Slansky JE and Farnham PJ (1996) Transcriptional control of cell growth: the E2F gene family. Curr. Top. Microbiol.and Immunol. (1996) 208: 1–30.

Smith EJ, Leone G, DeGregori J, Jakoi L, Nevins JR (1996) The accumulation of an E2F-p130 transcriptional repressor distinguishes a G0 cell state from a G1 cell state. Mol. Cell. Biol. 16: 6965–6976

Sorensen TS, Girling R, Lee CW, Gannon J, Bandara LR, La Thangue NB (1996) Functional interaction between DP-1 and p53. Mol. Cell. Biol. 16: 5888–5895

Souza RF, Yin J, Smolinski KN, Zou T-T, Wang S, Shi Y-Q, Rhyu M-G, Cottrell J, Abraham J M, Biden K, Simms L, Leggett B, Bova G S, Frank T, Powell SM, Sugimura H, Young J, Harpaz N, Shimizu K, Matsubara N, Meltzer SJ (1997) Frequent mutation of the E2F-4 cell cycle gene in primary human gastrointestinal tumors. Cancer Res. 57: 2350–2353

Studbal H, Zalvide J, DeCaprio JA (1996) Simian Virus 40 Large T antigen alters the phosphorylation state of the RB-related proteins p130 and p107. J. Virol. 70: 2781–2788

Suzuki-Takahasi I, Kitagawa M, Saijo M, Higashi H, Ogino H, Matsumoto H, Taya Y, Nishimura S, Okuyama A (1995) The interactions of E2F with pRB and with p107 are regulated via the phosphorylation of pRB and p107 by a cyclin-dependent kinase. Oncogene 10: 1691–1698

Tao Y, Kassatly RF, Cress WD, Horowitz JM (1997) Subunit composition determines E2F DNA-binding site specificity. Mol. Cell. Biol. 17 (12): 6994–7007

Tevosian SG, Paulson KE, Bronson R, Yee AS (1996) Expression of the E2F-1/DP-1 transcription factor in murine development. Cell Growth Diff. 7: 43–52

Thalmeier K, Synovzik H, Mertz R, Winnacker E-L, Lipp M (1989) Nuclear factor E2F mediates basic transcription and trans-activation by E1a of the human MYC promoter. Genes Dev. 3: 527–536

Thelander M, Thelander L (1989) Molecular cloning and expression of the functional gene encoding the M2 subunit of mouse ribonucleotide reductase: a new dominant marker gene. EMBO J. 8: 2475–2479

Thut CJ, Goodrich JA, Tjian R (1997) Repression of p53-mediated transcription by MDM2: a dual mechanism. Genes Dev. 11: 1974–1986

Trouche D, Kouzarides T (1996) E2F1 and E1A 12S have a homologous activation domain regulated by RB and CBP. Proc. Natl. Acad. Sci. USA 93: 1439–1442

Vairo G, Livingston DM, Ginsberg D (1995) Functional interaction between E2F-4 and p130: evidence for distinct mechanisms underlying growth suppression by different retinoblastoma protein family members. Genes Dev. 9: 869–881

Verona R, Moberg K, Estes S, Starz M, Vernon JP, Lees JA (1997) E2F Activity is regulated by cell-cycle-dependent changes in subcellular localization. Mol. Cell Biol. 17: 7268–7282

Weintraub SJ, Prater CA, Dean DC (1992) Retinoblastoma protein switches the E2F site from positive to negative element. Nature 358: 259–261

Weintraub SJ, Chow KNB, Luo RX, Zhang SH, He S, Dean DC (1995) Mechanism of active transcriptional repression by the retinoblastoma protein. Nature 375: 812–815

Wells J, Held P, Illenye S, Heintz NH (1996) Protein-DNA interactions at the major and minor promoters of the divergently transcribed dhfr and rep3 genes during the chinese hamster ovary cell cycle. Mol. Cell. Biol. 16: 634–647

Wu C-L, Zukerberg LR, Ngwu C, Harlow E, Lees JA (1995) In vivo association of E2F and DP family proteins. Mol. Cell. Biol. 15: 2536–2546

Wu C-L, Classon M, Dyson N, Harlow E (1996) Expression of dominant-negative mutant DP-1 blocks cell cycle progression in G1. Mol. Cell. Biol. 16: 3698–3706

Wu X, Levine AJ (1994) p53 and E2F-1 cooperate to mediate apoptosis. Proc. Natl. Acad. Sci. USA 91: 3602–3606

Xu G, Livingston DM, Krek W (1995) Multiple members of the E2F transcription factor family are the products of oncogenes. Proc. Natl. Acad. Sci. USA 92: 1357–1361

Xu M, Sheppard KA, Peng CY, Yee AS, Piwnica-Worms H (1994) Cyclin A/CDK2 binds directly to E2F-1 and inhibits the DNA-binding activity of E2F-1/DP-1 by phosphorylation. Mol. Cell. Biol. 14: 8420–8431

Yagi H, Kato T, Nagata T, Habu T, Nozaki M, Matsushiro A, Nishimune Y, Morita T (1995) Regulation of the mouse histone H2A.X gene promoter by the transcription factor E2F and CCAAT binding protein. J. Biol. Chem. 270: 18759–18765

Yamaguchi M, Hayashi Y, Hirose F, Matsuoka S, Shiroki K, Matsukage A (1992) Activation of the mouse proliferating cell nuclear antigen gene promoter by adenovirus type 12 E1A proteins. Japan J. Canc. Res. 83: 609–617

Yamaguchi M, Hayashi Y, Matsukage A (1995) Essential role of E2F recognition sites in regulation of the proliferating cell nuclear antigen gene promoter during Drosophila development. J. Biol. Chem. 270: 25159–25165

Yamasaki L, Jacks T, Bronson R, Goillot E, Harlow E, Dyson N (1996) Tumor induction and tissue atrophy in mice lacking E2F-1. Cell 85: 537–548

Yamasaki L, Bronson R, Williams BO, Dyson NJ, Harlow E, Jacks T (1998) Loss of E2F-1 reduces tumorigenesis and extends lifespan of Rb(+/-) mice. Nature Genetics (in press)

Zhang Y, Chellappan SP (1995) Cloning and characterization of human DP2, a novel dimerization partner of E2F. Oncogene 10: 2085–2093

Zhu L, Harlow E, Dynlacht BD (1995a) p107 uses a p21CIP1-related domain to bind cyclin/cdk2 and regulate interactions with E2F. Genes Dev. 9: 1740–1752

Zhu L, Zhu L, Xie E, Chang L-S (1995b) Differential roles of two tandem E2F sites in repression of the human p107 promoter by retinoblastoma and other proteins. Mol. Cell. Biol. 15: 3552–3562

Zwicker J, Muller R (1997) Cell-cycle regulation of gene expression by transcriptional repression. Trends Genet. 13: 3–6

Subject Index

Printing: Saladruck, Berlin
Binding: Buchbinderei Lüderitz & Bauer, Berlin